MITIGATION OF HAZARDOUS COMETS AND ASTEROIDS

It is known that large asteroids and comets can collide with the Earth with severe consequences. Although the chances of a collision in a person's lifetime are small, collisions are a random process and could occur at any time. This book collects the latest thoughts and ideas of scientists concerned with mitigating the threat of hazardous asteroids and comets. It reviews current knowledge of the population of potential colliders, including their numbers, locations, orbits, and how warning times might be improved. The structural properties and composition of their interiors and surfaces are reviewed, and their orbital response to the application of pulses of energy is discussed. Difficulties of operating in space near, or on the surface of, very low mass objects are examined. The book concludes with a discussion of the problems faced in communicating the nature of the impact hazard to the public.

MICHAEL BELTON is Emeritus Astronomer at the National Optical Astronomy Observatory, Tucson, Arizona, and President of Belton Space Exploration Initiatives, LLC.

THOMAS MORGAN is a Discipline Scientist in the Office of Space Science at the National Aeronautics and Space Administration, Washington, District of Columbia.

NALIN SAMARASINHA is an Assistant Scientist at the National Optical Astronomy Observatory, Tucson, Arizona.

DONALD YEOMANS manages NASA's Near-Earth Object Program Office and JPL's Solar System Dynamics group at the Jet Propulsion Laboratory of the California Institute of Technology, Pasadena, California.

MITIGATION OF HAZARDOUS COMETS AND ASTEROIDS

edited by

MICHAEL J. S. BELTON
Belton Space Exploration Initiatives, LLC

THOMAS H. MORGAN
National Aeronautics and Space Administration

NALIN H. SAMARASINHA
National Optical Astronomy Observatory

DONALD K. YEOMANS
Jet Propulsion Laboratory, California Institute of Technology

CAMBRIDGE
UNIVERSITY PRESS

CAMBRIDGE UNIVERSITY PRESS
Cambridge, New York, Melbourne, Madrid, Cape Town, Singapore,
São Paulo, Delhi, Dubai, Tokyo, Mexico City

Cambridge University Press
The Edinburgh Building, Cambridge CB2 8RU, UK

Published in the United States of America by Cambridge University Press, New York

www.cambridge.org
Information on this title: www.cambridge.org/9780521173322

© Cambridge University Press 2004

First published 2004
First paperback edition 2011

A catalogue record for this publication is available from the British Library

Library of Congress Cataloguing in Publication data
Mitigation of hazardous impacts due to asteroids and comets / Michael J.S. Belton . . . [et al.], editors.
p.   cm.
Includes bibliographical references and index.
ISBN 0 521 82764 7
1. Near-earth asteroids.   2. Asteroids – Collisions with Earth – Prevention.
3. Comets – Collisions with Earth – Prevention.   1. Belton, M. J. S.
QB651.M57   2004
523.44 – dc22   2003065427

ISBN 978-0-521-82764-5 Hardback
ISBN 978-0-521-17332-2 Paperback

Contents

Contributors

Erik Asphaug
Earth Science Department
University of California at Santa Cruz
Santa Cruz, CA 95064, USA

A. J. Ball
Planetary and Space Sciences Research
  Institute
The Open University
Walton Hall
Milton Keynes, MK7 6AA, UK

Michael J. S. Belton
Belton Space Exploration Initiatives,
  LLC
430 S. Randolph Way
Tucson, AZ 85716, USA

Paul Benoit
Arkansas–Oklahoma Center for Space
  and Planetary Sciences
Chemistry and Biochemistry
  Department
University of Arkansas
Fayetteville, AR 72701, USA

Richard P. Binzel
Massachusetts Institute of Technology
54-410
Cambridge, MA 02139, USA

William F. Bottke, Jr.
Department of Space Studies
Southwest Research Institute
1050 Walnut Street
Boulder, CO 80302, USA

Alberto Cellino
INAF-Osservatorio Astronomico
  di Torino
Strada Osservatorio 20
I-10025 Pino Torinese (TO), Italy

Clark R. Chapman
Department of Space Studies
Southwest Research Institute
1050 Walnut Street
Boulder, CO 80302, USA

Steven R. Chesley
Jet Propulsion Laboratory
MS 301-150
4800 Oak Grove Drive
Pasadena, CA 91109, USA

Bruce A. Conway
Department of Aeronautical and
  Astronautical Engineering
306 Talbot Laboratory
MC-236

University of Illinois
Urbana, IL 61801, USA

Mario Di Martino
INAF-Osservatorio Astronomico
  di Torino
Strada Osservatorio 20
I-10025 Pino Torinese (TO), Italy

Melissa Franzen
Arkansas–Oklahoma Center for Space
  and Planetary Sciences
Chemistry and Biochemistry
  Department
University of Arkansas
Fayetteville, AR 72701, USA

Jon D. Giorgini
Jet Propulsion Laboratory
MS 301-150
4800 Oak Grove Drive
Pasadena, CA 91109, USA

Christian Gritzner
Institute for Aerospace Engineering
Dresden University of Technology
D-01062 Dresden, Germany

Alan W. Harris
DLR Institute of Planetary Research
Rutherfordstrasse 2
D-12489 Berlin, Germany

Keith A. Holsapple
University of Washington
Box 352400
Seattle, WA 98195, USA

Walter F. Huebner
Southwest Research Institute
P. O. Drawer 28510
San Antonio, TX 78228, USA

Robert Jedicke
University of Hawai'i
Institute for Astronomy,
2680 Woodlawn Drive
Honolulu, HI 96822, USA

Ralph Kahle
DLR Institute of Planetary Research
  Rutherfordstrasse 2
D-12489 Berlin, Germany

Mikhail Kareev
Arkansas–Oklahoma Center for Space
  and Planetary Sciences
Chemistry and Biochemistry
  Department
University of Arkansas
Fayetteville, AR 72701, USA

W. Kofman
Laboratoire de Planétologie
Bâtiment D de Physique
B.P. 53
F-38041 Grenoble Cedex 9,
  France

P. Lognonné
Départment des Etudes Spatiales
Institut de Physique du Globe de Paris
  4 Avenue de Neptune
F-94100 Saint Maur des Fossés Cedex,
  France

Shauntae Moore
Arkansas–Oklahoma Center for Space
  and Planetary Sciences
Chemistry and Biochemistry
  Department
University of Arkansas
Fayetteville, AR 72701, USA

Alessandro Morbidelli
Centre National des Recherches
    Scientifiques
Observatoire de la Côte d'Azur
BP 4229
F-06304 Nice Cedex 4, France

Thomas H. Morgan
National Aeronautics and Space
    Administration Headquarters
Code SE
300 E Street, SW
Washington, DC 20546, USA

David Morrison
National Aeronautics and space
    Administration
Ames Research Center
MS 200-7
Moffett Field, CA 94035, USA

Shawn Nichols
Arkansas–Oklahoma Center
    for Space and Planetary
    Sciences
Chemistry and Biochemistry
    Department
University of Arkansas
Fayetteville
AR 72701, USA

Steven J. Ostro
Jet Propulsion Laboratory
MS 300-233
4800 Oak Grove Drive
Pasadena, CA 91109, USA

M. Pätzold
Institut für Geophysik und
    Meteorologie
Universität zu Köln

Albertus-Magnus-Platz
D-50923 Köla, Germany

A. Safaeinili
Jet Propulsion Laboratory
MS 300-319
4800 Oak Grove Drive
Pasadena, CA 91109, USA

Nalin Samarasinha
National Optical Astronomy
    Observatory
950 North Cherry Avenue
Tucson, AZ 85719, USA

Daniel J. Scheeres
Department of Aerospace Engineering
The University of Michigan
Ann Arbor, MI 48109, USA

Derek Sears
Arkansas–Oklahoma Center for Space
    and Planetary Sciences
Chemistry and Biochemistry
    Department
University of Arkansas
Fayetteville, AR 72701, USA

K. Seiferlin
Institut für Planetologie
Westfälische Wilhelms-Universität
    Münster
Wilhelm-Klemm-Straße 10
D-48149 Münster, Germany

Timothy B. Spahr
Center for Astrophysics
Harvard University
60 Garden Street
Cambridge, MA 02138, USA

T. Spohn
Institut für Planetologie

Westfälische Wilhelms-Universität
  Münster
Wilhelm-Klemm-Straße 10
D-48149 Münster, Germany

Duncan Steel
Joule Physics Laboratory
University of Salford
Salford, M5 4WT, UK

James D. Walker
Southwest Research Institute
P. O. Drawer 28510
San Antonio, TX 78228, USA.

Donald K. Yeomans
Jet Propulsion Laboratory
MS 301-150
4800 Oak Grove Drive
Pasadena, CA 91109, USA

Preface

The chapters in this book are based on a series of invited lectures given at the "Workshop on Scientific Requirements for Mitigation of Hazardous Comets and Asteroids," which was held in Arlington, Virginia, on September 3–6, 2002. The focus of the workshop was to determine what needs to be done to ensure that an adequate base of scientific knowledge can be created that will allow efficient development of a reliable, but as yet undefined, collision mitigation system when it is needed in the future.

To achieve this goal essentially all aspects of near-Earth objects were discussed at the workshop, including the completeness of our knowledge about the population of potential impactors, their physical and compositional characteristics, the properties of surveys that need to be done to find hazardous objects smaller than 1 km in size, our theoretical understanding of impact phenomena, new laboratory results on the impact process, the need for space missions of specific types, education of the public, public responsibility for dealing with the threat, and the possible roles in the United States of the National Aeronautics and Space Administration (NASA), the military, and other government agencies in mitigating the threat.

Most of these topics are, we believe, well covered by the material contained within this volume and so it should serve both as a snapshot of the state of the collision hazard issue in the United States in late 2002 and also a useful sourcebook for reference into the associated technical literature. In addition, other opinions and insights into these and other topics discussed at the workshop can be found in a companion volume of extended abstracts that is available on the World Wide Web at http://www.noao.edu/meetings/mitigation/eav.html.

During the workshop it became clear that the prime impediment to further advances in this field is the lack of any assigned responsibility to any national or international governmental organization to prepare for a disruptive collision and the absence of any authority to act in preparation for some future collision mitigation attempt. In addition, some 18 major conclusions were formulated that provided the basis for five recommendations. The full report of the workshop can be downloaded from http://www.beltonspace.com but, in brief, the essence of the recommendations is:

- That, in the United States, NASA be assigned the responsibility to advance this field.
- That a new and adequately funded program be instituted at NASA to create, through space missions and allied research, the specialized knowledge base needed to respond to a future threat of a collision from an asteroid or comet nucleus.
- That the United States sponsored Spaceguard Survey be extended to find the hazardous part of the population of possible impactors down to 200 m in size.
- That the military more rapidly communicate surveillance data on natural airbursts.
- That governmental policy-makers formulate a chain of responsibility for action in the event a threat to the Earth becomes known.

It is our earnest hope that through the publication of this volume interested scientists, engineers, politicians, and governmental managers will find an adequate source of reliable and quantitative information to underwrite the work that needs to be done to achieve the sense of the above recommendations.

Acknowledgments

We would like to thank E. J. Weiler and C. Hartman at the Office of Space Science of the National Aeronautics and Space Administration for providing the primary support for the "Workshop on Scientific Requirements for Mitigation of Hazardous Comets and Asteroids" and underwriting the publication of this book. Other essential support was also received from Science Applications International Corporation, Ball Aerospace and Technologies Corporation, Lockheed Martin Space Systems Company, the National Optical Astronomy Observatory, and the University of Maryland. Thanks are also due to our colleagues R. Greeley, S. Mitton, J. Garget, and E. Yossarian at Cambridge University Press for helping us with the production of this book and to M. A'Hearn and L. Diamond and their colleagues at the University of Maryland who provided logistical support during the workshop. We also thank D. Isbell at the National Optical Astronomy Observatory for organizing press and public outreach activities surrounding the workshop. M. Newhouse, D. Bell, and D. Gasson helped maintain the website and the online abstract submission procedure for the workshop. M. Hanna, P. Marenfeld, and A. Heinrichs helped us with the graphics. S. Ostro and R. Binzel kindly provided the figures for the front and back covers.

All of the chapters were subjected to peer review and we thank the following for providing timely and critical reviews: E. Asphaug, L. Benner, S. Chesley, P. Chodas, E. Clark, D. Crawford, D. R. Davis, D. Dearborn, R. Farquhar, P. Geissler, S. Gulkis, A. W. Harris, W. Huebner, P. Jenniskens, T. Lay, J. Melosh, G. Stokes, P. Thomas, and R. Williams.

Glossary

a   *See* semimajor axis.

ablation   Removal of molten material from a meteor due to heat generated by friction as it travels through the atmosphere.

absolute magnitude   Magnitude (a logarithmic measure of brightness typically in the V spectral band) corrected to a heliocentric and geocentric distance of 1 AU and to a solar phase angle of zero degrees. Note: this definition is different to the common usage of "absolute magnitude" by non-planetary astronomers. *See also magnitude*.

achondrites   A class of stony meteorites without chondrules. They may have non-solar composition.

albedo   Ratio of reflected light to the amount incident. Shiny white objects have high albedos whereas rough black surfaces have low albedos.

aphelion   Furthest point in a heliocentric orbit from the Sun.

aphelion distance   Distance from the Sun to the aphelion, generally denoted by Q.

Apollo asteroids   Asteroids with semimajor axis ≥ 1.0 AU and perihelion distance ≤ 1.0167 AU.

argument of perihelion   Angle along the orbit of an object measured from the ascending node to perihelion.

Amor asteroids   Asteroids with perihelion distance, q, given by 1.0167 AU $< q \leq 1.3$ AU.

ascending node   One of the two points of intersection between an orbit and the ecliptic plane. Of these two, the ascending node corresponds to the point where the object moves north of the ecliptic.

Astronomical Unit (AU)   Mean distance between the Earth and the Sun (149.6 million km).

Aten asteroids   Asteroids with semimajor axis < 1.0 AU and aphelion distance >0.9833 AU.

asteroids   Small rocky objects that generally have orbits between Mars and Jupiter. They are the primary source of meteorites. The largest asteroids are few hundred kilometers in size, but most are much smaller.

binary asteroids   Two asteroids gravitationally bound to each other that orbit around the common center of gravity.

bolides   Bright meteors. Typically brighter than the brightest stars, they can be as bright as the full moon.

C-class asteroids   A taxonomic class of asteroids common in the outer part of the main belt. They are often presumed to have a similar surface composition to carbonaceous chondrites.

carbonaceous chondrites   Chondritic meteorites with abundant carbon. They appear to be the most primitive type of meteorite.

charge-coupled device (CCD)   Detector used to capture images in visible light. A CCD is a solid state electronic device made out of a thin wafer of silicon. Its active area is organized into an array of pixels (picture elements).

Chicxulub crater   A large crater situated near Mexico's Yucatan peninsula. Widely thought to be the impact crater associated with the K/T boundary mass extinction.

chondrites   A class of stony meteorites containing condrules. With the exception of volatile elements their composition is similar to that of the Sun.

chondrules   Roughly spherical objects found in the interior of chondritic meteorites. Chondrules are thought to have originated as molten droplets floating freely in the solar nebula.

comet   A solar system object that is distinguished by a diffuse appearance and often a tail of dust and gas. As a comet comes close to the Sun, increased sublimation of volatiles from the nucleus forms a gravitationally unbound atmosphere (coma) rapidly increasing its brightness.

comet nucleus   The solid part of a comet. A leftover icy planetesimal from the early solar system. The size of a typical nucleus is few kilometers across.

comet Shoemaker-Levy 9   A comet whose nucleus was broken into several pieces due to tides raised by Jupiter's gravitational field. It crashed into the atmosphere of Jupiter in July 1994 generating long-lasting disturbances in the cloud and stratospheric layers.

co-orbital   Objects sharing the same or similar orbits.

Cretaceous/Tertiary boundary (K/T boundary)   The abrupt change in Earth's fossil history in rocks laid down some 65 million years ago. It separates the Cretaceous (K) from the Tertiary (T) geologic periods, and is associated with a great mass extinction of many species including the dinosaurs.

debiased population   Any population (such as the near-Earth objects) corrected for observational biases.

delay-Doppler image   A two-dimensional image constructed using the time delay information in radar reflections from different parts of a solar system object.

Delta v   Change of velocity needed for a specific alteration to the trajectory of a spacecraft.

descending node   One of the two points of intersection between the orbit and the ecliptic plane. The descending node corresponds to the point were the object moves south of the ecliptic.

dielectric constant   Property of a material that describes how it interacts with electromagnetic fields.

Doppler effect   Changes in the wavelength as seen by an observer due to the radial motion of the observer or the source. A source moving away from the observer results in a redshift while a motion towards the observer causes a blueshift.

e   *See* eccentricity.

Earth-crossing asteroid (ECA)   An asteroid in an orbit that crosses the orbit of the Earth.

eccentricity   Measure of the elongation of an ellipse. Eccentricity of a circle is 0 while that of a parabola is 1.

ecliptic comet population   The class of comets with small orbital inclinations and Tisserand parameter larger than 2. These are thought to have originated in the Kuiper belt which exists beyond Neptune's orbit.

equation of state   Relationship that describes the behavior between the physical properties (e.g., pressure, density, and temperature) of a substance.

ESA   European Space Agency.

escape velocity   Minimum velocity required to escape the gravitational field of an object. This is also the minimum impact velocity of an impactor on a direct approach from afar. Earth's escape velocity is about $11.2\ \mathrm{km\,s^{-1}}$.

fireball   *See* bolides.

gravity regime   The situation in which the formation of a crater is controlled by the gravity of the target rather than its material strength.

H   *See* absolute magnitude.

Halley-type comets   Comets in isotropic orbits with orbital periods less than about 200 years and Tisserand parameter less than 2. These are thought to have originated in the Oort cloud.

Hohmann transfer   Most fuel efficient intermediate transfer orbit when moving from one orbit to another.

hopping   Movement with a light jump or skip. A spacecraft maneuver proposed for low-gravity environments.

hovering   The act of remaining or lingering above the surface. A spacecraft maneuver proposed for low-gravity environments.

hypersonic   Equal to or larger than five times the speed of sound in air. Supersonic refers to speeds greater than that of sound in air.

IAU   International Astronomical Union.

i   *See* inclination.

inclination   Angle between the orbital plane and the ecliptic plane.

interior-to-the-Earth objects (IEOs)   Objects with aphelion distances less than 0.983 AU.

interplanetary dust particles (IDPs)   Dust particles in interplanetary space. The origins of IDPs include collisions among asteroids and Kuiper belt objects (KBOs), cometary activity, and dust grains from the interstellar medium.

Jupiter-family comets (JFC)   A subset of ecliptic comet population influenced by the gravity of Jupiter. Tisserand parameter is between 2 and 3. Thought to be of Kuiper belt origin.

K/T boundary   *See* Cretaceous/Tertiary boundary.

Kuiper belt   A large reservoir of cometary nuclei beyond the orbit of Neptune. The first Kuiper belt object (KBO) was discovered in 1992.

longitude of the ascending node   Angle from the first point of Aries (an astronomical reference point) to the ascending node of an orbit.

LINEAR   The Lincoln Near-Earth Asteroid Research program of the Lincoln Laboratory at the Massachusetts Institute of Technology.

lossy medium   Medium in which electromagnetic or elastic wave attenuation occurs.

magnitude   Number expressed on a logarithmic scale to indicate the brightness of a celestial object. A magnitude difference of 5 equals a brightness difference of 100. Brighter objects have lower numerical values for the magnitude. Often magnitude measurements are made with a spectral filter (e.g., U, B, V, R, I) that corresponds to a particular wavelength band of light. *See also* absolute magnitude.

main belt asteroids   Asteroids whose orbits lie between the orbits of Mars and Jupiter.

mass driver   A machine proposed to alter the orbit of asteroids. Reaction forces due to ejection of surface material by the mass driver cause the orbit to change.

meteor   A glowing object in the atmosphere caused as a meteoroid enters the atmosphere and is strongly heated by friction.

meteor shower   An event in which many meteors appear to radiate from the same point in the sky. They occur when the Earth passes through a cometary dust trail.

meteorite   An object of extraterrestrial origin which survives atmospheric entry and reaches the surface of the Earth largely intact.

meteoroid   A small solid extraterrestrial body that becomes a meteor as it enters the Earth's atmosphere.

microgravity conditions   An environment in which the net gravitational force is very small (e.g., on the surfaces of NEOs). Microgravity conditions can be simulated when free-falling.

minimum orbital intersection distance (MOID)   Closest possible approach distance between the osculating orbits of two objects. However, the objects themselves need not be at the closest orbital locations at the same instance.

mitigation   The act of alleviating, e.g., alleviating hazards posed by asteroids and comets that may be on a collision course with Earth.

monolithic asteroid   An asteroid whose structure is that of a single rigid body.

Monte Carlo simulations   Numerical technique that uses random samples and statistical methods to solve physical or mathematical problems.

NASA   National Aeronautics and Space Administration.

near-Earth objects (NEOs)   Those asteroids or comets whose orbits cause them to pass close to Earth or to cross the Earth's orbit. Near-Earth asteroid (NEA) refers specifically to an asteroid with such an orbit.

NEA   *See* near-Earth objects.

nearly isotropic comet population   A sub-population of comets with orbits that are randomly inclined to the ecliptic plane. Tisserand parameter is less than 2. They are thought to be of Oort cloud origin.

NEAT   Near-Earth Asteroid Tracking program at the Jet Propulsion Laboratory in Pasadena, California.

non-gravitational forces   Forces that are not of gravitational origin which act on asteroids and comets and that can sometimes significantly alter their orbits. Examples include reaction forces due to outgassing of volatile materials and the Yarkovsky and Poynting–Robertson forces.

Oort cloud   A large reservoir of cometary nuclei extending from few thousand AU to nearly 100 000 AU from the Sun. The outer part of the Oort cloud is approximately spherical. The Oort cloud has never been observed but its existence is inferred from the orbits of comets with long orbital periods or with nearly parabolic trajectories.

outgassing   The phenomenon of the loss of volatile materials from a solar system body, e.g., the sublimation of water and other gases from cometary nuclei.

Palermo Scale (also called the Palermo Technical Impact Hazard Scale)   A scale developed to enable NEO scientists to categorize and prioritize potential impact risks. It is a logarithmic scale which compares the likelihood of the detected potential impact with the average risk posed by objects of the same size or larger over the years until the date of the potential impact.

perihelion   Closest point on a heliocentric orbit to the Sun.

perihelion distance   Distance from the Sun to the perihelion. Generally denoted by q.

permittivity   A parameter that measures how a material interacts with electro-magnetic fields.

planetesimals   Small rocky or icy bodies formed in the solar nebula and from which larger bodies such as the planets were formed. Some asteroids and cometary nuclei are planetesimals or fragments of planetesimals.

ponds   A localized flat region on an asteroid devoid of craters and rocks (e.g., on asteroid Eros as revealed by the images from NEAR–Shoemaker mission).

porosity   A measure of the empty space in a solid. A high porosity object is less dense.

potentially hazardous objects (PHOs)   Near-Earth objects with minimum orbital intersection distance to Earth of 0.05 AU or less. PHAs refer to PHOs that are specifically asteroids.

Poynting–Robertson effect   A deacceleration and spiraling inward of small dust particles orbiting the Sun due to their interaction with solar radiation.

radar   Acronym for radio detection and ranging. A technique that uses the reflection of transmitted radio waves for detecting and characterizing a remote object.

radiation pressure   Pressure (i.e., force per unit area) exerted by electromagnetic radiation.

regolith   A layer of broken rocks and soil at the surface of an asteroid or a comet generated by collision with objects in space.

resonance   Amplification of the response of a system to an external stimulus that has the same (or commensurable) frequency as the natural frequency of the system.

Roche limit   The maximum distance from a planet where a strengthless satellite will be pulled apart by gravitational tides. If the density of the satellite is the same as the planet, the Roche limit is 2.4 times the radius of the planet.

rubble pile hypothesis   Hypothesis that some asteroids are made up of a loosely held aggregate of collision fragments of different sizes. The aggregate is held together by gravity rather than by cohesive forces.

S-class asteroids   A specific taxonomic class of asteroids. A common asteroid class found in the inner main belt.

scattered disk   Component of the Kuiper disk containing objects with highly eccentric orbits.

seismology   The science of studying sub-surface structures through the propagation of elastic waves.

semimajor axis   Half the distance between perihelion and aphelion. The mean distance between the Sun and an object in a heliocentric orbit.

silicates   Any rock or mineral containing silica (SiO_4).

solar concentrator   A reflector to concentrate solar radiation on to an NEO and vaporize the surface material at the heated spot. The resultant thrust caused by outgassing causes the NEO orbit to change.

solar electric propulsion   A technology that involves the acceleration of an ionized gas (e.g., xenon) to propel a spacecraft. It provides low thrust over extended periods and is sometimes preferred over conventional chemical propulsion in missions of high complexity. Also known as ion propulsion.

solar phase angle   Angle subtended at an object by the directions to the Sun and the Earth. Sometimes called the phase angle.

solar sail   A large ultralight sail that uses solar radiation pressure to provide an acceleration.

Spaceguard goal   The goal of discovering 90% of NEOs larger than 1 k in diameter by 2008.

strength regime   The situation in which the formation of a crater is dependent on the strength of the target rather than on its gravity.

synchronous orbit   An orbit in which an orbiting body has an orbital period equal to the mean rotation period of the body being orbited (with the same direction as of rotation).

taxonomy   Classification system based on observed properties; e.g., for asteroids their colors, spectral properties, and albedos are used.

thermal conductivity   Measure of the ability of a material to conduct heat.

thermal inertia   Measure of the ability of a material to conduct and store heat.

thermal-IR   Region of the electromagnetic spectrum that corresponds to the range of dominant wavelengths of a room-temperature black body. The range of wavelengths cover approximately from 3 to 100 μm.

three-sigma error (or uncertainty)   Statistical expression of error; e.g., for a Gaussian distribution, the probability of the value of a parameter being within its three-sigma error is 99.7%.

thrusters   A propulsion device.

tides   Surface deformations caused by the differential gravitational forces acting on a celestial body.

Tisserand parameter   A conserved quantity in the restricted circular three-body problem. Tisserand parameter with respect to Jupiter is widely used to classify asteroids and comets from a dynamical perspective.

tomography   A technique by which the interior structure is determined by the refection or transmission of elastic or electromagnetic waves.

Torino Scale   A scale for categorizing the Earth impact hazard due to asteroids and comets. This is intended as a tool to communicate with the public in contrast to the more technical Palermo Scale.

TNT   Unit of energy associated with atomic explosions. A 1 megaton TNT explosion is equivalent to 4.18×10^{15} J.

tsunami   A huge sea wave caused by a sudden disturbance such as an earthquake, a landslide, or the collision of an asteroid into the ocean.

Tunguska event   An explosion over Tunguska, Siberia, in 1908 that was caused by the collision and disruption of a small asteroid or a comet nucleus in the atmosphere.

volatiles   Chemical species that evaporate at low temperatures.

Yarkovsky effect   Acceleration due to recoil force caused by the thermal reradiation from an irregular object. This can cause substantial changes to asteroidal orbits and also affect their rotational states.

1

Recent progress in interpreting the nature of the near-Earth object population

William F. Bottke, Jr.
Southwest Research Institute

Alessandro Morbidelli
Observatoire de la Côte d'Azur

Robert Jedicke
University of Hawai'i

1 Introduction

Over the last several decades, evidence has steadily mounted that asteroids and comets have impacted the Earth over solar system history. This population is commonly referred to as "near-Earth objects" (NEOs). By convention, NEOs have perihelion distances $q \leq 1.3$ AU and aphelion distances $Q \geq 0.983$ AU (e.g., Rabinowitz et al. 1994). Sub-categories of the NEO population include the Apollos ($a \geq 1.0$ AU; $q \leq 1.0167$ AU) and Atens ($a < 1.0$ AU; $Q \geq 0.983$ AU), which are on Earth-crossing orbits, and the Amors (1.0167 AU $< q \leq 1.3$ AU) which are on nearly-Earth-crossing orbits and can become Earth-crossers over relatively short timescales. Another group of related objects that are not yet been considered part of the "formal" NEO population are the IEOs, or those objects located inside Earth's orbit ($Q < 0.983$ AU). To avoid confusion with standard conventions, we treat the IEOs here as a population distinct from the NEOs. The combined NEO and IEO populations are comprised of bodies ranging in size from dust-sized fragments to objects tens of kilometers in diameter (Shoemaker 1983).

It is now generally accepted that impacts of large NEOs represent a hazard to human civilization. This issue was brought into focus by the pioneering work of Alvarez et al. (1980), who showed that the extinction of numerous species at the Cretaceous–Tertiary geologic boundary was almost certainly caused by the impact of a massive asteroid (at a site later identified with the Chicxulub crater in the Yucatan peninsula) (Hildebrand et al. 1991). Today, the United Nations, the US Congress, the European Council, the UK Parliament, the IAU, NASA, and ESA have all made official statements that describe the importance of studying

---

Mitigation of Hazardous Comets and Asteroids, ed. M. J. S. Belton, T. H. Morgan, N. H. Samarasinha, and D. K. Yeomans. Published by Cambridge University Press. © Cambridge University Press 2004.

and understanding the NEO population. In fact, among all worldwide dangers that threaten humanity, the NEO hazard may be the easiest to cope with, provided adequate resources are allocated to identify all NEOs of relevant size. Once we can forecast potential collisions between dangerous NEOs and Earth, action can be taken to mitigate the potential consequences.

In this chapter, we review the progress that has been made over the last several years by our team to understand the NEO population, to quantify the collision hazard, and to determine the possibility that existing or near-future surveys may detect/catalog all dangerous NEOs. As such, we employ theoretical and numerical models that can be used to estimate the NEO orbital and size distributions. Our model results are constrained by the observational efforts of numerous NEO surveys, which constantly scan the skies for as yet unknown objects. The work presented here is primarily based on three papers: Bottke *et al.* (2002a), Morbidelli *et al.* (2002a), and Jedicke *et al.* (2003). The interested reader should also examine Morbidelli *et al.* (2002b).

2 Dynamical origin of NEOs

2.1 The near-Earth asteroid population

Dynamical studies over the last several decades have shown that asteroids located in the main belt between the orbits of Mars and Jupiter can reach planet-crossing orbits by increasing their orbital eccentricity under the action of a variety of resonant phenomena (e.g., J. G. Williams, see Wetherill 1979; Wisdom 1983). (Main belt asteroids are believed to enter resonances via the thermal drag forces called the Yarkovsky effect; for a review, see Bottke *et al.* 2002b). Here we classify resonances according to two categories: "powerful resonances" and "diffusive resonances," with the former distinguished from the latter by the existence of associated gaps in the main belt asteroid semimajor axis, eccentricity, and inclination (a, e, i) distribution. A gap is formed when the timescale over which a resonance is replenished with asteroidal material is far longer than the timescale which resonant asteroids are transported to the NEO region. The most notable resonances in the "powerful" class are the v_6 secular resonance at the inner edge of the asteroid belt and several mean motion resonances with Jupiter (e.g., 3:1, 5:2, and 2:1 at 2.5, 2.8, and 3.2 AU respectively). Because the 5:2 and 2:1 resonances push material onto Jupiter-crossing orbits, where they are quickly ejected from the inner solar system by a close encounter with Jupiter, numerical results suggest that only the first two resonances are important delivery pathways for NEOs (e.g., Bottke *et al.* 2000, 2002a). For this reason, we focus our attention here on the properties of the v_6 and 3:1 resonances.

2.2 The υ_6 resonance

The υ_6 secular resonance occurs when the precession frequency of the asteroid's longitude of perihelion is equal to the sixth secular frequency of the planetary system. The latter can be identified with the mean precession frequency of Saturn's longitude of perihelion, but it is also relevant in the secular oscillation of Jupiter's eccentricity (see Morbidelli 2001: ch. 7). The υ_6 resonance marks the inner edge of the main belt. In this region, asteroids have their eccentricity increased enough to reach planet-crossing orbits. The median time required to become an Earth-crosser, starting from a quasi-circular orbit, is about 0.5 My. Accounting for their subsequent evolution in the NEO region, the median lifetime of bodies started in the υ_6 resonance is $\sim$2 My, with typical end-states being collision with the Sun (80% of the cases) and ejection onto hyperbolic orbit via a close encounter with Jupiter (12%) (Gladman *et al.* 1997). The mean time spent in the NEO region is 6.5 My, longer than the median time because υ_6 bodies often reach $a < 2$ AU orbits where they can reside for tens of millions of years (Bottke *et al.* 2002a). The mean collision probability of objects from the υ_6 resonance with Earth, integrated over their lifetime in the Earth-crossing region, is $\sim$0.01 (Morbidelli and Gladman 1998).

2.3 The 3:1 resonance

The 3:1 mean motion resonance with Jupiter occurs at $\sim$2.5 AU, where the orbital period of the asteroid is one-third of that of the giant planet. The resonance width is an increasing function of the eccentricity (about 0.02 AU at $e = 0.1$ and 0.04 AU at $e = 0.2$), while it does not vary appreciably with the inclination. Inside the resonance, one can distinguish two regions: a narrow central region where the asteroid eccentricity has regular oscillations that bring them to periodically cross the orbit of Mars, and a larger border region where the evolution of the eccentricity is wildly chaotic and unbounded, so that the bodies can rapidly reach Earth-crossing and even Sun-grazing orbits. Under the effect of Martian encounters, bodies in the central region can easily transit to the border region and be rapidly boosted into the NEO space (see Morbidelli: 2001 ch. 11). For a population initially uniformly distributed inside the resonance, the median time required to cross the orbit of the Earth is $\sim$1 My, and the median lifetime is $\sim$2 My. Typical end-states for test bodies including colliding with the Sun (70%) and being ejected onto hyperbolic orbits (28%) (Gladman *et al.* 1997). The mean time spent in the NEO region is 2.2 My (Bottke *et al.* 2002a), and the mean collision probability with the Earth is $\sim$0.002 (Morbidelli and Gladman 1998).

The diffusive resonances are so numerous that they cannot be effectively enumerated. Therefore, we only discuss their generic dynamical effects below.

2.4 Diffusive resonances

In addition to the few wide mean motion resonances with Jupiter described above, the main belt is also crisscrossed by hundreds of thin resonances: high-order mean motion resonances with Jupiter (where the orbital frequencies are in a ratio of large integer numbers), three-body resonances with Jupiter and Saturn (where an integer combination of the orbital frequencies of the asteroid, Jupiter, and Saturn is equal to zero) (Nesvorny *et al.* 2002), and mean motion resonances with Mars (Morbidelli and Nesvorny 1999). The typical width of each of these resonances is of the order of a few 10^{-4}–10^{-3} AU.

Because of these resonances, many, if not most, main belt asteroids are chaotic (e.g., Nesvorny *et al.* 2002). The effect of this chaoticity is very weak, with an asteroid's eccentricity and inclination slowly changing in a secular fashion over time. The time required to reach a planet-crossing orbit (Mars-crossing in the inner belt, Jupiter-crossing in the outer belt) ranges from several 10^7 years to billions of years, depending on the resonances and the starting eccentricity. Integrating real objects in the inner belt ($2 < a < 2.5$ AU) for 100 My, Morbidelli and Nesvorny (1999) showed that chaotic diffusion drives many main belt asteroids into the Mars-crossing region. The flux of escaping asteroids is particularly high in the region adjacent to the υ_6 resonance, where effects from this resonance combine with the effects from numerous Martian mean motion resonances.

It has been shown that the population of asteroids solely on Mars-crossing orbits, which is roughly four times the size of the NEO population, is predominately resupplied by diffusive resonances in the main belt (Migliorini *et al.* 1998; Morbidelli and Nesvorny 1999; Michel *et al.* 2000; Bottke *et al.* 2002a). We call this region the "intermediate-source Mars-crossing region," or IMC for short. To reach an Earth-crossing orbit, Mars-crossing asteroids random walk in semimajor axis under the effect of Martian encounters until they enter a resonance that is strong enough to further decrease their perihelion distance below 1.3 AU. The mean time spent in the NEO region is 3.75 My (Bottke *et al.* 2002a).

The paucity of observed Mars-crossing asteroids with $a > 2.8$ AU is not due to the inefficiency of chaotic diffusion in the outer asteroid belt, but is rather a consequence of shorter dynamical lifetimes within the vicinity of Jupiter. For example, Morbidelli and Nesvorny (1999) showed that the outer asteroid belt – more specifically the region between 3.1 and 3.25 AU – contains numerous high-order mean motion resonances with Jupiter and three-body resonances with Jupiter and Saturn, such that the dynamics are chaotic for $e > 0.25$. To investigate this, Bottke *et al.* (2002a) integrated nearly 2000 observed main belt asteroids with $2.8 < a < 3.5$ AU, $i < 15°$, and $q < 2.6$ AU for 100 My. They found that ~20% of them entered the NEO region. Accordingly, they predicted that, in a steady-state scenario, the outer main belt region could provide ~600 new NEOs per My, but the mean time that these bodies spend in the NEO region was only ~0.15 My.

2.5 Near-Earth comets

Comets also contribute to the NEO population. They can be divided into two groups: those coming from the trans-Neptunian region (the Kuiper belt or, more likely, the scattered disk) (Levison and Duncan 1994; Duncan and Levison 1997) and those coming from the Oort cloud (e.g., Weissman *et al.* 2002). Some NEOs with comet-like properties may also come from the Trojan population as well, though it is believed their contribution is small compared to those coming from the trans-Neptunian region and the Oort cloud (Levison and Duncan 1997). The Tisserand parameter T, the pseudo-energy of the Jacobi integral that must be conserved in the restricted circular three-body problem, has been used in the past to classify different comet populations (e.g., Carusi *et al.* 1987). Writing T with respect to Jupiter, we get (Kresak 1979):

$$T = (a_{JUP}/a) + 2\cos(i)\,[(a/a_{JUP})(1 - e^2)]^{1/2} \qquad (1.1)$$

where a_{JUP} is the semimajor axis of Jupiter. Adopting the nomenclature provided by Levison (1996), we refer to $T > 2$ bodies as ecliptic comets, since they tend to have small inclinations, and $T < 2$ bodies as nearly isotropic comets, since they tend to have high inclinations.

Numerical simulations suggest that comets residing in particular parts of the trans-Neptunian region are dynamically unstable over the lifetime of the solar system (e.g., Levison and Duncan 1997; Duncan and Levison 1997). Those ecliptic comets that fall under the gravitational sway of Jupiter ($2 < T < 3$) are called Jupiter-family comets (JFCs). These bodies frequently experience low-velocity encounters with Jupiter. Though most model JFCs are readily thrown out of the inner solar system via a close encounter with Jupiter (i.e., over a timescale of ~0.1 My), a small component of this population achieves NEO status (Levison and Duncan 1997). The orbital distribution of the ecliptic comets has been well characterized using numerical integrations by Levison and Duncan (1997), who find that most JFCs are confined to a region above $a = 2.5$ AU. To create comets with orbits like 2P Encke, which have $T > 3$, it may be necessary to invoke non-gravitational forces.

Nearly isotropic comets, comprised of the long-period comets and the Halley-type comets, come from the Oort cloud (Weissman *et al.* 2002) and possibly the trans-Neptunian region (Levison and Duncan 1997; Duncan and Levison 1997). Numerical work has shown that nearly isotropic comets can be thrown into the inner solar system by a combination of stellar and galactic perturbations (Duncan *et al.* 1987). At this time, however, we lack a complete understanding of their dynamical source region (e.g., Levison *et al.* 2001).

To understand the population of ecliptic comets and nearly isotropic comets, we need to understand more than cometary dynamics. Comets undergo physical evolution as they orbit close to the Sun. In some cases, active comets evolve into dormant, asteroidal-appearing objects, with their icy surfaces covered by a lag deposit of non-volatile dust grains, organics, and/or radiation processed material which prevents volatiles from sputtering away (e.g., Weissman *et al.* 2002). Accordingly, if a $T < 3$ object shows no signs of cometary activity, it is often assumed to be a dormant or possibly extinct comet. In other cases, comets self-destruct and totally disintegrate (e.g., comet C/1999 S4 (LINEAR)). The fraction of comets that become dormant or disintegrate among the ecliptic and nearly isotropic comet populations must be understood to gauge the true impact hazard to the Earth. These issues will be discussed in greater detail below.

2.6 Evolution in NEO space

The dynamics of bodies in NEO space is strongly influenced by a complicated interplay between close encounters with the planets and resonant dynamics. Encounters provide an impulse velocity to the body's trajectory, causing the semimajor axis, eccentricity, and inclination to change by a quantity that depends on both the geometry of the encounter and the mass of the planet. Resonances, on the other hand, keep the semimajor axis constant while changing a body's eccentricity and/or inclination.

In general, NEOs with $a \sim 2.5$ AU or smaller do not approach Jupiter even at $e \sim 1$, so that they end their evolution preferentially by an impact with the Sun. Particles that are transported to low semimajor axes ($a < 2$ AU) and eccentricities have dynamical lifetimes that are tens of millions of years long (Gladman *et al.* 1997) because there are no statistically significant dynamical mechanisms to pump up eccentricities to Sun-grazing values. To be dynamically eliminated, the bodies in the evolved region must either collide with a terrestrial planet (rare), or be driven back to $a > 2$ AU, where powerful resonances can push them into the Sun. Bodies that become NEOs with $a > 2.5$ AU, on the other hand, are preferentially transported to the outer solar system or are ejected onto hyperbolic orbit by close encounters with Jupiter. This shorter lifetime is compensated by the fact that these objects are constantly resupplied by fresh main belt material and newly arriving Jupiter-family comets.

3 Quantitative modeling of the NEO population

The observed orbital distribution of NEOs is not representative of the real distribution, because strong biases exist against the discovery of objects on some types of orbits. Given the pointing history of a NEO survey, the observational bias for a body with a given orbit and absolute magnitude can be computed as the probability of being in the field of view of the survey with an apparent magnitude brighter than the limit of detection (Jedicke 1996; Jedicke and Metcalfe 1998; see review in Jedicke *et al.* 2002). Assuming random angular orbital elements of NEOs, the bias is a function $B(a, e, i, H)$, dependent on semimajor axis, eccentricity, inclination, and on the absolute magnitude H. Each NEO survey has its own bias. Once the bias is known, in principle the real number of objects N can be estimated as

$$N(a, e, i, H) = n(a, e, i, H)/B(a, e, i, H) \qquad (1.2)$$

where n is the number of objects detected by the survey. The problem, however, is that we rarely have enough observations to obtain more than a coarse understanding of the debiased NEO population (i.e., the number of bins in a four-dimensional orbital-magnitude space can grow quite large), though such modeling efforts can lead to useful insights (Rabinowitz 1994; Rabinowitz *et al.* 1994; Stuart 2001).

An alternative way to construct a model of the real distribution of NEOs relies on dynamics. Using numerical integration results, it is possible to estimate the steady-state orbital distribution of NEOs coming from each of the main source regions defined above. Here we describe the method used by Bottke *et al.* (2002a). First, a statistically significant number of particles, initially placed in each source region, is tracked across a network of (a, e, i) cells in NEO space until they are dynamically eliminated. The mean time spent by these particles in those cells, called their residence time, is then computed. The resultant residence time distribution shows where the bodies from the source statistically spend their time in the NEO region. As it is well known in statistical mechanics, in a steady-state scenario, the residence time distribution is equal to the relative orbital distribution of the NEOs that originated from the source. This allowed Bottke *et al.* (2002a) to obtain steady-state orbital distributions for NEOs coming from five prominent NEO sources: the υ_6 resonance, the 3:1 resonance, the IMC population (which is a clearinghouse for all of the diffusive resonances in the main belt up to $a = 2.8$ AU), the outer main belt, which includes numerous powerful and diffusive resonances between 2.8 and 3.5 AU, and the Jupiter-family comets. The overall NEO orbital distribution was then constructed as a linear combination of these five distributions, with the contribution of each source dependent on a weighting function. The nearly isotropic comet population was excluded in this model, but its contribution will be discussed below.

The NEO magnitude distribution, assumed to be source-independent, was constructed so its shape could be manipulated using an additional parameter. Combining the resulting NEO orbital-magnitude distribution with the observational biases associated with the Spacewatch survey (Jedicke 1996), Bottke *et al.* (2002a) obtained a model distribution that could be fit to the orbits and magnitudes of the NEOs discovered or accidentally re-discovered by Spacewatch. A visual comparison showed that the best-fit model adequately matched the orbital-magnitude distribution of the observed NEOs. The fitting procedure for the determination of the parameters was improved by considering additional constraints on the ratios among the populations in the NEO region, in the IMC region, and in the considered portion of the outer belt. The resulting best-fit model nicely matches the distribution of the NEOs observed by Spacewatch (see Bottke *et al.* 2002a: Fig. 10).

Note that once the values of the parameters of the model are computed by best-fitting the observations of *one* survey, the steady state orbital-magnitude distribution of the *entire* NEO population is determined. This distribution is valid also in regions of orbital space that have never been sampled by any survey because of extreme observational biases. This underlines the power of the dynamical approach for debiasing the NEO population.

4 The debiased NEO population

This section is strongly based on the results of the modeling effort by Bottke *et al.* (2002a). Unless explicitly stated, all numbers reported below are taken from that work.

The total NEO population contains 960 ± 120 objects with absolute magnitude $H < 18$ (roughly 1 km in diameter) and with $a < 7.4$ AU. These results are consistent with other recent estimates (Rabinowitz *et al.*, 2000; D'Abramo *et al.*, 2001; Stuart, 2001). Current observational completeness of this population is ~ 55–60%. The NEO absolute magnitude distribution in the range $13 < H < 22$ is $N(H) = 13.9 \times 10^{0.35(H-13)} dH$, implying $24\,500 \pm 3000$ NEOs with $H < 22$; the error bar on the exponent is ± 0.02. Assuming that the albedo distribution is not dependent on H, this magnitude distribution implies a cumulative size distribution with exponent -1.75 ± 0.1. This distribution agrees with estimates obtained by Rabinowitz *et al.* (2000), who directly debiased the magnitude distribution observed by the NEAT (New Earth Asteroid Tracking) survey. Also, it is consistent with the crater size distributions of young surfaces on Venus, Earth, Mars, and the Moon.

The Bottke *et al.* (2002a) model implies that $37 \pm 8\%$ of the NEOs come from the υ_6 resonance, $25 \pm 3\%$ from the IMC population, $23 \pm 9\%$ from the 3:1 resonance, $8 \pm 1\%$ from the outer belt population, and $6 \pm 4\%$ from the Transneptunian region. Thus, the long-debated cometary contribution to the NEO population from the Jupiter-family comets does not exceed 10%. Note that the Bottke *et al.* model was constrained in the JFC region by several objects that are almost certainly dormant comets. For this reason, factors that have complicated discussions of previous JFC population estimates (e.g., issues of converting cometary magnitude to nucleus diameters, etc.) are avoided. Note, however, that the Bottke *et al.* model does not

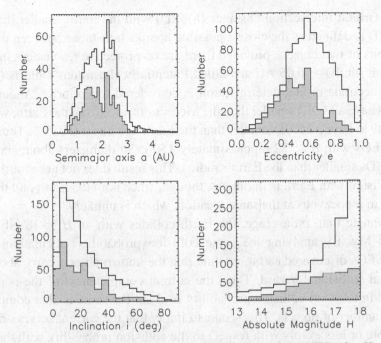

Figure 1.1 The debiased orbital and size distribution of the NEOs for $H < 18$. The predicted NEO distribution (dark solid line) is normalized to 960 NEOs. It is compared with the 426 known NEOs (as of December 2000) from all surveys (shaded histogram). NEO observational completeness was ~44% at the time this plot was created. Most discovered objects have low e and i. From Bottke *et al.* (2002a).

account for the contribution of comets of Oort cloud origin. This issue will be discussed below.

The debiased (a, e, i, H) distribution of the NEOs with $H < 18$ is shown in Fig. 1.1 as a series of four one-dimensional plots (see Bottke *et al.* (2002a) for other representations of this data). For comparison, the figure also reports the distribution of the objects discovered up to $H < 18$, all surveys combined. As one sees, most of the NEOs that are still undiscovered have H larger than 16, e larger than 0.4, a in the range 1–3 AU, and i between 5 and 40 degrees. The populations with $i > 40°$ and $a < 1$ AU or $a > 3$ AU have a larger relative incompleteness, but contain a much more limited number of undiscovered bodies. Of the total NEOs, $32 \pm 1\%$ are Amors, $62 \pm 1\%$ are Apollos, and $6 \pm 1\%$ are Atens; and $49 \pm 4\%$ of the NEOs should be in the evolved region ($a < 2$ AU), where the dynamical lifetime is strongly enhanced. As far as the objects inside Earth's orbit, or IEOs, the ratio between the IEO and the NEO populations is about 2%. Thus, there are only about 20 IEOs with $H < 18$.

With this orbital distribution, and assuming random values for the argument of perihelion and the longitude of node, about 21% of the NEOs turn out to have a

Minimal Orbital Intersection Distance (MOID) with the Earth smaller than 0.05. The MOID is defined as the closest possible approach distance between the osculating orbits of two objects, provided there are no protective resonances in action. NEOs with MOID < 0.05 AU are called Potentially Hazardous Objects (PHOs), and their accurate orbital determination is considered top priority. About 1% of the NEOs have a MOID smaller than the Moon's distance from the Earth, while the probability to have a MOID smaller than the Earth's radius is 0.025%. Thus, of the 24 500 NEOs with $H < 22$ (approximately 150 m in diameter) about six should have MOIDs smaller than the Earth's radius. This result does not necessarily imply that a collision with Earth is imminent, though, since both the Earth and the NEO still need to rendezvous at the same location, which is unlikely.

We estimate that, on average, the Earth collides with an $H < 18$ NEO once every 0.5 My. By applying the same collision probability calculations to the $H < 18$ NEOs discovered so far, we find that the known objects carry about 47% of the total collisional hazard. Thus, the current completeness of the population computed in terms of collision probability is about the same as that computed in terms of number of objects. This seems to imply that the current surveys discover NEOs more or less evenly with respect to the collision probability with the Earth. The most dangerous and still largely undetected NEO sub-population is that with $0.8 < a < 2.2$ AU and $i < 15°$, with little dependence on the eccentricity.

5 Nearly isotropic comets

We now come to the issue of the contribution of nearly isotropic comets (NICs) to the NEO population (and the terrestrial impact hazard). Dynamical explorations of the orbital distribution of the nearly isotropic comets (Wiegert and Tremaine 1999; Levison *et al.* 2001) indicate that, in order to explain the orbital distribution of the observed population, NICs need to rapidly "fade" (i.e., become essentially unobservable). In other words, physical processes are needed to hide some fraction of the returning NICs from view. One possible solution to this so-called "fading problem" would be to turn bright active comets into dormant, asteroidal-appearing objects with low albedos. If most NICs become dormant, the potential hazard from these objects could be significant. An alternative solution would be for cometary splitting events to break comets into smaller (and harder-to-see) components. If most returning NICs disrupt, the hazard to the Earth from the NIC population would almost certainly be smaller than that from the near-Earth asteroid population.

To explore this issue, Levison *et al.* (2002) took several established comet dynamical evolution models of the NIC population (Wiegert and Tremaine 1999; Levison *et al.* 2001), created fake populations of dormant NICs from these models, and ran these fake objects through a NEO survey simulator that accurately mimics the performance of various NEO surveys (e.g., LINEAR, NEAT) over a time period stretching from 1996 to 2001 (Jedicke *et al.* 2003). Levison *et al.* (2001) then compared their model results to the observed population of dormant comets found over the same time period. For example, the survey simulator discovered one out of every 22 000 dormant NICs with orbital periods > 200 years, $H < 18$, and perihelion $q < 3$ AU. This result, combined with the fact that only two dormant objects with comparable parameters had been discovered between 1996 and 2001, led them to predict that there are total of 44 000 ± 31 000 dormant NICs with orbital periods $P > 200$ years, $H < 18$, and perihelion $q < 3$ AU.

Levison *et al.* (2002) then used these values to address the fading problem by comparing the total number of fake dormant NICs discovered between 1996 and 2001 to the observed number. The results indicated that dynamical models that fail to destroy comets over time produce ∼100 times more dormant NICs than can be explained by current NEO survey observations. Hence, to resolve this paradox, Levison *et al.* (2002) concluded that, as comets evolve inward from the Oort cloud, the vast majority of them must physically disrupt.

Assuming there are 44 000 dormant comets with $P > 200$ years, $H < 18$, and perihelion $q < 3$ AU, Levison *et al.* (2002) estimated that they should strike the Earth once per 370 My. In contrast, the rate that active comets with $P > 200$ years strike the Earth (both new and returning) is roughly once per 32 My (Weissman 1990; Morbidelli 2001). For comets with $P < 200$ years, commonly called Halley-type comets (HTCs), Levison *et al.* (2002) estimate there are 780 ± 260 dormant objects with $H < 18$ and $q < 2.5$ AU. This corresponds to an Earth impact rate of once per 840 My. Active HTCs strike even less frequently, with a rate corresponding to once per 3500 My (Levison *et al.* 2001, 2002). Hence, since all of these impact rates are much smaller than that estimated for $H < 18$ NEOs (one impact per 0.5 My), we conclude that NICs currently represent a tiny fraction of the total impact hazard.

In summary, the consequences of the results described above are that asteroids rather than comets provide most of the present-day impact hazard. The only cautionary note that we raise on this point is that the modeling described above does not yet account for comet showers, which are presumably caused when perturbations from passing stars disturb the Oort cloud and send numerous impactors toward the inner solar system. We believe that comet showers may have provided a significant number of large terrestrial impactors over solar system history, but that comet showers are probably not relevant for the study of the impact hazard over at least the next few hundred years.

6 The albedo distribution of the NEOs

Like the main belt asteroids, NEOs show a variety of taxonomic classes, which depend on their chemical and physical composition. Some of them (mostly S-class) have high albedos, while others (mostly C-class) have a low reflectivity. We can categorize the NEOs as "bright" and "dark" bodies, according to their albedo being higher or lower than 0.089, a threshold defined by Tedesco (1994) as the minimum of the bimodal IRAS asteroid albedo histogram, which effectively separates C-class from S-class asteroids. In terms of taxonomic data, following the work of Bus (2002), we consider that the bodies of class A, K, L, O, Q, R, S, V, Xe, Xk, are "bright" while the bodies of class B, C, D, T, Xc are "dark."

At this time, the debiased proportion of dark and bright bodies in the NEO population is unknown. Of the ∼1500 NEOs discovered so far, taxonomic data have only been published for ∼300 (Binzel *et al.* 2002). The albedo or taxonomic distribution observed in this small sample is almost certainly biased enough that it cannot be considered representative of the distribution of the whole population (Morbidelli *et al.* 2002a). Determining the debiased dark/bright ratio, however, is not just a matter of curiosity. Understanding the distribution of NEO albedos would allow us to convert this H distribution into a size distribution, and ultimately into a mass distribution. This last conversion is now possible thanks to the sharp correlation between bulk density and taxonomic class (for a review of asteroid bulk density measurements, see Britt *et al.* 2002). In essence, the bulk density of bright S-class asteroids is ∼2.7 g cm^{-3}, while that of dark C-class bodies is ∼1.3 g cm^{-3}. Ultimately, the coupled knowledge of the NEOs' mass distribution and orbital distribution would allow us to compute the collision rate as a function of collision energy, on the Earth and on the other planets. This work is essential to interpreting the cratering record of planetary surfaces as well as to quantify the hazard represented by the NEO population for our civilization.

To debias the albedo distribution of the NEOs, Morbidelli *et al.* (2002a) took advantage of the NEO orbital-magnitude distribution described above. Recall that this NEO model uses five intermediate sources to characterize the NEO population: the ν_6 resonance, the 3:1 resonance, the outer portion of the main belt (i.e., 2.8-3.5 AU), the Mars-crossing population adjacent to the main belt, and the Jupiter-family comet population. The model also establishes the relative contribution of these sources to the NEO population. By computing the albedo distribution of the bodies in/near each of the five sources, Morbidelli *et al.* (2002a) deduced the albedo distribution of the NEO population as a function of semimajor axis, eccentricity, and inclination. A problem with this strategy, however, is that the albedo distribution of main belt asteroids over the same size range as the observed NEOs (diameter $D <$ 10 km) is not currently known. For this reason, Morbidelli *et al.* (2002a) determined the albedo distribution of large asteroids in and/or near each NEO source region and used these results to deduce the albedo distribution of smaller asteroids in the same regions. This method required them to make some assumptions about the absolute magnitude distributions of both asteroid families and background asteroids. For example, to understand the contribution of various families to the NEO population, Morbidelli *et al.* (2002a) extrapolated the observed absolute magnitude distributions of the families up to some threshold value H_{thr}, beyond which they assumed that the families' absolute magnitude distributions were background-like.

Morbidelli *et al.* (2002a) found that $H_{thr} = 14.5$ provides the best match to the color vs. heliocentric distance distribution observed by the Sloan Digital Sky Survey (SDSS) (Ivezic *et al.*, 2001). With this value of H_{thr}, their model predicts that the debiased ratio between dark and bright (albedo smaller or larger than 0.089) objects in any *absolute magnitude*-limited sample of the NEO population is 0.25 ± 0.02. Once the observational biases are properly taken into account, this agrees very well with the observed dark/bright NEO ratio (0.165 for $H < 20$). Note that this ratio is skewed towards bright bodies because, for a given H value, there are many more bright asteroids than dark asteroids. The dark/bright ratio of NEOs changes to 0.87 ± 0.05, however, if a *size-limited* sample is considered, because bright asteroids convert to small sizes while dark asteroids convert to large sizes. Morbidelli *et al.* (2002a) estimated that the total number of NEOs larger than 1 km is 855 ± 110 which, when compared to the total number of NEOs with $H < 18$ (960 ± 120) (Bottke *et al.* 2002a), shows that the usually assumed conversion that a $H = 18$ body is approximately 1 km in diameter slightly overestimates the number of kilometer-size objects.

Combining this orbital distribution model with the new albedo distribution model, and assuming that the density of bright and dark bodies is 2.7 and $1.3 \, \mathrm{g \, cm}^{-3}$, respectively, Morbidelli *et al.* (2002a) estimated that the Earth should undergo a 1000-megaton collision every $63\,000 \pm 8000$ years. On average, the bodies capable of producing 1000 megaton of impact energy are those with $H < 20.6$. The NEOs discovered so far carry only $18 \pm 2\%$ of this collision probability.

7 The Spaceguard goal

The realization that globally devastating asteroid or comet impacts occur on timescales relevant to human beings has recently spurred the development of survey systems capable of detecting NEOs which threaten Earth. In this section, we examine the prospects for discovering 90% of all NEOs with absolute magnitude $H < 18$ (roughly $D > 1$ km bodies) by 2008. In May 1998, NASA committed to achieving this level of completion among the 1-km-sized NEOs within 10 years. This target will hereafter be referred to as the "Spaceguard goal" (e.g., Morrison 1992; Morrison *et al.* 2002). The work described here is based on that of Jedicke *et al.* (2003).

Our primary motivation here is to determine whether existing NEO surveys can meet the Spaceguard goal. If they cannot meet this target then it is important to ask to what depth must the surveys scan the sky to achieve the goal in a reasonable time frame? What is "reasonable"? Simple linear extrapolations of NEO discovery rates to the expected total population predict that the Spaceguard goal is achievable; but this assumes that the undiscovered NEOs are as likely to be discovered as the known sample. In the process of answering these questions we will also touch upon the benefits of space-based surveys and examine the merits of multi-hemispheric NEO surveying observatories.

7.1 Modeling the detection of the NEO population

There are two main elements required for simulating the discovery of NEOs: (1) a realistic model of the distribution of their orbital elements and absolute magnitudes and (2) a simulation of the search strategy.

To get the first element, we used the NEO orbital and absolute magnitude model constructed by Bottke *et al.* (2002a). From this model, we generated 15 independent sets of 961 NEOs and IEOs with $H < 18$. Each of the 15 NEO models were generated with different contributions from each of the five source regions consistent with the statistical errors in the Bottke *et al.* (2002a) NEO model. Individual NEOs were generated according to the expected distribution for that particular model. In this way, our ensemble of models samples both the statistical variation of the models and the statistical variation within the models themselves. The three angular elements for the orbit were generated randomly in the interval $(0, 2\pi)$. The slope parameter (G) was fixed at 0.22 which is the number-weighted mean between bright ($G = 0.25$) and dark ($G = 0.15$) NEOs according to Morbidelli *et al.* (2002a).

For the second element, we built a survey simulator that was designed to mimic the behavior of a telescopic survey (with preselected characteristics) that searches the night sky for asteroids and comets. To run the code, we input a set of objects that includes (1) their Keplerian orbital elements, (2) their absolute magnitude, and (3) their slope parameter. We also input characteristics corresponding to the nature of the survey under consideration (e.g., the fields of sky it covers at a given time and the limiting magnitude of the system). In some cases the survey might cover the entire sky at one instant while in others it might observe many small rectangular fields during the course of a night. Similarly, the limiting magnitude of the system may be abrupt (100% to some magnitude and then 0%) or gentle (100% to some magnitude and then decreasing linearly to 0% in a fixed range of magnitudes). The position of objects on the sky at any time is provided by an ephemeris calculator. Each of the 15 sets of generated NEOs were run through our survey simulator in various configurations.

If an object lies in a survey field the code calculates the object's magnitude and the detection efficiency (ε) at that magnitude in order to determine whether a detection is possible. An object is considered detected if a random number r in the range $[0,1)$ is such that $r \leq \varepsilon$. When an object is detected for the first time the simulator outputs a list of discovery characteristics (e.g., rates of motion, apparent magnitude, location with respect to opposition). If the object is detected but has already been discovered in an earlier field the detection is recorded as a re-discovery. We filter potential NEOs from our survey simulator discovery output using criteria similar to those used by contemporary NEO surveys (like thresholds on the candidate's rates of motion). Objects may be missed by the survey for any number of reasons (e.g., objects are far from Earth, they do not reflect enough light, they pass so quickly through the search volume that they streak across multiple pixels of a CCD, reducing their peak-pixel signal and chance of detection). Other code details can be found in Jedicke *et al.* (2003).

Our most realistic simulator was tuned to match the operating performance of an existing wide-field survey (LINEAR) and, in this way, implicitly incorporates all the efficiency-dependent parameters relevant to that survey. To test our code, we examined the survey code's performance against actual LINEAR results. Prior to January 1, 1999, 273 NEOs with $H < 18$ were known. Starting from that date, LINEAR discovered 176 new $H < 18$ NEOs over the next 2 years. We find that our survey simulator takes 2.3 years to increase the number of detected objects from 273 to 449, nearly the same amount of time. The rate of discovery and the absolute magnitude of the detected and accidentally re-discovered objects was also reasonable. For a second test, we compared the orbital and absolute magnitude values of the 544 NEOs values with $H < 18$ that were discovered as of January 1, 2002 with the distribution of the first 544 objects found by our pseudo-LINEAR simulation. A visual comparison of the known and simulated distributions suggest a reasonable agreement between the two distributions (see Jedicke *et al.* 2003).

7.2 Prospects for achieving the Spaceguard goal

Using our survey simulator, we can investigate the prospects for achieving the Spaceguard goal, i.e., discovering 90% of the NEOs with $H < 18$ within the next 6 years (Fig. 1.2). Our results suggest that the LINEAR detector system (as it performed in 1999 and 2000) cannot achieve the Spaceguard goal; because of the difficulty in detecting some hard-to-find objects (e.g., IEOs), this system would not achieve 90% completion to $H = 18$ until the year 2035 ± 5. Over the next 6 years, LINEAR as of 1999–2000 would only reach 70% completeness. It is important to point out, however, that our pseudo-LINEAR survey is *not* the actual LINEAR survey which steadily improves its performance in both area coverage and limiting magnitude. We predict that a survey with limiting magnitude $V = 21.5$ starting today would be needed to achieve the Spaceguard goal.

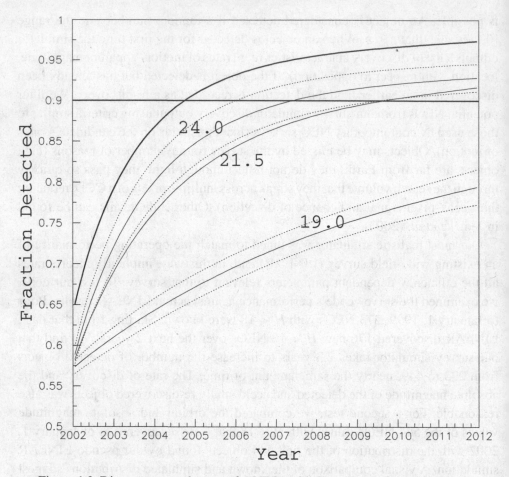

Figure 1.2 Discovery completeness for NEOs with $H \leq 18$ as a function of time for pseudo-LINEAR surveys with limiting magnitudes of $V = 19.0, 21.5$, and 24.0. The dotted lines enveloping the solid curves show the root mean square error in the fraction detected as a function of time. The error on the completion fraction for the $V = 21.5$ line are not shown for the sake of clarity. They are of the same order as the errors on the other two lines. The thick solid line shows the results for an all-sky survey to $V = 21.5$. The solid horizontal line is at 90% completeness. From Jedicke *et al.* (2003).

If time is of the essence, it might be worthwhile to locate the observing platform for NEOs in space. A satellite survey system on a heliocentric orbit interior to Earth could detect IEOs and dangerous Aten class NEOs (with $a < 1.0$ AU and aphelion $Q \geq 0.983$ AU) that are very difficult to detect from Earth. It would also avoid Earth-based disadvantages such as daylight, poor weather, and the inability to look close to the Sun. The primary disadvantage would be its high cost.

Using our survey simulator, we examined what a $V = 19.0$ space-based platform could do, assuming that a band of sky $\pm 20°$ around the ecliptic could be searched

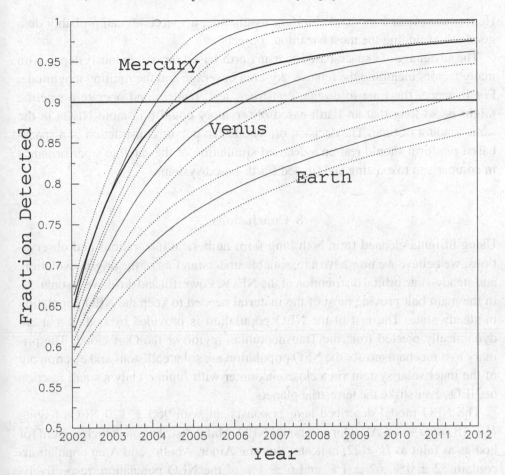

Figure 1.3 Discovery completeness for NEOs with $H \leq 18$ as a function of time for space-based surveys with $V = 19.0$. The surveys take place in a circular orbit at the semimajor axis of each of the three innermost planets and cover the entire night sky each day to within 45° of the Sun. The dotted lines enveloping the solid curves show the root mean square error in the fraction detected as a function of time. The thick solid line shows the results for an all-sky survey to $V = 21.5$. The solid horizontal line is at 90% completeness. From Jedicke *et al.* (2003).

every day, with the exception of a ±40° cone around the Sun. The results for surveys from Earth-, Venus-, and Mercury-like orbits are shown in Fig. 1.3. We find that the key parameter that determined the efficiency of the space survey is the orbit of the satellite, with the survey becoming more effective as the spacecraft approaches the Sun. Interestingly, a survey from inside Earth's orbit, unlike a geocentric or ground-based survey, would preferentially discover the most dangerous NEOs, because these objects, by definition, must pass very close to the ecliptic at 1 AU. If the spacecraft is located inside 1 AU, it can detect these objects close to opposition.

Hence, even with a moderate limiting magnitude, a spacecraft can probably do a good job of finding the most hazardous objects.

The advantage of a space-based over an Earth-based survey obviously depends on many factors including the relative sky-area coverage and the limiting magnitude. For instance, the time to design, construct, test, launch, and operate a satellite might be so long that an Earth-based observatory could find more NEOs in the same amount of time. The decision on whether to pursue construction of a space-based platform should rest on a detailed simulation of the satellite's performance in comparison to existing and funded Earth-based systems.

8 Conclusions

Using insights gleaned from both long-term numerical integrations and observations, we believe we now have a reasonable understanding of the origin, evolution, and steady-state orbital distribution of the NEOs. Powerful and diffusive resonances in the main belt provide most of the material needed to keep the NEO population in steady state. The rest of the NEO population is provided by bodies that are dynamically ejected from the Transneptunian region or the Oort cloud. The primary loss mechanisms for the NEO population are solar collisions and ejection out of the inner solar system via a close encounter with Jupiter. Only a small fraction of NEOs ever strike the terrestrial planets.

The NEO model described here is consistent with 960 ± 120 NEOs having $H < 18$ and $a < 7.4$ AU. Our computed NEO orbital distribution, which is valid for bodies as faint as $H < 22$, indicates that the Amor, Apollo, and Aten populations contain $32 \pm 1\%$, $62 \pm 1\%$, and $6 \pm 1\%$ of the NEO population, respectively. We estimate that the population of objects completely inside Earth's orbit (IEOs) arising from our source regions is 2% the size of the NEO population. Overall, our model predicts that $\sim 61\%$ of the NEO population comes from the inner main belt ($a < 2.5$ AU), $\sim 24\%$ comes from the central main belt ($2.5 < a < 2.8$ AU), $\sim 8\%$ comes from the outer main belt ($a > 2.8$ AU), and $\sim 6\%$ comes from the Jupiter-family comet region ($2 < T < 3$).

Using the data and tools described above, we have estimated the albedo distribution of the NEO population and the approximate collisional hazard for the Earth (i.e., frequency of collisions as a function of the impact energy). These results indicate the Earth should undergo a 1000-megaton collision every $63\,000 \pm 8000$ years. Only a small fraction of all the bodies than can produce such blasts have yet been found. We have also used our model of the orbit and absolute magnitude distribution for NEOs to make realistic simulations of asteroid surveys. We find it unlikely that the Spaceguard goal (discovery of 90% of the $H < 18$ objects by 2008) can be achieved by existing systems (as of 1999–2000). We predict that a survey with limiting magnitude $V = 21.5$ starting in mid-2002 would be needed to achieve the Spaceguard goal. Recent advances in technology for existing surveys and newly proposed observatories, however, suggest the Spaceguard goal may be reached only a few years after 2008.

References

Alvarez, L. W., Asaro, F., and Michel, H. V. 1980. Extraterrestrial cause for the Cretaceous–Tertiary extinction. *Science* **208**, 1095–1099.

Binzel, R. P., Lupishko, D., Di Martino, M., *et al.* 2002. Physical properties of near-Earth objects. In *Asteroids III*, eds. W. F. Bottke, A. Cellino, P. Paolicchi, and R. P. Binzel, pp. 255–271. Tucson, AZ: University of Arizona Press.

Bottke, W. F., Jedicke, R., Morbidelli, A., *et al.* 2000. Understanding the distribution of near-Earth asteroids. *Science* **288**, 2190–2194.

Bottke, W. F., Morbidelli, A., Jedicke, R., *et al.* 2002a. Debiased orbital and absolute magnitude distribution of the near-Earth objects. *Icarus* **156**, 399–433.

Bottke, W. F., Vokrouhlicky, D., Rubincam, D. P., *et al.* 2002b. The effect of Yarkovsky thermal forces on the dynamical evolution of asteroids and meteoroids. In *Asteroids III*, eds. W. F. Bottke, A. Cellino, P. Paolicchi, and R. P. Binzel, pp. 395–408. Tucson, AZ: University of Arizona Press.

Britt, D. T., Yeomans, D., Housen, K., *et al.* 2002. Asteroid density, porosity, and structure. In *Asteroids III*, eds. W. F. Bottke, A. Cellino, P. Paolicchi, and R. P. Binzel, pp. 485–500. Tucson, AZ: University of Arizona Press.

Bus, S. J., Vilas, F., and Barucci, M. A. 2002. Visible-wavelength spectroscopy of asteroids. In *Asteroids III*, eds. W. F. Bottke, A. Cellino, P. Paolicchi, and R. P. Binzel, pp. 169–182. Tucson, AZ: University of Arizona Press.

Carusi, A., Kresak, L., Perozzi, E., *et al.* 1987. High-order librations of Halley-type comets. *Astron. Astrophys.* **187**, 899–905.

D'Abramo, G., Harris, A. W., Boattini, A., *et al.* 2001. A simple probabilistic model to estimate the population of near-Earth asteroids. *Icarus* **153**, 214–217.

Duncan, M. J. and Levison, H. F. 1997. A scattered comet disk and the origin of Jupiter family comets. *Science* **276**, 1670–1672.

Duncan, M., Quinn, T., and Tremaine, S. 1987. The formation and extent of the solar system comet cloud. *Astron. J.* **94**, 1330–1338.

Gladman, B. J., Migliorini, F., Morbidelli, A., *et al.* 1997. Dynamical lifetimes of objects injected into asteroid belt resonances. *Science* **277**, 197–201.

Hildebrand, A. R., Penfield, G. T., Kring, D. A., *et al.* 1991. Chicxulub crater: a possible Cretaceous–Tertiary boundary impact crater on the Yucatan Peninsula. *Geology* **19**, 867–871.

Ivezic, Z. Tabachnik, S., Rafikov, R., *et al.* 2001. Solar system objects observed in the Sloan Digital Sky Survey commissioning data. *Astron. J.* **122**, 2749–2784.

Jedicke, R. 1996. Detection of near-Earth asteroids based upon their rates of motion. *Astron. J.* **111**, 970–982.

Jedicke, R. and Metcalfe, T. S. 1998. The orbital and absolute magnitude distributions of main belt asteroids. *Icarus* **131**, 245–260.

Jedicke, R., Larsen, J., and Spahr, T. 2002. Observational selection effects in asteroid surveys. In *Asteroids III*, eds. W. F. Bottke, A. Cellino, P. Paolicchi, and R. P. Binzel, pp. 71–87, Tucson, AZ: University of Arizona Press.

Jedicke, R., Morbidelli, A., Petit, J.-M., *et al.* 2003. Earth and space-based NEO survey simulations: prospects for achieving the Spaceguard goal. *Icarus* **161**, 17–33.

Kresak, L. 1979. Dynamical interrelations among comets and asteroids. In *Asteroids*, ed. T. Gehrels, pp. 289–309. Tucson, AZ: University of Arizona Press.

Levison, H. F. 1996. Comet taxonomy. In *Completing the Inventory of the Solar System*, eds. T. W. Rettig and J. M. Hahn, ASP Conference Series **107**, 173–191.

Levison, H. F. and Duncan, M. J. 1994. The long-term dynamical behavior of short-period comets. *Icarus* **108**, 18–36.

   1997. From the Kuiper belt to Jupiter-family comets: the spatial distribution of ecliptic comets. *Icarus* **127**, 13–32.

Levison, H. F., Dones, L., and Duncan, M. J. 2001. The origin of Halley-type comets: probing the inner Oort cloud. *Astron. J.* **121**, 2253–2267.

Levison, H. F., Morbidelli, A., Dones, L., *et al.* 2002. The mass disruption of Oort cloud comets. *Science* **296**, 2212–2215.

Michel, P., Migliorini, F., Morbidelli, A., *et al.* 2000. The population of Mars-crossers: classification and dynamical evolution. *Icarus* **145**, 332–347.

Migliorini, F., Michel, P., Morbidelli, A., *et al.* 1998. Origin of Earth-crossing asteroids: a quantitative simulation. *Science* **281**, 2022–2024.

Morbidelli, A. 2001. *Modern Celestial Mechanics: Aspects of Solar System Dynamics*. London, UK: Gordon and Breach.

Morbidelli, A. and Gladman, B. 1998. Orbital and temporal distributions of meteorites originating in the asteroid belt. *Meteor. and Planet. Sci.* **33**, 999–1016.

Morbidelli, A. and Nesvorny, D. 1999. Numerous weak resonances drive asteroids toward terrestrial planets' orbits. *Icarus* **139**, 295–308.

Morbidelli, A., Jedicke, R., Bottke, W. F., *et al.* 2002a. From magnitudes to diameters: the albedo distribution of near-Earth objects and the Earth collision hazard. *Icarus* **158**, 329–342.

Morbidelli, A., Bottke, W. F., Froeschle, C., *et al.* 2002b. Origin and evolution of near-Earth objects. In *Asteroids III*, eds. W. F. Bottke, A. Cellino, P. Paolicchi, and R. P. Binzel, pp. 409–422. Tucson, AZ: University of Arizona Press.

Morrison, D. 1992. *The Spaceguard Survey: Report of the NASA International Near-Earth-Object Detection Workshop*. Washington, DC: NASA.

Morrison, D., Harris, A. W., Sommer, G., *et al.* 2002. Dealing with the impact hazard. In *Asteroids III*, eds. W. F. Bottke, A. Cellino, P. Paolicchi, and R. P. Binzel, pp. 739–754. Tucson, AZ: University of Arizona Press.

Nesvorny, D., Ferraz-Mello, S., Holman, M., *et al.* 2002. Regular and chaotic dynamics in the mean motion resonances: implications for the structure and evolution of the main belt. In *Asteroids III*, eds. W. F. Bottke, A. Cellino, P. Paolicchi, and R. P. Binzel, pp. 379–394. Tucson, AZ: University of Arizona Press.

Rabinowitz, D. L. 1994. The size and shape of the near-Earth asteroid belt. *Icarus* **111**, 364–377.

Rabinowitz, D. L., Bowell, E., Shoemaker, E. M., *et al.* 1994. The population of Earth-crossing asteroids. In *Hazards due to Comets and Asteroids*, ed. T. Gerhels, pp. 285–312. Tucson, AZ: University of Arizona Press.

Rabinowitz, D. L., Helin, E., Lawrence, K., *et al.* 2000. A reduced estimate of the number of kilometre-sized near-Earth asteroids. *Nature* **403**, 165–166.

Shoemaker, E. M. 1983. Asteroid and comet bombardment of the Earth. *Ann. Rev. Earth and Planet. Sci.* **11**, 461–494.

Stuart, J. S. 2001. A near-Earth asteroid population estimate from the LINEAR survey. *Science* **294**, 1691–1693.

Tedesco, E. F. 1994. Asteroid albedos and diameters. In *Asteroids, Comets, Meteorites*, eds. A. Milani, M. DiMartino, and A. Cellino, pp. 55–74. London, UK : Kluwer.

Weissman, P. R. 1990. The cometary impactor flux at Earth. In *Global Catastrophes in Earth History*, eds. V. L. Sharpton and P. D. Ward, pp. 171–180. Boulder, CO: Geological Society of America.

Weissman, P. R., Bottke, W. F., and Levison, H., 2002. Evolution of comets into asteroids. In *Asteroids III*, eds. W. F. Bottke, A. Cellino, P. Paolicchi, R. Binzel, pp. 669–686. Tucson, AZ: University of Arizona Press.

Wetherill, G. W. 1979. Steady state populations of Apollo–Amor objects. *Icarus* **37**, 96–112.

Wiegert, P. and Tremaine, S. 1999. The evolution of long-period comets. *Icarus* **137**, 84–121.

Wisdom, J. 1983. Chaotic behavior and the origin of the 3/1 Kirkwood gap. *Icarus* **56**, 51–74.

2

Earth impactors: orbital characteristics and warning times

Steven R. Chesley

Jet Propulsion Laboratory, California Institute of Technology

Timothy B. Spahr

Center for Astrophysics, Cambridge, MA

1 Introduction

The most important requirement, scientific or otherwise, for any impact mitigation is the recognition of the hazard, since, in the absence of a perceived impact risk, there is neither the incentive nor the capability to address the threat. Therefore, the success of any potential mitigation effort will rely heavily upon our ability to discover, track, and analyze threatening objects. In this chapter we will consider the effectiveness of the present surveying and monitoring capabilities by bombarding the Earth with a large set of simulated asteroids that is statistically similar to the impacting population.

Our objective is to determine where on the sky impactors may most readily be detected by search instruments and to evaluate current search techniques for their effectiveness at detecting asteroids on impact trajectories. We also consider the likelihood that existing survey efforts would find previously undiscovered impactors with just weeks to months of warning time. We discuss the factors that affect whether an impactor detection is actually recognized as a near-Earth asteroid (NEA) discovery and announced to the community for further analysis, including impact monitoring. We close with an example demonstrating how automatic impact monitoring can detect a distant impending impact immediately after discovery, when the impact probability is very low, and how the threat gradually grows more severe during the discovery apparition. In many cases the threat will not be alarming until the object is re-detected at a subsequent apparition. This can substantially diminish the effective warning time, and hence shorten the time available to mitigate the impact.

Mitigation of Hazardous Comets and Asteroids, ed. M. J. S. Belton, T. H. Morgan, N. H. Samarasinha, and D. K. Yeomans. Published by Cambridge University Press. © Cambridge University Press 2004.

2 Derivation of synthetic impactors

We wish to form a large set of "typical" impactors, and for this purpose we begin with the debiased NEA population model developed by Bottke *et al.* (2000). Starting with 10^6 values for semimajor axis a, eccentricity e, and inclination i that represent the Bottke *et al.* NEA distribution, we generated approximately 2×10^8 NEAs by adding uniformly distributed longitudes of ascending node Ω and arguments of perihelion ω.

This very large initial NEA set was first reduced to about 58 000 objects for which the minimum possible orbital separation, or Minimum Orbital Intersection Distance (MOID), is low enough to permit an impact. Specifically, the impact is possible if the MOID is less than the Earth's capture cross-section $r_\oplus V_{\mathrm{imp}}/V_\infty$, where $r_\oplus$ is the Earth radius, V_{imp} is the impact velocity at the Earth's surface, and V_∞ is the encounter or hyperbolic excess velocity. These objects represent the subset of orbits that could impact, but an actual impact also requires that the object's arrival at the minimum distance point be timed to coincide with the Earth's. In this development we call this very low-MOID subset of the NEAs the Potentially Hazardous Asteroids (PHAs), although this is a non-standard usage. According to this definition, the Bottke *et al.* model predicts about one PHA for every 4000 NEAs.

Because not all of the PHAs are equally likely to impact, we next sampled 1000 impactors from among the PHAs according to the fraction of their orbital period that they spend within the Earth-capture cross-section of the Earth's orbit. Deriving the impactors in this way allows for the more hazardous orbital classes, such as low-inclination, Earth-like, or tangential orbits, to have appropriately increased prominence among the simulated impactors. This *hazard fraction f* is computed by finding the mean anomaly values for which the object enters and exits the toroid formed by the capture cross-section surrounding the Earth's orbit. We used the actual Earth orbit, not a circular approximation, for these computations. Values for f in our PHA sample range from as much as a few percent for Earth-like orbits down to 10^{-9} for low-MOID cometary orbits. The hazard fraction is similar to the impact probability per node crossing, and is distinct from, for example, the Öpik (1951) or Wetherill (1967) impact probabilities, which average the impact probability over the precession cycle of the object. Our intent is to measure the hazard posed by a given PHA over a timescale short enough that orbital evolution is not significant, say on the order of a few decades, but even f is probably not ideal for this purpose across the entire range of NEA orbits. In particular, for objects with very low encounter velocities the time spent within the capture cross-section torus can be fairly long but this increased risk may be partially offset by a long synodic period. However, the objects for which this concern may be significant comprise only a few percent

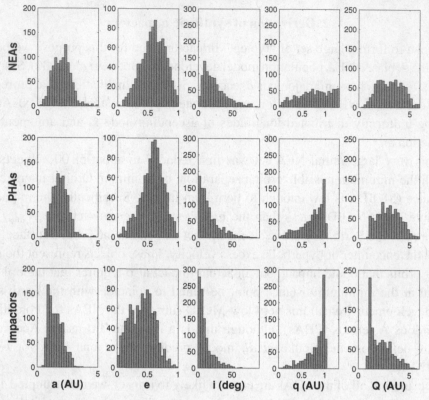

Figure 2.1 Orbital characteristics of modeled NEAs, PHAs, and impactors.

of our impacting population and we do not believe this plays a significant role in the overall results. Nonetheless, robust techniques for measuring a given orbit's mean impact rate over short timescales are not addressed in the literature and this is an avenue for further research.

3 Impactor orbital characteristics

It is instructive to compare the orbital characteristics of the entire modeled NEA population with those of the PHAs and the impactors. Figure 2.1 compares the distributions of a, e, i, q, and Q, where q and Q are the perihelion and aphelion distances, respectively. From this figure and from Table 2.1, it is clear that the impacting population has several distinct features. For instance, Earth-similar orbits are prominent among the impactors, as evidenced by the relative excess of impactors with $a \simeq 1$ AU, low e, and low i. Indeed, low-inclination orbits are strongly predominant among the PHAs, and even more so among the impactors. We also note that shallow-crossing orbits, i.e., those with either q or Q within

Table 2.1 *Orbital characteristics of NEA subpopulations*

	NEAs (%)	PHAs (%)	Impactors (%)
Shallow crossers			
Interior ($Q < 1.05$ AU)	1	1	11
Exterior ($q > 0.95$ AU)	8	22	38
Deep crossers	61	77	53
Atens ($a < 1$AU)	7	7	23
Low inclination ($<5°$)	6	25	38
Low V_∞ (<10 km s^{-1})	–	15	53

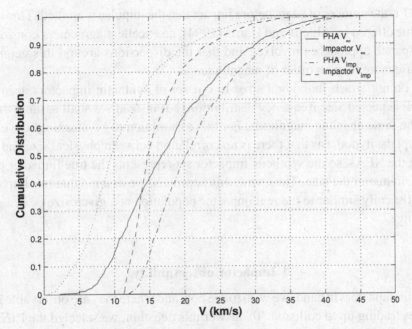

Figure 2.2 Cumulative distributions of impact velocity V_{imp} and encounter velocity V_∞ for the PHA and impactor sets. The impact velocity is a simple function of the Earth escape velocity ($V_e \simeq 11.2$ km s^{-1}) and the encounter velocity: $V_{imp}^2 = V_\infty^2 + V_e^2$.

0.05 AU of the Earth, have a substantially increased prominence among the impactors. The interior shallow crossers, in particular, have a ten-fold increase in the impacting population, when compared to the entire PHA population.

In general, the encounter velocity V_∞ and impact velocity V_{imp} are significantly greater for PHAs than for impactors. Figure 2.2 compares the cumulative distributions of impact and encounter velocities between the PHA and impactor sets. The impact velocity, which is a good indicator of the ΔV cost for spacecraft rendezvous (and hence impact mitigation), has a median value that is about 5 km s^{-1} less for impactors than for PHAs. This is due to the predominance of low-i shallow-crossing orbits for impactors.

Carusi *et al*. (2002) describe a simple formula for computing the velocity change required to deflect an asteroid by a given distance within a specified time interval. Using their approach, we can compute the required Δv to deflect each impactor by one Earth radius as a function of lead time. Taking the geometric mean of these values we find the relation

$$\Delta v = \frac{0.035 \text{ m s}^{-1}}{T}, \tag{2.1}$$

where T is the number of years before impact that the impulse is applied. This is half the value offered by Ahrens and Harris (1994), an excellent agreement considering the disparate approaches involved and that the dispersions around this geometric mean are more than an order of magnitude.

We do not attach individual sizes to our set of synthetic impactors according to some expected size frequency distribution, but instead we shall assume that all have the same absolute magnitude. For this assumption to be meaningful we have to accept the hypothesis that there is no correlation between physical size and orbit among the NEAs, so the synthetic impactor set represents the true impactor population, no matter the size. Thus, with this approach, we assume that our impactors are statistically similar to the real impactor population at a given size or in a given size range.

4 Impactor observability

With our impactors in hand, we wish to see how and where they are observable in the decades leading up to collision. To answer this question, we selected the LINEAR Experimental Test Site (ETS), near Socorro, New Mexico, as the observing location. We consider an object observable if it is situated at least 60° from the Sun, with proper motion in the range 0.05–10.0° per day and at no more than 3 air masses. Furthermore, it must be brighter than the detection limit of the survey telescope.

The question of whether an object is bright enough for detection depends upon both its assumed intrinsic brightness and the survey's assumed limiting magnitude, or depth. In preparing the ephemerides, we only need to calculate the difference between the apparent visual magnitude and the absolute magnitude, $S = V - H$. Then, given an assumed absolute magnitude H and survey limiting magnitude V_{lim}, the object is considered bright enough to be detected if $S + H < V_{\text{lim}}$, or equivalently if $S < V_{\text{lim}} - H$. For this report, we will vary the assumed H, while assuming a

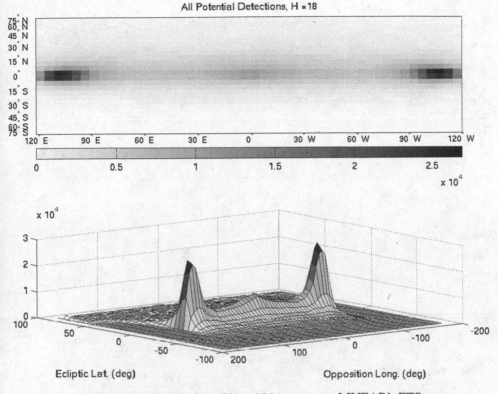

Figure 2.3 Sky-plane distribution of $H = 18$ impactors at LINEAR's ETS, assuming a survey depth of $V_{lim} = 20$. In the top diagram, the origin is the solar opposition point.

survey depth of $V_{lim} = 20$ throughout. This allows us to frame our results in terms of absolute magnitude, rather than the much less intuitive $V_{lim} - H$. It also means that any of these results can be applied to a different survey depth simply by incrementing or decrementing the quoted H value by the corresponding change from $V_{lim} = 20$.

Figure 2.3 indicates the sky-plane density of the 1000 impactors over the 100-year period leading up to impact. The plot indicates clearly that, for a $V_{lim} = 20$ survey, the most favorable region to search for $H = 18$ impactors is near the ecliptic between solar elongations of $60°$ and $90°$. There is also a modest concentration at opposition because objects are brightest at full phase. On the other hand, the peaks away from opposition at low elongations are present because the sky density of objects is much higher as we look through the "belt" of impactors. However, as the elongation becomes smaller the increasing density of objects is eventually overcome by the decreasing brightness due to phase losses. These phase losses are more significant for smaller objects, and the corresponding plot (not shown) for $H = 20$ shows the peaks at low elongations are less substantial than the opposition peak. For $H = 22$, these peaks are no longer significant.

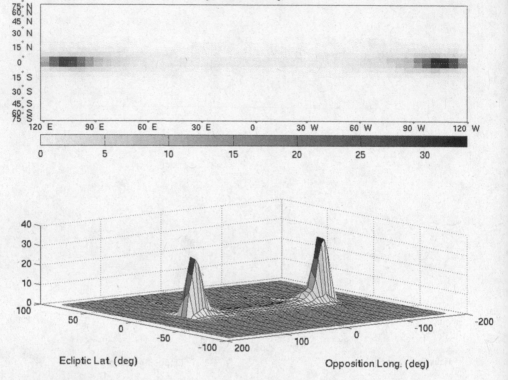

Figure 2.4 Sky-plane distribution of impactor hazard. This is the same as Fig. 2.3, except the residence times in each bin are weighted according to the associated hazard fraction f for that object.

Another approach to plotting the sky density of impactors is to weight them according to their likelihood of collision. In other words, instead of weighting all objects equally when accumulating the results, we can weight them according to the hazard fraction f described in Section 2. The result of this alternative approach, which can be thought of as mapping impact probability onto the sky, is presented in Fig. 2.4 for the same data set used in Fig. 2.3. The results indicate clearly that the *hazard*-density approach even further favors the low-elongation regions over the opposition region, as compared to the *object*-density plots. Indeed, using this approach, the prime discovery region for hazardous objects is situated predominantly at low elongations for sizes as small as $H = 22$.

For various reasons, searching the near-Sun region is more operationally challenging than searching near opposition, where present survey efforts have been concentrated. Also, current survey strategies are tailored towards fulfilling the Spaceguard goal of finding all NEAs larger than 1 km in diameter, and searches for NEAs, as opposed to impactors, are clearly the most productive around opposition since so many NEAs are of the Amor class, with orbits completely exterior to the Earth's.

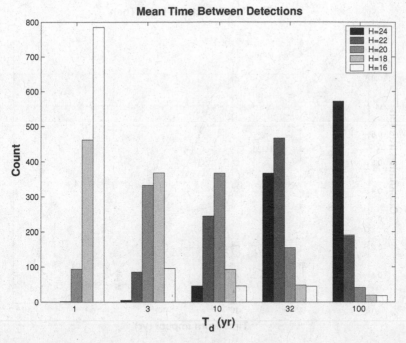

Figure 2.5 Histogram of mean time T_d between detections (not discoveries) for impactors at various absolute magnitudes, for the OPPOSITION survey. Objects that escaped detection over the entire simulation are included in the 100-year bin.

On the other hand, a search for large impactors is not expected to find such an object over human timescales, simply because large impacts are so rare. If a large NEA *is* set to impact the Earth in the next century it will most readily be detected far from opposition, as we show in the next section.

5 Survey simulations

Given our set of impactors, one can ask whether and when they would be discovered by various NEO surveys with differing sky coverages and limiting magnitudes. To consider these questions, we simulated two fictitious NEO surveys over the century prior to impact. The first survey, dubbed "OPPOSITION," is based loosely on the strategy and capability of the LINEAR system (Stokes *et al.*, 2000), the world's most prolific discoverer of NEOs. OPPOSITION covers the ecliptic and opposition regions heavily with modest coverage at higher ecliptic latitudes, but no coverage at solar elongations less than 90°. The survey model assumes 75% clear weather and some down time due to lunar interference. Figure 2.5 indicates the distribution of the mean time between detection by OPPOSITION for the synthetic impactors at various sizes. We note from this chart that the mean time between detection for most impactors larger than $H = 20$ will be a few decades or less.

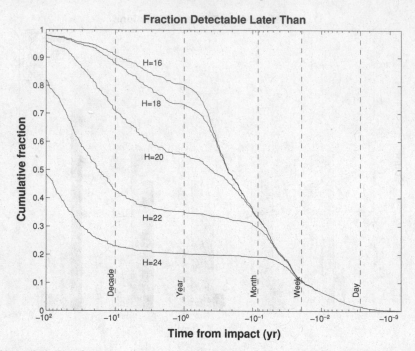

Figure 2.6 The fraction of impactors serendipitously detected later than a given time before impact, for the OPPOSITION survey.

Another interesting result that can be extracted from the data is the detection lead time. Figure 2.6 depicts the fraction of impactors that were serendipitously detected by OPPOSITION before impact as a function of the time until impact and as a function of absolute magnitude. For example, the plot indicates that 48% of $H = 24$ impactors will be detected sometime in the century leading up to impact, and that only 20% will be detected in the last year before impact. Similarly we can see that, for $H = 20$ objects, about one-third will be serendipitously detected in the last month before impact, whereas 12% will be detected only in the last week before impact. Differencing these two numbers tells us that roughly 20% of previously undiscovered $H = 20$ impactors will be detected by OPPOSITION with 1–4 weeks of warning time, which, in principle, could be sufficient to mount an effective evacuation of the impact region or threatened coastal regions.

Figure 2.6 also indicates that OPPOSITION would only detect 98% of objects, no matter the size. This is because the survey never looks beyond 90° from opposition and some shallow-crossing interior objects rarely enter the search region. To further consider this point we tested a second survey strategy, with equipment and location similar to OPPOSITION. However, this survey, which we call "NEAR-SUN," never searches near the opposition region, instead concentrating exclusively on the high-density regions indicated in Figs. 2.3 and 2.4. Specifically, NEAR-SUN trolled for impactors within 15° of the ecliptic and at solar elongations ranging from 60° to

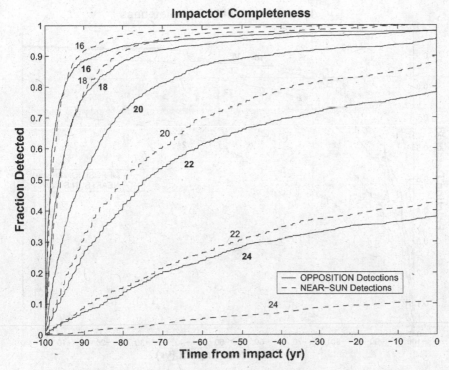

Figure 2.7 Comparison of OPPOSITION and NEAR-SUN survey performance, in terms of impactor completeness. Each curve is labeled with the assumed absolute magnitude.

110°. Figure 2.7 compares the two surveys' impactor completeness as a function of time prior to impact for several different impactor sizes. From that plot it is clear that NEAR-SUN beats OPPOSITION at detecting large impactors ($H \leq 18$), especially when the completeness exceeds 80%. But for smaller objects the phase losses prevent NEAR-SUN from discovering objects as rapidly.

We can measure survey completeness by the number of objects detected, as in Fig. 2.7, or by the percentage of the aggregate impact flux detected, in much the same way as Fig. 2.4 projected the risk onto the sky. The result of this approach is given in Fig. 2.8, where we can see that, despite its rather limited sky coverage, NEAR-SUN is markedly better at detecting the most hazardous large impactors, even excelling as faint as $H = 20$. Note also that the 2% of impactors that went undetected by OPPOSITION during the 100-year simulation actually comprised about 6% of the hazard. This is because the interior shallow-crossing objects that were missed hold a disproportionate share of the aggregate impact risk.

The obvious conclusion from these results is that the near-Sun region should not be neglected when searching for large impactors. Given, however, that this region of sky is only observable for a few hours each morning and evening, the remaining

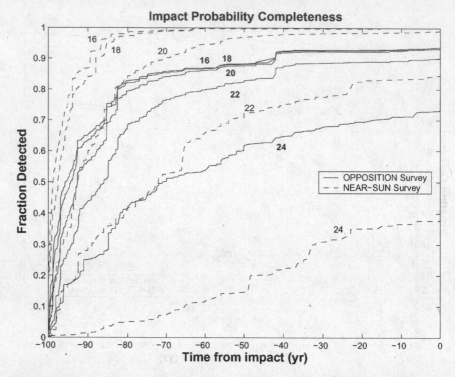

Figure 2.8 Comparison of OPPOSITION and NEAR-SUN survey performance, in terms of impact hazard completeness. Each curve is labeled with the assumed absolute magnitude.

telescope time could be used for an opposition-type survey, thus concentrating on the two most productive areas of sky to survey for the most hazardous objects. Note also that the size results given in this section are for a survey depth of $V_{lim} = 20$. For a fainter survey, the absolute magnitudes need to be adjusted fainter accordingly. For example, if we extend the survey to $V_{lim} = 24$, the proposed limit of the recently announced Pan-STARRS project (Kaiser and Pan-STARRS Team, 2002), then we can use the $H = 20$ curves of Fig. 2.8 (for $V_{lim} = 20$) to represent the case where $H = 24$ and $V_{lim} = 24$. Thus the figure indicates that for Pan-STARRS the NEAR-SUN strategy would be more effective than the OPPOSITION strategy for all sizes of interest.

6 Warning time: post-detection issues

The detection lead time is important in determining the time available for mitigation, but it is not the only factor. There are several hurdles to be crossed before a detection is announced to the community as a discovery. The first among these

Table 2.2 *Full precision orbital elements*
for simulated impactor

Quantity	Value
Epoch	2000-01-27.64180445 TDT
a	1.92551764232665 AU
e	0.506153768512869
Ω	337.350228048783°
ω	151.900629508322°
i	11.5131345762734°
M	−2.90435767774188°

is the recognition that an object has unusual motion (Jedicke, 1996). If a survey makes this determination it calls out the detection when forwarding the data to the Minor Planet Center (MPC). If the MPC staff agrees with the assessment then it is placed on the WWW NEO Confirmation Page for verification by other observers. If the object is followed and confirmed to be an NEA then the MPC will issue a discovery announcement. At any of these steps a potential impactor detection could be scrapped, and thus escape discovery. It is well known that this actually happens, but it is difficult to discern the extent to which these factors delay discovery of NEAs.

There is also some delay between the discovery of the asteroid and the recognition that it poses a threat worthy of mitigation. The idea of continually monitoring the ever-evolving asteroid orbit catalog for possibilities of impact is fairly new, and the first automatic collision monitoring system was fielded less than 3 years ago. Today there are two independent and parallel systems, Sentry[1] at the Jet Propulsion Laboratory in California and NEODyS[2] at the University of Pisa in Italy, that are operating continuously to scan for potential impacts. These efforts have been successful at detecting potentially hazardous future encounters for newly discovered asteroids and reporting the results to the NEO community. Follow-up observers have responded enthusiastically with observations that permit the hazard assessment to be refined and usually eliminated.

We close with a case study of the rate at which the probability of an impending impact increases after discovery, and we examine how this can affect the lead time for mitigation. We have selected a typical impactor on a shallow-crossing exterior orbit with moderate inclination (Table 2.2). It impacts the Earth on February 26, 2000. For this example we assume $H = 22$ and that the object is discovered at the second apparition before impact, in December 1983. With only the 3-day discovery data arc the actual impact is detected by Sentry (at a probability of 2×10^{-7}), along

---

[1] http://neo.jpl.nasa.gov/risk   [2] http://newton.dm.unipi.it/neodys

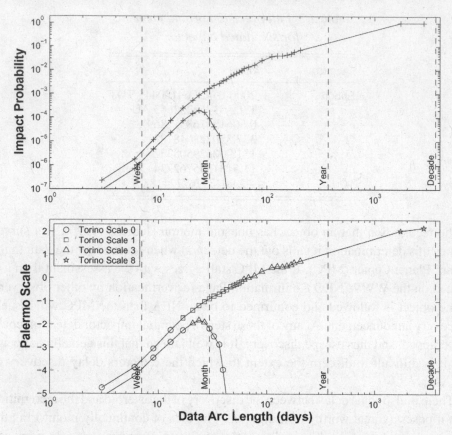

Figure 2.9 For the simulated impactor outlined in the text, with orbital elements given in Table 2.2, these plots present the computed impact probability (top) and Palermo/Torino Scale values (bottom) as a function of the data arc length used in the orbit determination process.

with many other spurious impact possibilities. (Here we use the term "spurious" to indicate a potential impact detection that will eventually be eliminated with more data, rather than to indicate that the detection is somehow incorrect or erroneous.) Among these early spurious potential impacts is one occurring on the same day as the true one. Such cases typically indicate a relationship, described by the theory of interrupted returns (Milani *et al.* 2002), between the two dynamically distinct potential impact trajectories.

Figure 2.9 depicts the impact probabilities for both of these potential impacts as a function of the observed arc length. The behavior of the spurious potential impact is typical of cases seen to date: the impact probability rises as long as the Earth remains within a collapsing uncertainty region, until eventually the uncertainty region shrinks to the point that it no longer encompasses the Earth, at which point the impact probability falls steeply. In this case, the probability reaches a peak of

2×10^{-4} with 3 weeks of data and is completely eliminated less than 3 weeks later. In contrast, the true impact's probability increases steadily during the discovery apparition until the object is lost in the daytime sky, with more than 6 months of observations. The impact probability stands at 7% at the end of the discovery apparition and the impact is less than 16 years away. The impact is confirmed 4 years later when the object is recovered at the next observing opportunity. At this point only 11 years remain until impact.

Two impact hazard scales are in regular use, the Torino Scale (Binzel 2000), which is intended for public communications, and the Palermo Technical Scale (Chesley *et al.* 2002), which is tailored to facilitate communication among specialists. The time history of these two scale values is also depicted in Fig. 2.9 for this simulation. We remark that the spurious impact detection did reach a Torino Scale value of 1 before beginning its decline, which is not unusual. We also note that the true impact event was rated positive on the Palermo Scale after only 68 days of observation. A positive Palermo Scale value indicates that the event is more threatening than the average impact flux for objects in the size range $H < 22$. As is typical of impacts occurring a few decades into the future, the Torino Scale rating exited the "1" region, in this case moving into the "3" region, at about the same time that the Palermo Scale became positive. During the 4 years that the simulated impactor was unobservable, the Palermo Scale stood at $+0.74$, which is greater than the highest real-world value computed to date ($+0.38$ for 2002 NT$_7$). This exercise raises the question as to what might be done to prepare for the potential impact during the years when the impact is not confirmed but the object is unobservable.

We face a similar but less severe situation even now, with the object 1997 XR$_2$. This asteroid, which at $H = 20.8$ is approximately 230 m in diameter, was heavily observed for 28 days at its discovery apparition in late 1997. With this data set the JPL Sentry system reports two potential impacts in the year 2101, each rating Torino Scale 1 and with impact probabilities around 5×10^{-5}. This asteroid is expected to be recovered at its next observing opportunity in 2006–2007, at which point the possibility of impact will either be confirmed or (more likely) eliminated. Incidentally, this object would almost certainly have been observed far longer at the discovery apparition in 1997–8 had the potential impacts been recognized at the time, but, unfortunately, the first automatic monitoring system was then still several years away.

7 Conclusions

This chapter describes our work to characterize the orbits and observability of Earth impacting asteroids and to understand the factors affecting the impact warning time. We summarize the major conclusions with the following remarks.

- The orbital distribution of the impacting population is noticeably different from the general low-MOID population, with an increased presence of low-i orbits, shallow-crossing orbits, and low-V_∞ orbits.
- Statistically, the impacting population will collide with Earth at lower relative velocities than the general PHA population. Since the impact velocity is a rough indicator of the spacecraft ΔV costs to effect a rendezvous, impactors should be more accessible for spacecraft mitigation attempts than objects in the general PHA population.
- Surveys will most readily detect impactors for which $H < V_{lim}$ in two fairly small "sweet spots" situated within 15° of the ecliptic and 90–120° from opposition. Fainter objects will be preferentially found in the opposition region.
- Among the impacting population, those most prone to impact are predominantly concentrated around the sweet spots if $H < V_{lim} + 2$. Surveys that concentrate on these regions reduce the risk of large impacts significantly faster than opposition-style surveys.
- A LINEAR-like survey would detect roughly 20% of previously undetected $H = 20$ impactors with 1–4 weeks of warning time, which in some cases may be enough time to mount an effective evacuation of affected areas.
- After its initial detection by a survey there are a series of hurdles that an impactor must cross before its hazard can be appreciated. If the object does make the leap from detection to discovery then automated impact monitoring systems can be expected to detect the impending impact very quickly. However, for impacts some decades in the future, the statistical hazard associated with the impact detection may remain fairly modest for months following the discovery. Depending upon the observing circumstances, this delay in recognition may trim several years from the time that could be dedicated to averting the impact.

Acknowledgments

The authors are grateful to Alessandro Morbidelli, who provided 1 million NEA orbits based upon the Bottke *et al.* (2000) population model, Alan Harris, who contributed important insight through various discussions and lectures, and Don Yeomans for his helpful review of this manuscript. This research was conducted in part at the Jet Propulsion Laboratory, California Institute of Technology, under a contract with the National Aeronautics and Space Administration (NASA).

References

Ahrens, T. J. and Harris, A. W. 1994. Deflection and fragmentation of near-Earth asteroids. In *Hazards due to Comets and Asteroids*, ed. T. Gehrels, pp. 897–927. Tucson, AZ: University Arizona Press.

Binzel, R. P. 2000. The Torino Impact Hazard Scale. *Planet. Space Sci.* **48**, 297–303.

Bottke, W. F., Jedicke, R., Morbidelli, A., *et al.* 2000. Understanding the distribution of near-Earth asteroids. *Science* **288**, 2190–2194.

Carusi, A., Valsecchi, G. B., D'Abroam, G., *et al.* 2002. Deflecting NEOs in route of collision with the earth. *Icarus* **159**, 417–422.

Chesley, S. R., Chodas, P. W., Milani, A., *et al.* 2002. Quantifying the risk posed by potential Earth impacts. *Icarus* **159**, 423–432.

Jedicke, R. 1996. Detection of near-Earth asteroids based upon their rates of motion. *Astron. J.* **111**, 970–982.

Kaiser, N. and Pan-STARRS Team. 2002. The Pan-STARRS optical survey telescope project. *Bull. Am. Astron. Soc.* **34**, 1304.

Milani, A., Chesley, S. R., Chodas, P. W., *et al.* 2002. Asteroid close approaches: analysis and potential impact detection. In *Asteroids III*, eds. W. F. Bottke, A. Cellino, P. Paolicchi, and R. P. Binzel, pp. 55–69. Tucson, AZ: University of Arizona Press.

Öpik, E. J. 1951. Collision probability with the planets and the distribution of planetary matter. *Proc. R. Irish Acad. Sect. A* **54**, 165–199.

Stokes, G. H., Evans, J. B., Viggh, H. E. M., *et al.* 2000. Lincoln Near-Earth Asteroid Program (LINEAR). *Icarus* **148**, 21–28.

Wetherill, G. W. 1967. Collisions in the asteroid belt. *J. Geophys. Res.* **72**, 2429–2444.

3

The role of radar in predicting and preventing asteroid and comet collisions with Earth

Steven J. Ostro

Jet Propulsion Laboratory, California Institute of Technology

Jon D. Giorgini

Jet Propulsion Laboratory, California Institute of Technology

1 Introduction

The current Spaceguard Survey classifies each known near-Earth asteroid (NEA) as either non-threatening or deserving of additional astrometric attention. For any possibly threatening object, the dominant issues are the uncertainty in its trajectory and physical nature as well as what can be done to reduce that uncertainty. Morrison *et al.* (2002) note that

From the standpoint of an allocator of society's resources, an uncertain threat calls for adaptive policies, delaying potentially costly action but informing later decision by investing in uncertainty-reduction measures. In the context of the NEO impact hazard, this means avoiding the costs of standing organizational structures and capital expenditures until a threat materializes. . . .

Thus reduction in uncertainty is tantamount to ensuring that unnecessary costs are avoided and that necessary actions are undertaken with adequate warning.

Ground-based radar is a knowledge-gathering tool that is uniquely able to shrink uncertainty in NEO trajectories and physical properties. The power of radar stems largely from the precision of its measurements (Table 3.1). The resolution of echoes in time delay (range) and Doppler frequency (radial velocity) is often of order 1/100 the extent of a kilometer-sized target, so several thousand radar image pixels can be placed on the target. Delay-Doppler positional measurements often have a fractional precision finer than 1/10 000 000, comparable to sub-milliarcsecond optical astrometry.

The single-date signal-to-noise ratio (SNR) of echoes, a measure of the number of useful imaging pixels placed on a target by a given radar data set, depends primarily on the object's distance and size. Figure 3.1 shows nominal values of

Mitigation of Hazardous Comets and Asteroids, ed. M. J. S. Belton, T. H. Morgan, N. H. Samarasinha, and D. K. Yeomans. Published by Cambridge University Press. © Cambridge University Press 2004.

Table 3.1 *Fractional precision of NEA radar measurements[a]*

	Range (m)	Radial velocity (m s^{-1})
Best radar resolution	$\sim$10	$\sim$0.000 1
Echo dispersion	$\sim$1 000	0.01 to 10
Astrometric "location"	$\sim$10 000 000	$\sim$10 000

[a] The optimal resolution of radar measurements of the distribution of echo power in time delay (range) and Doppler frequency (radial velocity) for observations of a large NEA is compared with the scale of the object's delay-Doppler extent and location.

SNR for Arecibo and Goldstone. Notwithstanding the heroic efforts by Zaitsev and colleagues in Russia and several intercontinental asteroid radar demonstrations involving Goldstone or Arecibo transmissions with reception of asteroid echoes in Japan, Spain, and Italy, the world's only effective NEO radars are at Arecibo and Goldstone, whose declination windows are $-1°$ to $38°$ and $> -40°$, respectively. However, given the historical funding difficulties experienced by those two systems (Beatty 2002), the future of radar astronomy cannot be taken for granted. Time will tell whether the USA will opt to maintain, much less improve, the current Arecibo and Goldstone radar telescopes.

In this chapter, we examine how our current radar capabilities might help at each stage of detecting and mitigating an impact hazard encountered during this century. See Ostro (1994) for a discussion of radar's role in hazard mitigation written a decade ago, Ostro *et al.* (2002) for a review of asteroid radar astronomy, and Harmon *et al.* (1999) for a review of comet radar astronomy.

2 Post-discovery astrometric follow-up

Once an asteroid is discovered, its orbital motion must be followed well enough to permit reliable prediction and recovery at the next favorable apparition. As of April 2004, 41% of the 595 identified Potentially Hazardous Asteroids (PHAs) are lost in the sense that the three-standard-deviation uncertainty in the time of the next close-approach exceeds $\pm$10 days, corresponding roughly to a plane-of-sky angular uncertainty greater than $90°$. (A PHA is defined by NASA's Jet Propulsion Laboratory as an object having a minimum orbit intersection distance with the Earth $\leq$0.05 AU and an absolute visual magnitude H $\leq$ 22.)

The first asteroid radar astrometry was obtained in 1968 (for 1566 Icarus) (Goldstein 1968; Pettengill *et al.* 1969). Prior to those observations, from simulations designed to evaluate the usefulness of optical and radar astrometry of Icarus in disclosing relativistic effects, Shapiro *et al.* (1968) concluded that radar data would

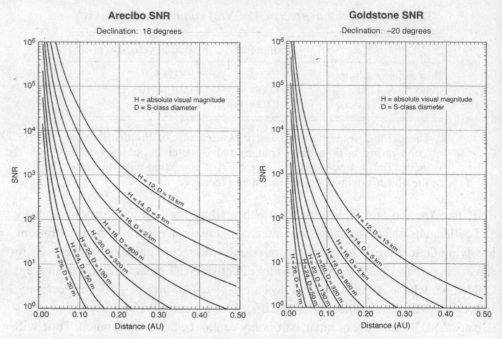

Figure 3.1 Predictions of the single-date signal-to-noise ratio (SNR): (a) for Arecibo echoes from asteroids at declination 18°, and (b) for Goldstone echoes from asteroids at declination −20°, as a function of the target's distance and absolute visual magnitude (converted to diameter by assuming an S-class optical albedo of 0.16) (Zellner 1979). Other assumptions include a 10% radar albedo, an equatorial view, a 4-h rotation period, and optimal values for system parameters. Plots for other declinations and distances are on the Internet (Ostro 2004a).

make an impressive contribution to the improvement of estimates of that asteroid's orbit. Two decades later, the potential of delay-Doppler measurements for small-body orbit refinement was examined comprehensively in a series of Monte Carlo simulations carried out by Yeomans et al. (1987). They showed that a single radar detection of a newly discovered NEA shrinks the instantaneous positional uncertainty at the object's next close approach by orders of magnitude with respect to an optical-only orbit, thereby preventing "loss" of the object. These conclusions have been substantiated quantitatively in the years since through comparison of radar+optical with optical-only positional predictions for recoveries of NEAs during the past decade (Table 3.2): The radar-based predictions historically have had pointing errors that average 310 times smaller than their optical-only counterparts, dramatically facilitating recovery.

Furthermore, radar astrometry can significantly reduce ephemeris uncertainties even for an object whose optical astrometry spans many decades. For example, radar measurements of 1862 Apollo (Ostro et al. 1991) at Arecibo showed that the object's optical-astrometry-based orbit, although based on 49 years of data, had

Table 3.2 *Residuals for past NEA recoveries*[a]

Object	Recovery date	O	R	O/R
1989 PB (4769 Castalia)	May 1990	24″	0.4″	60
1991 AQ	Sep 1994	57″	0.1″	380
1986 DA (6178)	Oct 1994	56″	0.9″	60
1991 JX (6489 Golevka)	Mar 1995	3600″	4.6″	780
1989 JA (7335)	Oct 1996	196″	99.3″	2
1986 JK (14827 Hypnos)	Apr 2000	114″	0.1″	910
1998 ML14	Nov 2002	125″	0.5″	250
1990 OS	Jun 2003	50477″	3200″	16

[a] Here O represents the positional offset (the observed position at recovery minus the predicted position) for a pre-recovery orbit solution incorporating only optical astrometry. R represents the residual for a pre-recovery orbit solution using radar combined with optical. O/R is the ratio of residuals for the two cases and is a measure of the relative reduction in position error when radar astrometry is included in the orbit solution.

a range error of 3750 ± 2 km. (See also the discussion of 1950 DA below.) The reduced uncertainties of a radar orbit can also aid recovery at fainter magnitudes. For example, 1998 ML14 was recovered (Table 3.2) at magnitude 21.2, only 0.5 arcseconds from the position predicted from a radar+optical orbit. The optical-only orbit would have suggested a search for the faint object 125 arcseconds away from the actual position.

For NEAs observed only during their discovery apparition, one can predict the uncertainty in the location during the next opportunity for optical observation, and hence the area of the sky for a search having a given likelihood of success. Table 3.3 lists the total sky area for the three-standard-deviation orbit-determination uncertainties mapped onto the sky at the next favorable Earth-based recovery date (which we define as the next time when the apparent visual magnitude is less than 20 during reasonable sky-brightness conditions) for both an optical-only orbit and a radar+optical orbit, for seven NEAs. Radar shrinks the required sky search area by an average factor of 2642, dramatically facilitating recovery. For six other objects in the table, the optical-only and radar+optical orbits are so different that the nominal recovery dates are months or years apart. Since the radar+optical solutions would be expected to be the more accurate, reliance on an optical-only solution would be unlikely to lead to recovery.

3 Window of predictability

A goal of the Spaceguard Survey is to provide as much warning as possible of any possibly dangerous approach of NEAs as large as 1 kilometer. However, since

Table 3.3 *Search areas for future NEA recoveries*

Object[a]	Most favorable Earth-based recovery date	Optical orbit recovery	Data span (days)	Astrometry Optical	Astrometry Doppler	Astrometry delay	Gap (yrs)	Search area (arcsec²) O	Search area (arcsec²) R	Search area (arcsec²) O/R
1996 JG	Nov 26 2003	~	20	265	3	3	7	1.5E5	310	484
2000 EH26	Jun 24 2005	~	140	47	4	2	5	47808	1055	45
2002 FD6	Sep 23 2006	~	15	556	2	4	4	3.9E5	146	2671
2002 BM26	Dec 20 2006	(2003)	87	218	2	2	4	–	294	–
2000 UG11	Oct 24 2008	(2025)	19	395	1	2	8	–	70	–
2001 AV43	Oct 21 2013	~	182	46	1	0	12	1.5E5	6.1E4	2
2002 FC	Jul 08 2021	~	137	191	3	1	19	14297	35	408
1998 KY26	May 31 2024	~	11	211	2	2	26	33534	786	43
2000 RD53	Oct 15 2031	~	102	322	5	4	31	89005	6	14834
2002 AV	Jan 25 2033	(2036)	39	210	3	2	31	–	816	–
2000 LF3	May 31 2046	(2090)	13	67	4	1	46	–	1.1E5	–
1999 TY2	Oct 03 2064	(2091)	5	115	1	0	65	–	2.3E6	–
2001 FR85	Mar 09 2081	(2082)	7	36	3	1	80	–	9.5E4	–

[a] The objects listed above were observed optically and with radar over a single apparition. We used least-squares to estimate optical-only and radar+optical orbits and then used the radar+optical orbit to determine a likely recovery date, defined as the next time when the apparent visual magnitude first drops below 20 under reasonable sky-brightness conditions, that is, with sufficient separation in the sky from the Moon and the Sun, taking into account elevation above the horizon as well as the fraction of the lunar surface that's illuminated. Column (R) lists the area of the three-sigma uncertainty ellipsoid projected in the plane of sky for the radar+optical orbit. Column (O) lists the optical-only orbit uncertainties at the time of recovery for the seven cases in which the optical-only solution predicted a recovery date within 1 month of the radar+optical prediction. The ratio O/R for those cases have a mean of 2642, providing some indication of how much larger the three-sigma search region would be with just optical data. For the six other cases, the optical-only orbit would not allow for recovery of the asteroid at all, at the time predicted; the year of most likely recovery indicated by such optical orbit solutions is in parentheses.

an orbit estimate is based on a least-squares fit to measurements of an asteroid's position over a small portion of its orbit, knowledge of the future trajectory generally is limited by statistical uncertainties that increase with the length of time from the interval spanned by astrometric measurements. Trajectory uncertainties are greatest and grow most rapidly during close planetary encounters, as the steeper gravity field gradient differentially affects the volume of space centered on the nominal orbit solution within which the asteroid is statistically located. Eventually the uncertainty region grows so large, typically within the orbit plane and along the direction of motion, that the prediction becomes meaningless.

Current ground-based optical astrometric measurements typically have angular uncertainties of between 0.2 and 1.0 arcsec (a standard deviation of 0.5 to 0.8 arcsec is common), corresponding to tens or hundreds or thousands of kilometers of uncertainty for any given measurement, depending on the asteroid's distance. Radar can provide astrometry referenced to the asteroid's center of mass, with uncertainties as small as ~ 10 m in range and ~ 1 mm s^{-1} in range rate. Since those measurements are orthogonal to plane-of-sky angular measurements and have relatively fine fractional precision, they offer substantial leverage on an orbit solution and normally extend NEO trajectory predictability intervals far beyond what is possible with optical data alone.

Let us define the window of predictability as the interval over which an object's Earth close approaches can be reliably known at the three-sigma level of confidence. Table 3.4 lists optical-only and radar+optical predictability windows for all radar-detected PHAs. For objects observed only during their discovery apparition, radar has enlarged the total window of predictability (past and future) by an average factor of eight, from 145 years for solutions based only on optical data to 1196 years when radar was included in the orbit solution. On average, radar has added 367 more years to the window of accurate future predictions.

When radar astrometry is excluded from the 29 single-apparition PHA radar+optical orbit solutions, 41% cannot have their next close approach predicted within the adopted confidence level using only the single apparition of optical data. This is the same percentage seen in the total population of PHAs. Radar astrometry obtained in these cases adds an average of 500 years of statistical confidence to their Earth encounter predictions, preventing them from being lost. For 2000 RD53 and 1999 FN53, the extension is through the end of this millennium.

We see that a discovery–apparition orbit solution containing radar astrometry can be compared favorably to a multiple-apparition, optical-only solution. As an example, 1998 ML14 is listed in both single and multiple apparition categories to show the effect of including the first six optical observations from the November 2002 recovery, which lengthened the data arc from 7 months to 5 years and lengthened the optical-only total knowledge window from 364 to 1721 years. By comparison,

Table 3.4 *Earth close-approach prediction intervals for radar-detected PHAs[a]*

Object	Astrometry			Days	Earth close-approach prediction intervals				R-O	Re-Oe	R/O
	Optical	Doppler	Delay		Optical-only		Radar+optical				
					Span	Years	Span	Years			
Single-apparition objects											
2000 CE59	163	2	3	210	1609–2601	992	1547–2703	1156	164	102	1.2
2002 SY50	522	2	5	72	1896–2051	155	1862–2071	209	54	20	1.3
1999 RQ36	210	1	3	208	1895–2060	165	1848–2080	232	67	20	1.4
1998 KY26	211	2	2	11	1959–2024	65	1959–2099	140	75	75	2.2
2000 QW7	850	1	0	121	1902–2087	185	1755–2185	430	245	98	2.3
2000 EW70	286	6	3	16	1971–2069	98	1929–2209	280	182	140	2.9
2000 DP107	395	1	9	250	1847–2286	439	1066–2392	1326	887	106	3.0
2000 YF29	156	2	1	207	1932–2083	151	1642–2136	494	343	53	3.3
1998 ML14	243	6	6	214	1874–2238	364	1100–2283	1183	819	45	3.3
2002 NY40	1441	5	2	35	1997–2049	52	1849–2081	232	180	32	3.5
2001 JV1	129	2	1	134	1874–2168	294	1266–2382	1116	822	214	3.8
2002 BG25	107	1	1	198	1667–2363	696	5BC–3167	3173	3172	804	4.6
2001 GQ2	323	2	1	15	1997–2084	87	1626–2100	474	387	16	5.5
2002 AY1	34	1	2	7	1848–2167	319	428–3034	2606	1842	867	8.2
2001 CP36	126	2	2	8	1972–2004	32	1628–2280	652	620	276	20.4
2000 ED14	57	1	0	16	1926–2000	74	29BC–3134	3164	3090	1134	41.8
2002 VE68	196	4	0	15	1994–2010	16	448–2653	2205	2189	643	137.8
1998 BY7	109	2	0	30	1998–1998	1	1995–1998	3	2	0	(3)
1990 OS	26	2	0	13	1990–1990	1	1966–2212	246	245	222	(246)
2002 FD6	277	2	4	15	2002–2002	1	1862–2161	299	298	159	(299)

Single-apparition objects (cont.)

Object	Optical	Doppler	Delay	Years	Span	Years	Span	Years	R-O	Re-Oe	R/O
2000 EH26	47	4	2	140	2000–2000	1	1806–2106	300	299	106	(300)
1996 JG	265	3	3	20	1996–1996	1	1851–2180	329	328	184	(329)
2000 UG11	395	1	2	19	2000–2000	1	1812–2142	330	329	142	(330)
2000 LF3	67	6	1	14	2000–2000	1	1583–2046	463	462	46	(463)
2002 BM26	218	2	2	87	2002–2002	1	1757–2312	555	554	310	(555)
2002 AV	210	3	2	39	2002–2002	1	1626–2702	1076	1075	700	(1076)
2000 RD53	322	5	4	102	2000–2000	1	1756–3023	1267	1267	1023	(1267)
2002 FC	191	3	1	137	2002–2002	1	73BC–2415	2489	2488	413	(2489)
1999 FN53	98	3	1	81	1999–1999	1	3570BC–4686	8257	8256	2687	(8256)
Mean (29)				145		1196			+1060	+367	

Object	Astrometry				Earth close-approach prediction intervals						
	Optical	Doppler	Delay	Years	Optical-only		Radar+optical		R-O	Re-Oe	R/O
					Span	Years	Span	Years			
Multiple-apparition objects											
4769 Castalia	122	7	7	13	1101–2837	1736	1043–2516	1474	−262	−321	0.8
1620 Geographos	1548	4	3	51	944–3188	2244	915–2900	1985	−259	−288	0.9
35396 (1997 XF11)	428	0	5	13	1627–2155	528	1627–2102	475	−53	−53	0.9
5604 (1992 FE)	162	0	3	17	1418–2184	766	1488–2156	668	−98	−28	0.9
6489 Golevka	686	30	26	9	1518–2706	1188	1621–2706	1085	−103	0	0.9
5189 (1990 UQ)	37	1	0	12	87BC–2660	2748	70BC–2660	2731	−17	0	1.0
49114 1998 ST27	287	1	3	3	1713–3775	2062	1690–3680	1990	−72	−95	1.0
25143 (1998 SF36)	628	6	10	3	1852–2170	318	1852–2170	318	0	0	1.0
4660 Nereus	371	2	11	21	1827–2166	339	1827–2166	339	0	0	1.0
7482 (1994 PC1)	268	2	0	28	1842–2361	519	1842–2361	519	0	0	1.0
2000 EE104	319	3	0	3	1638–2351	713	1638–2351	713	0	0	1.0
2201 Oljato	187	4	0	70	1666–2392	726	1666–2392	726	0	0	1.0
2002 HK12	516	1	0	17	1504–2299	795	1504–2299	795	0	0	1.0
4486 Mithra	233	9	8	16	1472–2255	783	1472–2281	809	26	26	1.0

Multiple-apparition objects (cont.)

33342	(1998 WT24)	736	1	6	3	1751–2675	924	1751–2675	924	0	0	1.0
	1991 AQ	84	3	5	10	1786–2731	945	1786–2731	945	0	0	1.0
13651	(1997 BR)	439	1	0	20	1693–2768	1075	1693–2768	1075	0	0	1.0
4183	Cuno	495	0	1	16	>1403–2481	1078	>1403–2481	1078	0	0	1.0
4034	(1986 PA)	180	1	0	16	1424–2682	1258	1424–2682	1258	0	0	1.0
23187	(2000 PN9)	295	2	1	12	>993–2325	1332	>993–2325	1332	0	0	1.0
6037	(1988 EG)	266	4	4	13	1412–2771	1359	1377–2771	1394	35	0	1.0
2101	Adonis	54	5	0	66	1244–2609	1365	1209–2609	1400	35	0	1.0
38071	(1999 GU3)	355	6	5	13	1196–2645	1449	1196–2645	1449	0	0	1.0
52761	(1998 ML14)	249	6	6	4	562–2283	1721	562–2283	1721	0	0	1.0
10115	(1992 SK)	217	2	8	46	932–2683	1751	932–2683	1751	0	0	1.0
9856	(1991 EE)	103	1	3	10	781–2567	1780	781–2567	1780	0	0	1.0
	1999 KW4	1624	0	2	3	1145–2929	1784	1127–2929	1802	18	0	1.0
7335	(1989 JA)	137	5	0	12	1362–3219	1857	1362–3219	1857	0	0	1.0
29075	(1950 DA)	223	5	8	51	>588–2880	2292	>588–2880	2292	0	0	1.0
22753	(1998 WT)	209	1	3	47	116–2562	2446	116–2565	2449	3	3	1.0
26663	(2000 XK47)	149	2	2	27	>71BC–2397	2469	>71BC–2397	2469	0	0	1.0
1566	Icarus	624	11	0	53	1206–3803	2597	1206–3803	2597	0	0	1.0
14827	Hypnos	159	11	0	14	249–2959	2710	249–2959	2710	0	0	1.0
3908	Nyx	605	1	6	17	77BC–2708	2786	77BC–2708	2786	0	0	1.0
53319	(1999 JM8)	408	5	3	13	811–3988	3177	811–3988	3177	0	0	1.0
7822	(1991 CS)	212	4	0	11	305BC–2840	3146	305BC–2884	3191	45	44	1.0
7341	(1991 VK)	398	1	1	11	.501–3797	3296	398–3797	3399	103	0	1.0
52387	(1993 OM7)	222	1	2	10	198BC–3192	3391	198BC–3207	3406	15	15	1.0
1999	FN53	133	3	1	1	3826BC–6789	10616	3826BC–6789	10616	0	0	1.0
1862	Apollo	283	8	4	71	1848–2351	503	1788–2362	574	71	11	1.1

Multiple-apparition objects (cont.)

8014	(1990 MF)	60	10	6	8	1568–2313	745	1568–2371	803	58	58	1.1
4179	Toutatis	1105	27	19	68	1221–2069	848	1117–2069	952	104	0	1.1
4953	(1990 MU)	95	2	0	27	>1519–3123	1604	>1519–3271	1752	148	148	1.1
3757	(1982 XB)	85	2	0	20	1184–2673	1489	1005–2673	1668	179	0	1.1
1981	Midas	96	1	0	25	1237–3122	1885	1011–3122	2111	226	0	1.1
Mean (45)							1803		1808	+4	−11	

[a] For each asteroid we give the time-span over which a numerically integrated orbit solution (along with its variational partial derivatives) based only on optical data can predict Earth close approaches when compared to an independent solution that also includes radar astrometry. Prediction intervals are bounded *either* by the first Earth approach by the first Earth approach ≤ 0.1 AU for which the three-sigma linearized uncertainty in the time of closest approach exceeds ±10 days *or* by the first Earth approach for which the three-sigma approach distance uncertainty at the nominal encounter time exceeds 0.1 AU, whichever occurs first. Objects with a '>' symbol did not have any Earth approaches closer than 0.1 AU between the given year and 1000 BC, the earliest date checked. Our uncertainties are based on a mapping of the measurement covariance matrix in which the higher-order non-linear terms in the integrated variational partials are neglected. Thus, in a few cases, non-linearities due to a particularly close approach may not be immediately detected. This table includes radar-detected asteroids (Ostro 2004b) for which radar astrometry has been reported (Giorgini 2004) and which are designated PHAs by JPL or the MPC, as well as three interesting cases that are not designated PHAs (1998 KY26, 2001 CP36, and 26663 (2000 XK47)). The first four columns give the numbers of optical, Doppler, and delay measurements, and the span of time they cover. Optical-only (O) and radar+optical (R) reliable prediction date intervals are given (the actual date range as well as the number of years spanned). R–O is the difference between the radar+optical and optical-only intervals. Re–Oe is the difference in the final year of the interval; it indicates how many additional years into the future radar can predict close approaches accurately. R/O is the ratio of the total span of years for the two solutions. If the optical-only interval is unity, we place R/O in parentheses. Integrations were performed using the DE406/408 planetary ephemeris and include relativistic perturbations due to the Sun, planets, and Moon as well as asteroids Ceres, Pallas, and Vesta. Whereas this table indicates the relative effect of radar astrometry, the limits of predictability for objects having multiple planetary encounters over centuries will normally be affected by additional factors such as radiation pressure, Yarkovsky acceleration, planetary mass uncertainties, and asteroid perturbations. These factors are not included here, since the precise models are unknown and key parameters are unmeasured.

during the discovery apparition, radar astrometry combined with optical data provided an interval of 1183 years. It required the recovery of 1998 ML14 before an optical-only solution yielded a prediction interval comparable to the discovery apparition combined with radar. Similarly, for 1999 FN53, discovery apparition radar indicated an 8257-year window not possible with optical data alone until the object was recovered about half an orbit period later.

For multiple-apparition objects, radar does not significantly extend the interval, which often is terminated centuries from the present era by one or more close planetary approaches whose detailed geometry simply cannot be discerned by any present-day data type. Nevertheless, radar improves the accuracy of multi-apparition orbits. A prime example is 1950 DA: the same upper-limit of AD 2880 exists whether or not radar is included in the multiple-apparition solution. However, including radar revealed a non-negligible impact potential in 2880 not apparent in optical solutions. This was because radar astrometry eliminated a bias in the optical data and reduced the 2880 uncertainty region by about 20% as compared to the optical only solution, resulting in the potential hazard detection (see Section 4.2).

In seven of the 45 multi-apparition cases, radar astrometry actually reduced the interval of prediction, while 17 cases were slightly extended. These disparate effects arise because the different nominal orbits for the optical and optical+radar solutions have slightly different planetary encounter circumstances, so their uncertainty regions increase in different ways. Thus the net effect of radar for these multi-apparition cases is to correct the length of the optically predicted interval, suggesting that if any optical-only orbit were to reveal a potentially hazardous close approach, it would be highly desirable to get radar astrometry to check the prediction.

4 Radar and collision probability prediction

For newly discovered NEOs, a collision probability is now routinely estimated (Milani *et al.* 2002) for close Earth approaches. This probability is combined with the asteroid's estimated diameter and the time until the approach to rate the relative degree of hazard using the Palermo Technical Scale (Chesley *et al.* 2002). The Jet Propulsion Laboratory's Sentry program maintains a "risk page" (Chesley 2004) which lists objects found to have a potential for impact within the next 100 years. However, for newly discovered objects, the limited number of initial astrometric observations typically does not permit accurate trajectory prediction. When an object's optical astrometric arc is only days or weeks long, the orbit is so uncertain that a potentially hazardous close approach cannot be distinguished from a harmless one or even a non-existent one. The object is placed on the Sentry page, then

Table 3.5 *Simulated impacting orbit*

ORBIT (heliocentric J2000.0 ecliptic elements):
Impacts Earth surface: 2028-Mar-30 15:51:38.5000 (CT)
Impact relative speed: 17.26 km/s
EPOCH = 1994-Mar-05 00:00:00.0000 = 2449416.5 JD (CT)
EC = 0.50990174495185
QR = 0.93177704136264 AU
IN = 15.587556441422 deg
OM = 10.5543473199928 deg
W = 215.77334777809 deg
TP = 2449468.8313169 JD
H = 19.0
G = 0.15

typically removed later, when additional optical astrometry is obtained and the span of observations is extended. However, almost as a rule, objects on the Sentry page have not been observed with radar.

4.1 A simulated impact scenario

If an asteroid is on collision course with Earth, this fact will be recognized much sooner with radar data than without it. To examine the possible progression of optical-only and radar+optical impact probability estimates prior to a collision, we constructed a simulation as follows.

First, from the initial set of statistically possible trajectories for a recently dis-covered asteroid, we selected an orbit that has a 2028 approach to within two Earth radii, a 1994 approach when it could have been discovered, and two post-discovery periods of visibility. That orbit was adjusted so as to change the 2028 close approach into an impact. We adopted an absolute visual magnitude of $H = 19$, which corresponds to an object with a diameter between 420 and 940 m and hav-ing a discovery-apparition peak brightness of magnitude 14. Thirteen years after discovery, the asteroid brightens to magnitude 19, so recovery would be possible. Subsequent additional observing opportunities exist, but are less favorable since the object does not again get brighter than 20th magnitude until 9 weeks before impact. Radar observations would be possible during the discovery apparition, but then not again until 2 weeks prior to impact. Table 3.5 gives the impacting orbit and Table 3.6 lists observing opportunities.

We then simulated optical astrometry using the impacting reference trajectory and a Gaussian residual noise model in which the residual mean and standard

Table 3.6 *Observing opportunities for the simulation*

Years since discovery		Date/time	Visual brightness (magnitude)	Radar SNR	Comments
0	1994	Mar 10	16.7	–	Optical discovery
		Mar 20	15.1	532	Arecibo start
		Mar 27	14.0	17791	Last day in Arecibo window
		Mar 28	14.3	1064	Goldstone start
		Mar 30	15.2	455	Goldstone stop
		Apr 30	19.5	–	Last optical data (no impact detection)
		Oct 14	22.0	–	Last optical data (if impact detection)
13	2007	Apr 19	20.0	–	Optical recovery
		Jul 17	19.0	–	Peak brightness
		Oct 15	22.0	–	Last optical data (impact detection)
20	2014	Dec 21	21.6	–	
21	2015	Feb 21	20.0	–	Peak brightness
		Oct 14	22.0	–	
34	2028	Jan 22	20.0	–	Optical recovery
		Mar 16	16.0	15	Goldstone detection possible
		Mar 30	9.5		
		12:49:56	6.0		Dark-sky naked eye visibility
		14:01:13	5.0		
		14:45:55	4.0		
		15:12:39	3.0		
		15:29:47	2.0		
		15:40:14	1.0		
		15:46:32	0.0		
		15:50:16	–1.0		
		15:51:38	–		Surface impact

deviation for each reporting site's astrometry was based on the actual observing results for 1994 AW7. We simulated radar data for Arecibo and Goldstone using the predicted SNRs to determine observing windows and potential measurement accuracy, adjusting the astrometry to emulate the residual statistics for previous radar campaigns.

Table 3.7 shows the impact probability that would be predicted for each of several cases with different amounts of discovery-apparition radar astrometry. A typical optical campaign at discovery (case B) does not show an unusual impact risk after 50 days of observations. However, if just two radar measurements are made 10 days after discovery (case C), the likelihood of a very close approach immediately becomes evident, along with a non-negligible impact probability. Comparison of

Table 3.7 *Simulation cases and results*[a]

Cases

Discovery apparition only

Description	Data span		Optical	Delay	Doppler	n-RMS
	1994	1994				
(A) Optical data only	Mar 10 – Mar 21		57	0	0	0.65
(B) Optical data only	Mar 10 – Apr 30		158	0	0	0.74
(C) Optical and initial radar	Mar 10 – Mar 21		57	1	1	0.65
(D) Optical and initial radar	Mar 10 – Apr 30		158	1	1	0.73
(E) Optical and all radar	Mar 10 – Apr 01		127	11	7	0.70
(F) Optical and all radar	Mar 10 – Apr 30		158	11	7	0.73
(G) Optical and all radar	Mar 10 – Oct 16		229	11	7	0.75

Discovery apparition plus recovery apparition 13 years later

Description	Data span		Optical	Delay	Doppler	n-RMS
	1994	2007				
(H) 2 appar optical only	Mar 10 – Oct 15		313	0	0	0.65
(I) 2 appar optical and radar	Mar 10 – Oct 15		313	11	7	0.66

Results

Table 3.7 (*cont.*)

Nominal date (Discovery + 34 years)	± (min)	NomDist (AU)	MinDist (AU)	MaxDist (AU)	N-sigs	Volume (km³)	Projected area (km²)	Impact, linear	Probability, non-linear
(A) May 23.57393	± 1.0E6	0.237 217	0.203 725	2.771 532	15.902	1.3E+17	6.0E+13	0.000 00	0.000 02
(B) Apr 10.07787	± 86217	0.081 599	0.010 467	1.364 941	266.000	1.8E+10	1.1E+09	0.000 00	0.000 27
(C) Mar 30.40767	± 671	0.002 550	0.000 000	0.007 037	1.6707	9.6E+08	6.0E+07	0.006 94	0.006 79
(D) Mar 30.73868	± 172	0.000 620	0.000 001	0.001 759	1.5278	1.6E+07	2.0E+06	0.029 19	0.027 63
(E) Mar 30.61238	± 107	0.000 554	0.000 001	0.001 267	2.1537	1.2E+06	1.6E+06	0.011 30	0.011 13
(F) Mar 30.65632	± 87	0.000 104	0.000 001	0.000 679	0.3228	6.2E+05	9.1E+05	0.191 10	0.194 30
(G) Mar 30.64146	± 60	0.000 261	0.000 001	0.000 659	1.6603	3.2E+05	6.1E+05	0.042 31	0.037 81
(H) Mar 30.66428	± 0.22	0.000 001	0.000 001	0.000 001	0.0000	26301	1750	1.000 00	1.000 00
(I) Mar 30.66426	± 0.20	0.000 001	0.000 001	0.000 001	0.0000	894	1433	1.000 00	1.000 00

[a] For each case in our simulation, the top part of the table indicates the number of optical and radar astrometric measurements, their date span, and the normalized root mean squares residual. In the "Results" section at the bottom, the first columns give the encounter time and its three-sigma uncertainty. NomDist is the solution's nominal (highest probability), numerically integrated Earth approach distance on the given date. MinDist and MaxDist are the minimum and maximum (three-sigma) approach distances from the linearized covariance mapping. N-sigs is the number of standard deviations required for the mapped covariance ellipsoid to intersect the surface of the Earth. The next columns give the volume of the three-sigma uncertainty region and the area it projects into a plane perpendicular to the impactor's velocity vector at encounter. The last columns give the impact probabilities computed by the linearized mapping method and by the non-linear method used by Sentry, JPL's automated hazard monitoring system.

cases A and C reveals that after the first two radar measurements, the volume of the uncertainty region is nine orders of magnitude smaller with the radar+optical orbit than with the optical-only orbit.

At the conclusion of case B's 50-day observing window, a 0.027% impact probability is indicated by the optical-only solution. This is noteworthy, but not unusual for single-apparition objects – there currently are four objects on the Sentry Risk Page with a comparable impact probability. However, with the radar astrometry (case F), a 19% impact probability is indicated at the same point in time. Radar reduces the volume of the uncertainty region at the encounter by five orders of magnitude compared to the optical-only case B. A 19% impact probability would attract additional resources and would extend the window of optical observability several months, down to at least magnitude 22 (case G). Due to marginally greater bias and noise in the simulated data as the target fades from view, the additional optical astrometry moves the solution's nominal close approach slightly further away from the Earth, decreasing the impact probability estimate.

If instead there is no radar data at the discovery apparition, recovery would probably still occur during the optically favorable apparition 13 years after discovery. If so, two such apparitions of optical data conclusively identify the impact event whether or not radar data is available (cases H and I), although the radar data reduces the volume of the uncertainty region by a factor of 29 compared to a solution based only on two apparitions of optical data. However, if the recovery does not occur, the next good opportunity to recover the object and clarify the impact risk, or perhaps to first become aware of it, would be 2 months prior to impact. Radar data during the discovery apparition guarantees the recovery by clearly indicating a high impact risk immediately, providing 34 years of warning instead of 21 years (or possibly only a few weeks).

4.2 Negative predictions, positive predictions, and warning time

To a great extent, the dominance of PHA trajectory uncertainties is a temporary one, an artifact of the current discovery phase. Predictions are made for single-apparition objects having a few days or weeks of measurements. The uncertainty region in such cases can encompass a large portion of the inner solar system, thereby generating small but finite impact probabilities that change rapidly as the data arc lengthens, or if high-precision radar delay and Doppler measurements can be made. Impact probabilities in such cases are effectively a statement that the motion of the asteroid is so poorly known that the Earth cannot avoid passing through the asteroid's large uncertainty region – hence the apparent impact "risk." As optical measurements are made, the region shrinks. The resulting change in impact probability, up or down, is effectively a statement about where the asteroid won't be – a "negative prediction" – rather than a "positive prediction" of where it will be. This is due to the modest positional precision of optical measurements.

In contrast, radar measurements naturally provide strong constraints on the motion and hence "positive predictions" about where an asteroid will be decades and often centuries into the future. *Thus radar measurements substantially open the time-window of positive predictability.* However, within a couple of decades, asteroids being found now (but unobserved by radar) will themselves have multiple optical apparitions and similarly be predictable in a positive way over centuries, as radar cases are now. In this way, orbit uncertainties for present-day radar cases illustrate what the situation will be by mid-century for most of the asteroids known today, and presumably for almost all PHAs as large as 1 kilometer.

Unless a significant impact is predicted to occur in the next few decades, mitigation will primarily be an issue for multi-apparition asteroids with decades of observations behind them (since most objects will typically have that much optical astrometry) or for radar-detected objects, with the impact predictions at least centuries in the future. Examination of the next 1000 years will be 30 times as likely to find an actual impactor than examination of the next few decades. This suggests that high-energy mitigation methods may be rendered obsolete before they could be implemented, not by technology, but by the changing nature of orbit predictability as the primary discovery phase ends and observations accumulate. Low-energy methods such as radiation pressure or Yarkovsky modification can potentially be implemented at lower cost, on timescales compatible with the actual hazard.

4.3 1950 DA

At this writing, there is only one known NEO with a potentially significant possibility of collision. For 29075 (1950 DA), integrations of the radar-refined orbit by Giorgini *et al.* (2002) revealed that in 2880 there could be a hazardous approach not indicated in the half-century arc of pre-radar optical data. The current nominal orbit represents a risk as large as 50% greater than that of the average background hazard due to all other asteroids from now through 2880, as defined by the Palermo Technical Scale (PTS value +0.17). 1950 DA is the only known asteroid whose danger could be above the background level. During the observations, a radar time-delay measurement corrected the optical ephemeris's prediction by 7.9 km, changing an optical-only prediction of a 2880 close approach to a nominal distance of 20 lunar distances into a radar-refined prediction of a nominal distance of 0.9 lunar distances.

The uncertainty in the closeness of 1950 DA's 2880 approach and hence in the probability of a collision (which could be as low as zero or as high as 1/300) is due to a combination of the factors in Table 3.8. The dominant factor is the Yarkovsky

Table 3.8 *Sources of uncertainty in 1950 DA's position during the 2880 close approach*[a]

Phenomenon	Relative maximum along-track effect
Solar particle wind	0.001
Galilean satellites	0.333
Galactic tide	0.833
Numerical integration error (64-bit)	1.000 (9900 km, 12 min)
Solar mass loss	1.333
Poynting–Robertson drag	2.333
Solar oblateness	4.083
Sun-barycenter relativistic shift	81.0
61 most perturbing "other" asteroids	144
Planetary mass uncertainty	156
Solar radiation pressure	1092
Yarkovsky effect	6924

[a] These factors are normally neglected in asteroid trajectory predictions spanning less than a century. From Giorgini *et al.* (2002).

acceleration, which is due to the anisotropic reradiation of absorbed sunlight as thermal energy and depends on the object's mass, size, shape, spin state, and global distribution of optical and thermal properties. Thus, unlike previous cases, predicting a potential 1950 DA impact with the Earth depends mostly on the asteroid's physical characteristics, not initial trajectory measurement. The accelerations are all small, but add up over time and are amplified by 15 close encounters with the Earth or Mars prior to 2880.

The 1950 DA example underscores the fundamental inseparability of the physical properties of NEAs and long-term prediction of their trajectories. The urgency of physically characterizing a threatening object naturally would increase as estimates of the collision probability rise and mitigation is transformed from a hypothetical possibility to an engineering requirement. If we take the hazard seriously, physical characterization of these objects deserves high priority.

5 Physical characterization

5.1 Images and physical models

With adequate orientational coverage, delay-Doppler images can be used to construct three-dimensional models (e.g., Hudson *et al.* 2000), to define the rotation state, and to constrain the internal density distribution. Even a single echo spectrum jointly constrains the target's size, rotation period, and sub-radar latitude. A series of Doppler-only echo spectra as a function of rotation phase can constrain the location of the center of mass with respect to a pole-on projection of the asteroid's convex envelope (e.g., Benner *et al.* 1999a). For objects in a non-principal-axis spin state, the hypothesis of uniform internal density can be tested directly (Hudson and Ostro 1995). Given a radar-derived model and the associated constraints on an object's internal density distribution, one can use a shape model to estimate the object's gravity field and hence its dynamical environment, as well as the distribution of gravitational slopes on the surface, which can constrain regolith depth and interior configuration.

For most NEAs, radar is the only Earth-based technique that can make images with useful spatial resolution. Therefore, although a sufficiently long, multi-apparition optical astrometric time base might provide about as much advance warning of a possibly dangerous close approach as a radar+optical data set, the only way to compensate for a lack of radar images is with a space mission.

5.2 *Extreme diversity*

As reviewed by Ostro *et al.* (2002), NEA radar has revealed both stony and metallic objects, principal-axis and complex rotators, very smooth and extraordinarily rough surfaces, objects that must be monolithic and objects that almost certainly are not, spheroids and highly elongated shapes, objects with complex topography and convex objects virtually devoid of topography, contact binaries, and binary systems. Figure 3.2 illustrates some of the diversity of NEAs. Obviously it is useless to talk about the physical characteristics of a "typical" PHA.

5.3 *Surface roughness and bulk density*

Porous, low-strength materials are very effective at absorbing energy (Asphaug *et al.* 1998). The apparently considerable macroporosity of many asteroids (Britt *et al.* 2002) has led Holsapple (2002) to claim that impact or explosive deflection methods may be ineffective, even for a non-porous asteroid if it has a low-porosity regolith only a few centimeters deep: "That leaves the low force, long time methods. However, even in those cases the problems of anchoring devices to the surface may make them very difficult."

The severity of surface roughness would be of concern to any reconnaissance mission designed to land or gather samples. The wavelengths used for NEAs at Arecibo (13 cm) and Goldstone (3.5 cm), along with the observer's control of the transmitted and received polarizations, make radar experiments sensitive to the surface's bulk density and to its roughness at centimeter to meter (cm-to-m) scales (e.g., Magri *et al.* 2001). An estimate of the surface bulk density offers a safe lower bound on the subsurface bulk density, and hence a lower bound on the

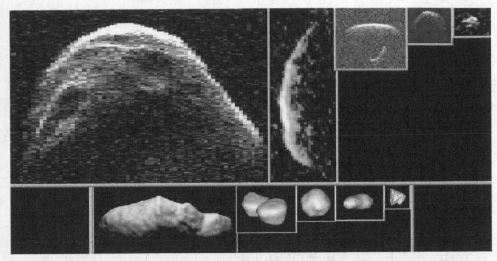

Figure 3.2 Radar delay-Doppler images and shape models. The top collage shows radar images of (left to right) 1999 JM8 (Benner *et al.* 2002a), Geographos (Ostro *et al.* 1996), the binary 1999 KW4 (Ostro *et al.* 2002), 1950 DA (Giorgini *et al.* 2002), and Golevka (Hudson *et al.* 2000). The bottom collage shows renderings of shape models of (left to right) Toutatis (Hudson *et al.* 2003), Castalia (Hudson and Ostro 1994), Nyx (Benner *et al.* 2002b), Bacchus (Benner *et al.* 1999b), and Golevka (Hudson *et al.* 2000). The relative scale of the images and models is approximately correct; Nyx is about 1 km in diameter.

asteroid's mass. Bulk density is a function of regolith porosity and grain density, so if an asteroid can confidently be associated with a meteorite type, then the average porosity of the surface can be estimated. Values of porosity estimated by Magri *et al.* (2001) for nine NEAs range from 0.28 to 0.78, with a mean and standard deviation of 0.53 ± 0.15. The current results suggest that most NEAs are covered by at least several centimeters of porous regolith, and therefore the above warning by Holsapple may be valid for virtually any object likely to threaten collision with Earth.

The fact that NEAs' circular polarization ratios (SC/OC) range from near zero to near unity (Fig. 3.3) means that the cm-to-m structure on these objects ranges from negligible to much more complex than any seen by the spacecraft that have landed on Eros (whose SC/OC is about 0.3, near the NEA average), the Moon, Venus, or Mars. 2101 Adonis and 1992 QN (Benner *et al.* 1997) and 2000 EE104 (Howell *et al.* 2001) are the extreme examples, with SC/OC near unity.

Ostro *et al.* (2002) claim that an asteroid's SC/OC can be taken as a crude estimate of the fraction of the surface area covered by roughly wavelength-sized rocks. To what extent might the surface rock coverage be representative of the structural configuration inside the object? NEA surfaces apparently can have rock coverages

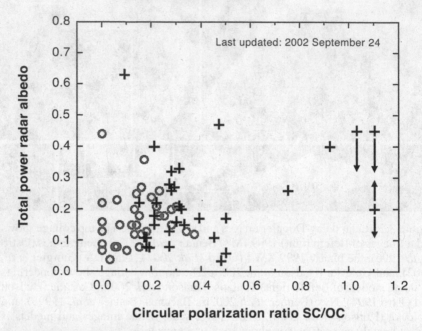

○ **Main-belt asteroids**

+ **Near-Earth asteroids**

Figure 3.3 Radar properties of NEAs and mainbelt asteroids. The two crosses with single arrows give upper bounds on the albedo; the cross with two arrows indicates an unknown albedo.

anywhere from negligible to total, and NEA interiors apparently can lie anywhere in the Richardson *et al.* (2002) relative-tensile-strength-vs.-porosity parameter space. Is there any relation between the two? If so, then an object's radar properties may indicate possibilities for its interior and hence for mitigation options. If not, then those properties still constrain options for spacecraft surface operations.

5.4 Binary NEAs: mass and density

The most basic physical properties of an asteroid are its mass, its size and shape, its spin state, and whether it is one object or two. The current Arecibo and Goldstone systems are able to identify binary NEAs unambiguously and at this writing have imaged 13 (Margot *et al.* 2002, and references therein; Nolan *et al.* 2002; references listed in Ostro 2004b), all of which are designated PHAs. Current detection statistics, including evidence from optical lightcurves (Pravec 2004) suggest that between 10% and 20% of PHAs are binary systems.

Analysis of echoes from these binaries is yielding our first measurements of PHA densities. Delay-Doppler images of 2000 DP107 (Margot *et al.* 2002) reveal a 800-m primary and a 300-m secondary. The orbital period of 1.767 d and semi-major axis of 2620 ± 160 m yield a bulk density of 1.7 ± 1.1 g cm^{-3} for the primary. DP107 and the five other radar binaries have spheroidal primaries spinning near the breakup-point for strengthless bodies. Whether binaries' components were mutually captured following a highly dispersive impact into a much larger body (Richardson *et al.* 2002, and references therein) or formed by tidal disruption of an object passing too close to an inner planet (Margot *et al.* 2002), it seems likely that the primaries are unconsolidated, gravitationally bound aggregates, so Holsapple's warning applies to them.

5.5 *Radar investigations, mission design, and spacecraft navigation*

Whether a PHA is single or binary, mitigation will involve spacecraft operations close to the object. Maneuvering near a small object is a non-trivial challenge, because of the weakness and complexity of the gravitational environment (Scheeres *et al.* 2000). Maneuvering close to either component of a binary system would be especially harrowing.

The instability of close orbits looms as such a serious unknown that unless we have detailed information about the object's shape and spin state, it would be virtually impossible to design a mission capable of autonomous navigation close to the object. Control of a spacecraft orbiting close to an asteroid requires knowledge of the asteroid's location, spin state, gravity field, size, shape and mass, as well as knowledge of any satellite bodies that could pose a risk to the spacecraft. Radar can provide information on all these parameters. Knowledge of the target's spin state as well as its shape (and hence nominal gravity harmonics under the assumption of uniform density) (Miller *et al.* 1999) would permit design of stable orbits immune to escape or unintended surface impact. (Upon its arrival at Eros, the NEAR Shoemaker spacecraft required almost 2 months to refine its estimate of the gravity field enough to ensure reliable close-approach operations.)

If it turns out to be necessary to have a sequence of missions beginning with physical reconnaissance and ending with a deflection, then a radar-derived physical model would speed up this process, reduce its cost, decrease complexity in the design and construction of the spacecraft, and improve the odds of successful mitigation. A reduced need for contingency fuel could be significant enough to allow a smaller launch vehicle for the mission. For example, the result might save $100 million via a switch from a Titan III launch vehicle to a Titan IIS, or $200 million for a switch from a Titan IV to a Titan III. The ability of prior radar reconnaissance to reduce mission cost, complexity and risk was embraced by the Department of Defense in their proposed Clementine II multiple-flyby mission (Hope *et al.* 1997), all of whose candidate targets either had already been observed with radar (Toutatis, Golevka) or were radar observable prior to encounter (1987 OA, 1989 UR).

Ironically, although PHAs include the lowest-ΔV rendezvous targets in the solar system, Japan's Hayabusa (MUSES-C) sample-return mission to 25143 Itokawa (1998 SF36) is the world's first rendezvous mission to a PHA. Results of radar imaging of that asteroid (Ostro *et al.* 2001) are being used by the Japanese Institute of Space and Astronautical Science in planning for the late 2005 rendezvous, and radar observations during the asteroid's mid-2004 close approach will be used for navigational assistance and to refine the model derived from the 2001 images. Radar-derived shape models of small NEAs have made it possible to explore the evolution and stability of close orbits (e.g., Scheeres *et al.* 1996, 1998), and this experience is currently being applied to Hayabusa.

Radar refinement of physical properties and radar refinement of orbits are very tightly coupled: shape modeling necessarily involves refinement of the delay-Doppler trajectory of the center of mass through the observing ephemerides. With very precise radar astrometry, a spacecraft lacking onboard optical navigation could be guided into orbit around, or collision course with, an asteroid. For example, consider how Goldstone observations shrunk the positional error ellipsoid of Geographos, an object already heavily observed by optical telescopes, just prior to a planned Clementine flyby of that target on August 31, 1994 (Ostro 1996). Before Goldstone ranging observations carried out during August 28–29, the overall dimension of the positional error ellipsoid was ~11 km. The radar astrometry collapsed the ellipsoid's size along the line of sight to several hundred meters, so its projection toward Clementine on its inbound leg would have been 11 × 2 km. Goldstone–VLA radar aperture synthesis angular astrometry (see discussions by de Pater *et al.* 1994, and Hudson *et al.* 2000), could have shrunk the error ellipsoid's longest dimension to about 1 km, about half of Geographos' shortest overall dimension. For less well-observed objects, the gains could be substantially more, as with 1862 Apollo's 3750 km radar range correction.

5.6 Modeling the efficiency of explosive deflection

Mitigation scenarios include the use of explosives to deflect the projectile (Ahrens and Harris 1992). However, as demonstrated by Asphaug *et al.* (1998), the outcome of explosive energy transfer to an asteroid or comet (via a bomb or a hypervelocity impact) is extremely sensitive to the pre-existing configuration of fractures and voids, and also to impact velocity. Just as porosity damps shock propagation, sheltering distant regions from impact effects while enhancing energy deposition at the impact point, parts of multi-component asteroids are preserved, because shock waves cannot bridge inter-lobe discontinuities. A radar-derived shape model would allow more realistic investigation (Asphaug *et al.* 1998) of the potential effectiveness of nuclear explosions in deflecting or destroying a hazardous asteroid.

5.7 Comets

The risk of a civilization-ending impact during this century is about the same as the risk of a civilization-ending impact by a long-period comet (LPC) during this millennium. At present, the maximum possible warning time for an LPC impact probably is between a few months and a few years. Comet trajectory prediction is hampered by optical obscuration of the nucleus and by uncertainties due to time-varying, non-gravitational forces. Comets are likely to be very porous aggregates, so concern about the ineffectiveness of explosive deflection is underscored in the case of comets.

Radar reconnaissance of an incoming comet would be the most reliable way to estimate the size of the nucleus (Harmon *et al.* 1999), could reveal the prevalence of centimeter-and-larger particles in the coma (Harmon *et al.* 1989, 1997), and would be valuable for determining the likelihood of a collision.

6 Recommendations

How much effort should be made to make radar observations of NEAs? For newly discovered objects, it is desirable to guarantee recovery and to ensure accurate prediction of close approaches well into the future, and at least throughout this century. Moreover, a target's discovery apparition often provides the most favorable radar opportunity for decades and hence a unique chance for physical characterization that otherwise would require a space mission. Similarly, even for NEAs that have already been detected, any opportunity offering a significant increment in echo strength and hence imaging resolution should be exploited. Binaries and non-principal-axis rotators, for which determination of dynamical and geophysical properties requires a long, preferably multi-apparition time base, should be observed extensively during any radar opportunity.

Construction of the proposed Large Synoptic Survey Telescope (LSST) has been endorsed (Belton *et al.* 2002), in part as a means to extend the Spaceguard Survey's

90% completeness goal for kilometer-sized objects down to 300-m objects. However, both Arecibo and Goldstone are already heavily oversubscribed, with only several percent of their time available for asteroid radar. Over the coming decades, it may become increasingly clear that most of the NEO radar reconnaissance that is technically achievable with Arecibo and Goldstone is precluded by the limited accessibility of those instruments, and that a dedicated NEO radar instrument is desirable.

An ideal NEO radar system (Ostro 1997) might consist of two antennas like the 100-m NRAO Greenbank Telescope (GBT, in West Virginia), one with a megawatt transmitter and one just for receiving, separated by a few tens of kilometers, operating at a wavelength of 0.9 cm (Ka band). Each antenna's gain could be 88 dB, compared to 73.5 dB for Arecibo. A two-antenna (bistatic) configuration would eliminate the frequent transmit/receive alternation and klystron power cycling required in single-antenna observations of NEOs and would double the available integration time. The antennas would be fully steerable, so any object could be tracked at least several times longer than at Arecibo. The combination of all these factors would make this dedicated NEO radar an order of magnitude more sensitive than the upgraded Arecibo telescope. The capital cost of building this system now, as calibrated by the GBT experience, would be within 10% of $180 million, comparable to the cost of a small Discovery mission and very close to the estimated cost of the LSST.

Acknowledgment

This research was conducted at the Jet Propulsion Laboratory, California Institute of Technology, under contract with the National Aeronautics and Space Administration (NASA). S. Ostro was partially supported by the Steven and Michele Kirsch Foundation.

References

Ahrens, T. J. and Harris, A. W. 1992. Deflection and fragmentation of near-Earth asteroids. *Nature* **360**, 429–433.

Asphaug, E., Ostro, S. J., Hudson, R. S., *et al.* 1998. Disruption of kilometre-sized asteroids by energetic collisions. *Nature* **393**, 437–440.

Beatty, J. K. 2002. Arecibo radar gets 11th-hour reprieve. Sky Publishing Corp. [cited April 23, 2004]. Available online at http://skyandtelescope.com/news/archive/article_285_1.asp.

Belton, M. J., Porco, C., A'Hearn, M., *et al.* (2002). *New Frontiers in the Solar System: An Integrated Exploration Strategy Solar System Exploration Survey*. Washington, DC: National Research Council.

Benner, L. A. M., Ostro, S. J., Giorgini, J. D., *et al.* 1997. Radar detection of near-Earth asteroids 2062 Aten, 2101 Adonis, 3103 Eger, 4544 Xanthus, and 1992 QN. *Icarus* **130**, 296–312.

Benner, L. A. M., Ostro, S. J., Rosema, K. D., *et al.* 1999a. Radar observations of asteroid 7822 (1991 CS). *Icarus* **137**, 247–259.

Benner, L. A. M., Hudson, R. S., Ostro, S. J., *et al.* 1999b. Radar observations of asteroid 2063 Bacchus. *Icarus* **139**, 309–327.

Benner, L. A. M., Ostro, S. J., Nolan, M. C., *et al.* 2002a. Radar observations of asteroid 1999 JM8. *Meteor. and Planet. Sci.* **37**, 779–792.

Benner, L. A. M., Ostro, S. J., Hudson, R. S., *et al.* 2002b. Radar observations of asteroid 3908 Nyx. *Icarus* **158**, 379–388.

Britt, D. T., Yeomans, D., Housen, K., *et al.* 2002. Asteroid density, porosity, and structure. In *Asteroids III*, eds. W. F. Bottke, A. Cellino, P. Paolicchi, and R. P. Binzel, pp. 485–500. Tucson, AZ: University of Arizona Press.

Chesley, S. R. 2004. Current impact risks. California Institute of Technology, Pasadena [cited April 23, 2004]. Available online at http://neo.jpl.nasa.gov/risks.

Chesley, S. R., Chodas, P. W., Milani, A., 2002. Quantifying the risk posed by potential Earth impacts. *Icarus* **159**, 423–432.

de Pater, I., Palmer, P., Mitchell, D. L., *et al.* 1994. Radar aperture synthesis observations of asteroids. *Icarus* **111**, 489–502.

Giorgini, J. D. 2004. Small-body astrometric radar observations. California Institute of Technology, Pasadena [cited April 23, 2004]. Available online at http://ssd.jpl.nasa.gov/radar_data.html.

Giorgini, J. D., Ostro, S. J., Benner, L. A. M., *et. al.* 2002. Asteroid 1950 DA's encounter with Earth in 2880: physical limits of collision probability prediction. *Science* **296**, 132–136.

Goldstein, R. M. 1968. Radar observations of Icarus. *Science* **162**, 903–904.

Harmon, J. K., Campbell, D. B., Hine, A. A., *et al.* 1989. Radar observations of comet IRAS-Araki-Alcock 1983d. *Astroph. J.* **338**, 1071–1093.

Harmon, J. K., Ostro, S. J., Benner, L. A. M., *et al.* 1997. Comet Hyakutake (C/1996 B2): radar detection of nucleus and coma. *Science* **278**, 1921–1924.

Harmon, J. K., Campbell, D. B., Ostro, S. J., *et al.* 1999. Radar observations of comets. *Planet. Space Sci.* **47**, 1409–1422.

Holsapple, K. A. (2002). The deflection of menacing rubble pile asteroids. In *Extended Abstracts from the NASA Workshop on Scientific Requirements for Mitigation of Hazardous Comets and Asteroids*, Arlington, VA, September 3–6, 2002, eds. E. Asphaug and N. Samarasinha. pp. 49–52.

Hope, A. S., Kaufman, B., Dasenbrock, R., *et al.* 1997. A Clementine II mission to the asteroids. In *Dynamics and Astrometry of Natural and Artificial Celestial Bodies, Proceedings of IAU Colloquium 165*, eds. I. M. Wytrzyszczak, J. H. Lieske, and R. A. Feldman, pp. 183–190. Dordrecht, The Netherlands: Kluwer.

Howell, E. S., Nolan, M. C., DeRemer, L., *et al.* 2001. Arecibo radar observations of near-Earth asteroid 2000 EE104. *Bull. Am. Astron. Soc.*, **33**, 1153.

Hudson, R. S. and Ostro, S. J. 1994. Shape of asteroid 4769 Castalia (1989 PB) from inversion of radar images. *Science* **263**, 940–943.

1995. Shape and non-principal-axis spin state of asteroid 4179 Toutatis. *Science* **270**, 84–86.

Hudson, R. S., Ostro, S. J., Jurgens, R. F., *et al.* 2000. Radar observations and physical modeling of asteroid 6489 Golevka. *Icarus* **148**, 37–51.

Hudson, R. S., Ostro, S. J., and Scheeres, D. J. 2003. High-resolution model of asteroid 4179 Toutatis. *Icarus* **161**, 348–357.

Magri, C., Consolmagno, G. J., Ostro, S. J., *et al.* 2001. Radar constraints on asteroid regolith compositions using 433 Eros as ground truth. *Meteor. and Planet. Sci.* **36**, 1697–1709.

Margot, J. L., Nolan, M. C., Benner, L. A. M., *et al.* 2002. Binary asteroids in the near-Earth object population. *Science* **296**, 1445–1448.

Milani, A., Chesley, S. R., Chodas, P. W., *et al.* 2002. Asteroid close approaches: analysis and potential impact detection. In *Asteroids III*, eds. W. F. Bottke, A. Cellino, P. Paolicchi, and R. P. Binzel, pp. 55–69. Tucson, AZ: University of Arizona Press.

Miller, J. K., Antreasian., P. J., Gaskell, R. W., *et al.* 1999. Determination of Eros physical parameters for near-Earth asteroid rendezvous orbit phase navigation. *Am. Astron. Soc.* Paper no. 99-463.

Morrison, D., Harris, A. W., Sommer, G., *et al.* 2002. Dealing with the impact hazard. In *Asteroids III*, eds. W. F. Bottke, A. Cellino, P. Paolicchi, and R. P. Binzel, pp. 739–754. Tucson, AZ: University of Arizona Press.

Nolan, M. C., Howell, E. S., Ostro, S. J., *et al.* 2002. 2002 KK_8. *IAU Circ. No.* **7921**.

Ostro, S. J. 1994. The role of groundbased radar in near-Earth object hazard identification and mitigation. In *Hazards due to Comets and Asteroids*, eds. T. Gehrels, pp. 259–282. Tucson, AZ: University of Arizona Press.

1996. Radar astrometry of asteroids, comets and planetary satellites. In *Dynamics and Ephemerides of the Solar System, Proceedings of IAU Symposium 172*, eds. S. Ferraz-Mello, B. Morando, and J.-E. Arlot, pp. 365–372. Dordrecht, The Netherlands: Kluwer.

1997. Radar reconnaissance of near-Earth objects at the dawn of the next millennium. *Ann. New York Acad. Sci.* **822**, 118–139.

2004a. Echo strength predictions. California Institute of Technology, Pasadena [cited April 23, 2004]. Available online at http://echo.jpl.nasa.gov/~ostro/snr/.

2004b. Radar-detected asteroids. California Institute of Technology, Pasadena [cited April 23, 2004]. Available online at http://echo.jpl.nasa.gov/ asteroids/index.html.

Ostro, S. J., Campbell, D. B., Chandler, J. F., *et al.* 1991. Asteroid radar astrometry. *Astron. J.* **102**, 1490–1502.

Ostro, S. J., Jurgens, R. F., Rosema, K. D., *et al.* 1996. Radar observations of asteroid 1620 Geographos. *Icarus* **121**, 44–66.

Ostro, S. J., Benner, L. A. M., Nolan, M. C., *et al.* 2001. Radar observations of asteroid 25143 (1998 SF36). *Bull. Am. Astron. Soc.* **33**, 1117.

Ostro, S. J., Hudson, R. S., Benner, L. A. M., *et al.* 2002. Asteroid radar astronomy. In *Asteroids III*, eds. W. F. Bottke, A. Cellino, P. Paolicchi, and R. P. Binzel, pp. 151–168. Tucson, AZ: University of Arizona Press.

Pettengill, G. H., Shapiro, I. I., Ash, M. E., *et al.* 1969. Radar observations of Icarus *Icarus*, **10**, 432–435.

Pravec, P. 2004. Binary near-Earth asteroids. Ondrejov Observatory [cited April 23, 2004]. Available online at http://www.asu.cas.cz/~asteroid/binneas.htm.

Richardson, D. C., Leinhardt, Z. M., Melosh, H. J., *et al.* 2002. Gravitational aggregates: evidence and evolution. In *Asteroids III*, eds. W. F. Bottke, A. Cellino, P. Paolicchi, and R. P. Binzel, pp. 501–515. Tucson, AZ: University of Arizona Press.

Scheeres, D. J., Ostro, S. J., Hudson, R. S., *et al.* 1996. Orbits close to asteroid 4769 Castalia. *Icarus* **121**, 67–87.

1998. Dynamics of orbits close to asteroid 4179 Toutatis. *Icarus* **132**, 53–79.

Scheeres, D. J., Williams, B. G., and Miller, J. K. 2000. Evaluation of the dynamic environment of an asteroid: applications to 433 Eros. *J. Guidance, Control and Dynamics* **23**, 466–475.

Shapiro, I. I., Ash, M. E., and Smith, W. B. 1968. Icarus: further confirmation of the relativistic perihelion precession. *Phys. Rev. Lett.* **20**, 1517–1518.

Yeomans, D. K., Ostro, S. J., and Chodas, P. W. 1987. Radar astrometry of near-Earth asteroids. *Astron. J.* **94**, 189–200.

Zellner, B. 1979. Asteroid taxonomy and the distribution of the compositional types. In *Asteroids*, ed. T. Gehrels, pp. 783–806. Tucson, AZ: University of Arizona Press.

4

Interior structures for asteroids and cometary nuclei

Erik Asphaug

University of California, Santa Cruz

1 Introduction

Mitigation of any hazard begins with a comprehensive understanding of the forces to be reckoned with. Reckoning – taking measure – is being done with great efficiency on one front: the first-order census of kilometer-scale near-Earth objects (NEOs) may be complete in the next few decades (Jedicke *et al.* 2003). But the most basic physical properties of these bodies remain unknown: how they are assembled, how they respond to tidal and impact stress, and how they will respond to the artificial perturbations that will one day be required.

This chapter provides an introduction to comets and asteroids and their geophysical evolution, and concludes with recommendations for theoretical and laboratory effort and spacecraft reconnaissance. Little is known for sure. Comprehensive introductions to the rapidly evolving science of comets and asteroids are found in the University of Arizona Press review volumes *Asteroids III* (Bottke *et al.* 2002) and *Comets II* (Festou *et al.* 2004).

1.1 Comets

Not long ago, interplanetary space near Earth was believed to be far emptier than it now appears. The only luminous entities besides the Moon were the passing comets, whose comae and tails can form some of the most extensive structures in the solar system, and which have been scrutinized since the dawn of astronomy. While these centuries of observation have led to an understanding of the dynamics and compositions of cometary envelopes, cometary nuclei – compact objects ranging from a few hundreds of meters to a few hundreds of kilometers diameter – remain a tight-wrapped mystery (Jewitt 1999; Meech *et al.* 2004).

Mitigation of Hazardous Comets and Asteroids, ed. M. J. S. Belton, T. H. Morgan, N. H. Samarasinha, and D. K. Yeomans. Published by Cambridge University Press. © Cambridge University Press 2004.

Despite the vivid displays for which comets are famous, nuclei themselves are typically darker than charcoal with albedos seldom above 0.05. From present observations, which are limited, there is nothing obvious to distinguish an inactive cometary nucleus from a primitive asteroid, and for this reason "asteroid" is sometimes used to describe any inactive small planetary body, regardless of presumed origin.

Active cometary nuclei are relatively pristine outer solar system objects emitting jets of volatiles during occasional (or long-term periodic) forays close to the Sun. Due to the swarm of particulates surrounding active nuclei, they are dangerous to approach at spacecraft flyby velocities. Only two comets (1 P/Halley and 19 P/Borrelly) have been successfully imaged (Keller *et al.* 1986; Soderblom *et al.* 2002) and at resolutions of ∼50 m per pixel at best. It is sometimes possible to use ground-based radar techniques to probe the nucleus of an active comet, and the diameters of active nuclei can be guessed through photometric subtraction. We have not yet obtained ground-based Doppler-delay radar images of cometary nuclei as we have for many near-Earth asteroids (NEAs) (see Chapter 3).

Compositions of cometary nuclei are inferred from the reflectance spectra of particulates found in their comae and from emission spectra of volatile fractionates in their comae and tails (e.g., Hayward *et al.* 2000). Nuclei – the objects of concern to mitigation – are by definition the residues of this active fractionation, so we don't know their composition first hand, but infer that they are relatively pristine reservoirs of icy outer solar system materials. The most hydrated meteorites and interplanetary dust particles (IDPs) may as far as we know be cometary or asteroidal, but these too have been highly processed relative to their parent bodies (see Chapter 8).

Densities of comets are believed to range between 0.2 and 1.0 $g\,cm^{-3}$ (Meech *et al.* 2004) based on analysis of P/Halley's non-gravitational motion, although allowable values for Halley formally range from 0.03 to 4.1 $g\,cm^{-3}$ (Peale 1989). A tighter but more model-dependent density constraint was derived by Asphaug and Benz (1994a) who demonstrated serious inconsistencies in all attempts using strength models to describe the tidal break-up of comet Shoemaker-Levy 9 around Jupiter into a "string of pearls" of ∼20 independent nuclei. They concluded that gravitational instability in a pulled-apart incohesive rubble pile is the only reasonable alternative, as discussed in detail below. Their density estimate of ∼0.5 to 1.0 $g\,cm^{-3}$ is tightly constrained by the rubble pile model, with lower or higher densities resulting in no clumping or a single massive clump, respectively, as confirmed by independent models (Solem 1994, 1995).

The validity of the rubble pile hypothesis is of singular importance to asteroid and comet mitigation, since a rubble pile does not respond at all like a monolith to impacts or explosions. Moreover, there exist gradations between incohesive sand piles, dry masses of rubble, bonded aggregates (snowballs or loose breccias), and

solid chunks of rock or ice. Each of these will have unique responses and requirements.

While the structural implications must be more directly addressed before we
anchor a mitigation technology to a simple model, it may well be that all comets
are rubble piles, for the Shoemaker-Levy 9 result is strongly corroborated by the
dozens of split comets that have imprinted themselves as crater chains on the icy
surfaces of Ganymede and Callisto (Schenk *et al.* 1996). Moreover, the rotational
periods of comets appear to be limited as one would expect for strengthless objects,
and their measured densities are consistent with ice-rich objects of high porosity.
A current review on comet structure, including a synopsis of the current evidence
for rubble piles, is provided by Weissman *et al.* (2004) in *Comets II*.

1.2 Asteroids

In comparison with comets, asteroids are newcomers to astronomy. But since the
discovery of the first asteroid in 1801, and the first near-Earth asteroid in 1898,
researchers have acquired far more data to work from, both astronomically and from
the recognition (see McSween 1999) that most meteorites are fragments of asteroids. Asteroids are typically better situated for spacecraft visitation than comets,
especially the thousands of near-Earth objects (NEOs) that are situated on relatively
easy trajectories and for which solar power for spacecraft operations is abundant.
Many NEOs are attainable at relatively low cost, providing us with ample opportunity to conduct geophysical exploration of these potentially hazardous objects
before one becomes an actual threat.

Asteroids are distinguished from comets by the absence of a coma. This enables
direct surface observation but also means we have less direct compositional information. Asteroid surfaces undergo a poorly understood "weathering" process (e.g.,
Clark *et al.* 2001) which masks their interior composition, even from proximal
observation. After a year of multispectral observation of asteroid 433 Eros from
orbit, Trombka *et al.* (2000) concluded it is an ordinary chondrite composition,
but only if one can explain away a large discrepancy in sulfur abundance without
invoking differentiation (to extract siderophiles including sulfur to the core of a
parent asteroid).

Asteroid compositions are represented by the meteorites found on Earth, although
the correlation is unclear. Meteorites are, in the words of McSween (1999), "samples
without geologic context." Establishing this context was, in fact, the primary science
goal of the NEAR mission, and it is of profound scientific importance to establish
clear links between NEOs, which are the immediate source for most meteorites,
and the many tons of exotic rocks sitting in collections here on Earth.

Asteroids and comets alike appear to be products of intense collisional bombardment (Chapman and Davis 1975; Stern and Weissman 2001). They are subplanetary in size, and not clearly dominated by either strength or gravity (Asphaug *et al.* 2002). The forces of evolution (such as tides and impacts) and the responses to that evolution (e.g., spin state, structure, and shape) may be the same for both classes of objects. Thus, a typical cometary nucleus may be structurally identical to a typical asteroid – the only difference being an enhanced ice fraction in a temperature realm where ice is mechanically not much different from rock. Alternatively cometary nuclei may turn out to be dramatically different from asteroids, with a different accretion history, a competent mantle, lag deposits, and structures welded by a mobile volatile inventory.

Asteroids derive from a primordial population of bodies which was, presumably, winnowed by a factor of ~1000 from the initial disk mass density[1] during the formation of terrestrial planets. They are evidently a residue of a chaotic dynamical process wherein planetesimals and planetary embryos, excited resonantly by Jupiter, were ejected in huge numbers. Those surviving – the parents of nearly all meteorites – need to be regarded as dynamically processed, perhaps highly selected specimens of primordial nebular material. Asteroids have, on average, been subject to a far more exciting history than most comets, although the most primitive asteroids appear to be bona fide relics from the protoplanetary nebula. It has been proposed that extinct comets contribute significantly to the asteroid population. While the contribution is now believed to be minor (see Weissman *et al.* 2002), it is recognized that compositional and structural similarities between primitive (C- and D-type) asteroids and evolved comets may be significant.

Clues to the interior structure of asteroids exist both in theory and in observation. Theory suggests that all asteroids larger than 1 kilometer are rubble piles – a result which was not generally accepted at first, but which is now backed up by rotation rate observations (see Pravec *et al.* 2002) and by observations that larger asteroids (all of the spacecraft-observed S-types) are crisscrossed by a network of fractures. Most compelling are the spacecraft and satellite-derived (Merline *et al.* 2002) measurements of asteroid density. Primitive C-type asteroids are less than half as dense as the closest meteorite analog, requiring porosities as high as 70% (compared with ~20% for sand), or else a very ice-rich interior. But Rivkin *et al.* (1997) found no evidence for hydrated mineral phases at C-type 253 Mathilde. The more highly evolved S-type asteroids such as 433 Eros (the class representing the majority of NEOs) also appear to have significant porosity (as high as ~40%

---

[1] If one distributes all the mass in the current planets, normalized to solar abundances, then the mass per unit area of the disk drops with distance from the Sun to approximately the $-3/2$ power, except for a 1000-fold mass deficit from Mars to the asteroid belt.

according to Consolmagno and Britt 1998). Rubble pile models are consistent with these high porosities, although it remains to be explained how large C-type asteroids can be much more porous than sand while sustaining central pressures exceeding 2×10^6 dyn cm^{-2} for Mathilde and 2×10^7 dyn cm^{-2} for Eugenia.

1.3 Satellite pairs

One similarity between comets and asteroids is particularly intriguing, though it may turn out to be coincidence. Binary systems (gravitationally bound pairs) are now known to be common among asteroid and comet populations alike. The first ground-based searches for main-belt and near-Earth asteroid satellites (Margot *et al.* 2002; Merline *et al.* 2002) found that 10–20% of all the large NEOs, and about the same fraction of large main-belt asteroids, are binaries. This agrees with cratering statistics on Earth, Moon, Mars, and Venus (Melosh and Stansberry 1991), which shows that ~15% of craters (produced by the impact of NEOs, of course) are doublets.

It has recently been discovered that binaries are fairly common in the trans-Neptunian population, and perhaps generally among comets (Noll *et al.* 2002). Discoveries are biased towards large semimajor axes and high mass ratios for these faint populations, and the first discovered pairs are almost at the limit where they become uncoupled by the Sun's tides. These could not be placed into orbit by impact unless far less massive than predicted by brightness models (Stern 2002). Weidenschilling (2002) proposes origin by disruptive capture, when two bodies collide inside a third object's Hill sphere leaving some fraction of bound debris. Goldreich *et al.* (2002) propose dissipative capture, when a swarm of smaller bodies exerts gravitational friction on two larger bodies already moving at low relative velocity. These pairs become bound and, until the swarm is dispersed, spiral inwards. A larger population of pairs with smaller semimajor axis is predicted by Goldreich *et al.* (2002) as well as a significant population of contact binaries.

These dynamical capture processes scale unfavorably to the inner solar system, where random velocities are higher, whereas collisional ejection of satellites scales favorably. Thus asteroid satellites are believed to form via collisions, either as an impact-ejected debris disk reaccumulating around a primary (as proposed for Earth–Moon formation) (Canup 2004), or as co-orbital ejecta fragments (Durda 1996). Alternatively, asteroid binaries may form during tidal disruption of cohesionless rubble piles (Bottke and Melosh 1996; see below). In either of these scenarios, the resulting asteroid binary system forms out of gravitationally reaccumulated debris, which is not the case for Goldreich *et al.*'s (2002) dissipative capture model for comets.

1.4 Distinctions

Comets and asteroids have distinct origins from distinct reservoirs, leading to significant differences in their original structure and also to the likely development of very different kinds of evolved structures. Comets, possessed of potentially mobile volatiles, may evolve into cohesive aggregates over time (e.g., Bridges *et al.* 1996), or form lag deposits and mantles, whereas a dry asteroid may become nearly strengthless if it evolves into a rubble pile (although very low cohesion can easily exceed gravitation as the binding force for small asteroids).

A second distinction is of particular importance to Earth-approaching comets: their intense activity when they enter the inner solar system. This activity is frequently catastrophic: a tenth of all comets disrupt during their first passage inside of 1 AU (Weissman 1980), sometimes without an identifiable trace. More commonly, cometary activity results in rapid mass wasting. If comets have undergone an impact evolution as intense as that of asteroids, as proposed by Stern and Weissman (2001), then the absence of craters on active nuclei (Keller *et al.* 1986; Soderblom *et al.* 2002) attests to widespread surface removal. By comparison, the most primitive asteroids appear to be saturated in relatively pristine giant crater forms (Thomas 1998).

A third distinction relates to the non-gravitational forces applied to active comets, which can excite their rotation state (Samarasinha and Belton 1995) possibly to the point of shape modification or disruption. There is growing evidence (Weissman *et al.* 2004) that cometary nuclei cannot sustain spin rates that exceed the threshold at which a rubble pile would fling itself apart (as is also the case for asteroids) (Harris 1996). Non-gravitational forces are now proposed for asteroids as well (Rubincam 2000), including forces which can excite their rotation from thermal radiation effects, although on a much longer timescale than is likely for an active comet.

2 Collisional evolution

The most significant geophysical forces acting on asteroids are planetary tides and mutual collisions; to comets one must add endogenic activity that can lead to mass wasting and disruption. While tidal disruption, discussed below, is not usually considered to be a common fate for small bodies, Asphaug and Benz (1996: Fig. 13) have shown that any *cohesionless* body suffering random encounters with a planet of twice its density (an asteroid near Earth; a comet near Jupiter) is about as likely to be destroyed by close tidal passage as it is to impact the planet. In other words, tidal disruption is as common as accretion for a population of rubble piles, or of bodies mantled in $\sim 1/5$ of a radius (half a sphere's mass) in regolith, or of molten bodies of radius R with viscosity lower than $\sim 10^{11}(R/100 \text{ km})^2$ poise (Asphaug *et al.* 2002).

Between tidal encounters, asteroids and comets are battered by impactors small and large, and comets are subject additionally to the pressure forces associated with ice sublimation and to jet-induced spin-up, so their geophysical evolution is complex.

2.1 Impactor populations

Depending on the size distribution of the impacting population (typically expressed as a cumulative power law $N_{>r} \propto r^{-a}$) impact evolution can either be dominated by a fusillade of small impacts ("sandblasting") or by a few major disruptive collisions. These evolutionary paths lead to radically different internal structure, and are thus of great importance from the perspective of hazard mitigation.

In terms of mass, momentum, and energy delivered by impact evolution, sandblasting dominates when $a > 3$ and large collisions dominate when $a < 3$. Since the size distributions of all measured populations appears to be shallower than 3, with typical values closer to 2, it is likely that a few major impact events have dominated the geophysical evolution of most small bodies. Korycansky and Asphaug (2003) found that the prolate shapes apparently common for comets and asteroids cannot be explained by sandblasting, but must be either original or cataclysmic. But their analysis is limited to gravity-scaled crater formation models. For a volatile-rich comet, a comparatively minor impact could have global or regional consequences due to sudden volatile expension, in which case pristine comets would suffer greatly in a modest collisional environment such as the inner solar system.

The small impacting population is not directly observable with present techniques, except as inferred from Earth-based meteor observations (see Chapter 8). Bell (2001) proposed a diminished population of small (meter-scale) impactors in the near-Earth asteroid population based on the absence of small craters and the persistence of blocks on the surface of asteroid 433 Eros. An absence of small impactors implies a removal mechanism for collisional debris. In the near-Earth population the Yarkovsky effect may selectively remove small particulates: this force results from asymmetric emission of blackbody radiation and is thus most effective at small sizes (sub-kilometer) and high emission temperatures (the inner solar system).

In the distant realm of comets, removal of fragmental material may also occur. While the Yarkovsky effect may be ineffective, mass wasting by sublimation may occur, which is another effect which scales as surface area divided by mass, or 1/size, so that the smallest comets may vanish. While observable Jupiter-family comets appear to follow a size distribution $a \approx 1.6$ (Weissman and Lowry 2004), the number of sub-kilometer comets with Earth-approaching periapses appears to be truncated in historical observation (Shoemaker and Wolfe 1982), with increasing telescope technology not resulting in a proportional discovery of small Earth-approaching comets. The cratering record in the Jupiter system (Zahnle *et al.* 2003) is also underrepresented by small comets. The break-up of comet LINEAR (D/1999 S4) led to water production rates measured by SOHO (Mäkinen *et al.* 2001) consistent with a fragment size distribution $a \approx 1.7$, suggesting that comets do break down to small sizes when they disrupt. If so, the smaller end of the population is somehow removed if comets follow a disruption cascade.

2.2 Response to impacts

However they bear the scars, typical comets may have suffered collisions as intense as those experienced by Eros and Mathilde – two well-studied asteroids whose geologies appear to be dominated by one or more impact craters on the verge of target disruption (Veverka *et al.* 1997; Thomas *et al.* 1999; Robinson *et al.* 2002; Sullivan *et al.* 2002). Rapidly rotating Eros comes close to intersecting its own Roche lobe (Miller *et al.* 2002) and is shaped by a large "saddle" probably caused by impact, while Mathilde has several craters which each span most of the asteroid's diameter.

Impact velocities in the outer solar system are slower than among the asteroids, and collision rates are lower, so the modern collisional evolution of comets is less intense than for asteroids. On the other hand, the ejection of comets from the accreting solar system might have led to frequent and intense collisions in their early history. Stern and Weissman (2001) calculate that Oort cloud comets are likely to have collided with a significant fraction of their own mass before dynamical ejection from the region of giant planets. If so, disruption was common.

Impacts into small comets might trigger the activity responsible for their rapid disintegration. It is known from seismic studies at Meteor Crater in Arizona (Ackerman *et al.* 1975) and from impact crater modeling (Asphaug *et al.* 1996) that significant fractures penetrate more than a crater diameter into the interior of a coherent geologic mass. A cratering event may therefore be expected to provide a direct pathway for the escape of interior volatiles, while also providing thermal energy in the form of a shock. Whether or not impact cratering triggers cometary activity, mass wasting appears to be a more rapid process than cratering on active nuclei, by the absence of identified craters on Halley and Borrelly.

3 Structural nomenclature

Richardson *et al.* (2002) established a common structural nomenclature for asteroids; for comets one must add the complexity of mantle development and perhaps melting of the insulated interior due to short-lived radionuclides (Prialnik and Podolak 1999), both of which can significantly alter the internal structure.

Monoliths are objects of low porosity and significant strength, and are good transmitters of elastic stress. Monoliths in an impact-evolved population must be smaller than the size that would accumulate its own impact ejecta. Escape velocity for a constant density body is proportional to its size (approximately 1 m s^{-1} per kilometer of diameter for spheres of ice), so that bodies larger than some transition size evolve into gravitational aggregates. Monoliths can be *fractured* by impact bombardment, in which case their tensile strength is compromised and may be reduced to zero, i.e., *shattered*. A fractured or shattered monolith might transmit a compressive stress wave fairly well, provided pore space has not been introduced between the major fragments. Tensile stress, however, is not supported across a fracture.

A *rubble pile* includes any shattered body whose pieces are furthermore translated and rotated into a loose packing. Stress waves of any sort are poorly transmitted across a rubble pile, although intense shocks may propagate by crushing and vapor expansion. Primordial rubble piles are objects that accreted as uncompacted cumulates to begin with, whether or not they have further evolved collisionally.

Differentiated asteroids (see Keil *et al*. 1994) are not of much interest to mitigation, except as progenitors of smaller silicate or iron NEAs, since all asteroids of this size range are known and none is a threat to Earth. These bodies (4 Vesta is believed to be an archetype) have a layered structure similar to that of terrestrial planets, with a core, mantle, and crust.

3.1 Kinds of aggregates

Rubble piles can have macro- or microporosity, leading to important distinctions in behavior. An asteroid or cometary nucleus consisting of quintillions of tiny grains might exhibit considerable cohesion and thus support high porosity. The total energy of contact bonds divided by the total mass of a granular asteroid (its bulk cohesion) is proportional to mass over contact area, therefore inversely proportional to grain diameter. A coarse aggregate is weaker than a fine one, if all else is equal. Moreover, a highly porous, finely comminuted body might accommodate significant compaction and thereby further resist impact disruption. Impact cratering on microporous bodies has been postulated to be a strange event involving crushing (Housen *et al*. 1999) rather than ejection.

A coarse rubble pile has far fewer contact surfaces distributed over the same total mass, and is therefore less cohesive, but similarly dissipative to impact energy. Asphaug *et al*. (1998) used a coarse rubble pile as a starting condition for impact studies, and demonstrated that the impact shock wave gets trapped in the impacted components, with few pathways of transmission to neighboring components. A coarse rubble pile thus experiences more localized cratering events, as is the case for microporous rubble piles. The difference is that energy confinement leads to most impact products escaping from the largest craters (Asphaug *et al.* 2002), as opposed to being compacted into the target. This distinction is of central relevance to mitigation, and is further addressed in the discussion of asteroid Mathilde, below.

Whether a comet or asteroid is macroporous or microporous, stress wave transmission is hindered due to the great attenuation of poorly consolidated ice and rock, making the survival of rubble piles more likely during impact, as demonstrated by the experiments of Ryan *et al.* (1991) and Love *et al.* (1993) and by numerical simulations.

3.2 Volatiles and cohesion

A volatile-rich aggregate (such as a cometary nucleus) is more cohesive than a dry aggregate (such as an asteroidal rubble pile) due to the facilitation of mechanical bonding, either directly (e.g. van der Waals forces) or indirectly during episodes of sublimation and frost deposition (Bridges *et al.* 1996). For gravity as low as on a typical cometary nucleus, frost or other fragile bonds can be critical to long-term survival during impact or tidal events. Comet Shoemaker-Levy 9, for example, could never have disrupted during its 1992 tidal passage near Jupiter, a nearly parabolic encounter with periapse of 1.3 jovian radii, had the tensile strength across the comet exceeded $\sim 10^3$ dyn cm^{-2} (Sekanina *et al.* 1994), weaker than snow. It thus matters greatly whether an asteroid or comet is truly strengthless or only extraordinarily weak.

Volatiles may help bind a comet, but as mentioned above their vapor expansion might help fuel a comet's disassembly during hypervelocity collisions. The energy of vaporization for ice is approximately ten times lower than that for rock, and the impact speed required to establish a shock wave is also lower. (On the other hand, impact speeds in the outer solar system are also lower.) The effect of supervolatiles such as CO, should they exist in sufficient quantities within the nucleus, would have an even more pronounced effect upon the expansion of impact ejecta (Durda *et al.* 2003). And so, while volatiles may provide some kind of structural integrity to an aggregate body, they may also reduce the size or speed of impactor required for catastrophic disruption.

Samarasinha (2001) offered expanding volatiles, propagating during solar approach from a warmer sub-surface layer to the interior of a coarsely porous comet, as an explanation for the disassembly of comet LINEAR (C/1999 S4) prior to perihelion. This brings up the final important effect of porosity, which is the tremendously efficient insulating property of granular media in vacuum. Weissman (1987) and Julian *et al.* (2000) showed that the surface thermal inertia for comet Halley was at least an order of magnitude less than that for solid water ice. If cometary nuclei have low thermal conductivities, then it is extremely difficult to transport energy during the course of a single perihelion to volatile reservoirs at depth. By extension (see Chapter 6), it may also be extremely difficult to couple the energies required of mitigation into a comet, unless technologies are developed to do so gradually.

4 The case for strength

The rubble pile concept has an interesting history, and it has met with resistance owing to the fact that a low-gravity body would seem to be quite fragile in the absence of appreciable strength. Davis *et al.* (1979) defined a threshold specific energy Q^*_D (impact kinetic energy per target mass) required to both shatter mechanical bonds and accelerate half an asteroid's mass to escaping trajectories. Shattering requires a lower specific energy $Q^*_s < Q^*_D$ to create fragments, none larger than half the target mass, not necessarily accelerating those fragments to dispersal. For small rocks $Q^*_D \rightarrow Q^*_s$ whereas for large bodies $Q^*_s/Q^*_D \rightarrow 0$. Whenever $Q^*_s << Q^*_D$, the probability of a shattering impact (which involves a smaller impactor) becomes far greater than the probability of dispersal, in which case a body might be expected to evolve into a pile of rubble, unless other effects (such as melting and compaction) were to dominate. The steeper the impactor population, the more likely it is that shattering will dominate dispersal, i.e., that the population evolves into rubble piles.

Davis *et al.* (1979) expressed impact strength as the sum of the shattering strength plus the gravitational binding energy of the target:

$$Q^*_D = Q^*_s + 4/5\pi\rho GR^2 \tag{4.1}$$

where R is the radius of a spherical target and ρ is its density. Equation 4.1 is called *energy scaling*; on a log–log graph of Q^*_D versus R it plots as a horizontal line ($Q^*_D \sim Q^*_s$ = constant), transitioning at a size of about 100 km (depending on Q^*_s) to a gravity-regime slope of 2. The size corresponding to this break in slope is known as the *strength–gravity transition* for catastrophic disruption.[2] Subsequent analysis has changed the slopes in both regimes (and the predicted transition size varies by orders of magnitude from model to model), but the concept was established that beyond some size, rubble piles might exist.

The quantity ρQ^*_s has dimensions of strength. It happens to be close to the corresponding static tensile strength of ice and rock in laboratory impact experiments (Fujiwara *et al.* 1989). Tensile strength was therefore used as an easily measured

---

[2] The strength–gravity transition for catastrophic disruption must be distinguished from the strength–gravity transition for planetary cratering. An object in the gravity regime for disruption (Earth is one) can certainly have strength-controlled craters.

proxy for ρQ^*s in early disruption theory. Upon this basis, it was concluded that primitive asteroids, comets, and early planetesimals (which are presumably of lower tensile strength than laboratory ice and rock) would be easily disrupted, and that survivors (the objects we see today) would be strong, intact bodies, and regolith would be thin or absent (Housen *et al.* 1979; Veverka *et al.* 1986). The ~100 km strength–gravity transition gained further support due to its consistency with two related notions: (1) the transition should occur when central pressure ~$2/3\pi G\rho^2 R^2$ equals rock strength; this transition occurs at about 100 km diameter for icy or rocky targets (rock being both stronger and denser), and (2) it should occur when gravitational binding energy per volume equals rock strength Y; neglecting constants this yields $R\rho = \sqrt{(Y/G)}$, and R of order several hundred kilometers, again whether for ice or rock.

With distinct ways of viewing small body structure converging upon a transition to the gravity regime at ~100 km sizes, the idea seemed safe that all but the largest comets and asteroids were monolithic. Certainly by the time of the spacecraft encounters with comet Halley, in 1986, the idea of structurally integral comets appeared to be on a solid foundation.

There were, however, some very serious problems with this conclusion. For one thing, the same collisional modeling predicted that bodies smaller than ~100 km would be unlikely to survive over billions of years, in contrast with their abundant population (see Chapman *et al.* 1989). An even more compelling argument against structural integrity was the manner in which comets come apart so effortlessly, as did P/Brooks 2 and, most recently, Shoemaker-Levy 9 (see below) and LINEAR. As we shall now see, the resolution to this dilemma appears to be that impact strength Q^*s and tensile strength Y are *not* simply related, and may even be *inversely* correlated. That is to say, structurally weak bodies are capable of absorbing large quantities of impact energy that would disrupt structurally strong bodies.

Experiments (Love *et al.* 1993) and modeling (Asphaug *et al.* 1998, 2002) have shown that loosely bonded aggregates can survive a projectile that would shatter and disperse an equal-mass monolith. It is now believed that some of the most fragile bodies in the solar system – porous aggregates with little or no cohesion – can be highly resistant to catastrophic disruption owing to their ability to dissipate and absorb impact energy. If so, there is no rationale for adopting tensile strength as a measure of an object's catastrophic disruption threshold. Equally important, ejection velocities from fragile bodies are correspondingly low, enabling them to hold on to their pieces. (In the strength regime, ejecta velocity scales with the square root of strength.) Like palm trees that bend in a storm, rubble piles may survive collisions that would shatter and disperse monolithic solids. They appear to be the natural end-state of comets and asteroids larger than a few hundred meters diameter.

5 Shoemaker-Levy 9

This comet was discovered in March 1993 as a chain of 21 discrete nuclei in orbit around Jupiter, and is nature's direct evidence for rubble pile structure. Dynamical integrations of the orbits of the nuclei backward in time (Chodas and Yeomans 1996) brought them together at a previous perijove passage on July 7.8, 1992 deep inside Jupiter's Roche limit.

The equilibrium tidal stress at the center of a homogeneous sphere is approximately

$$\sigma_T \sim GM_p\rho_c r_c^2 / R^3 \qquad\qquad (4.2)$$

where M_p is the mass of the planet (Jupiter), ρ_c and r_c are the density and radius of the comet, and R is the distance to the center of the planet; this is 1000 dyn cm^{-3} for a 1-km comet of density 0.6 g cm^{-3} at Shoemaker-Levy 9's periapse of 1.31 R_J. (Note also that tidal stress and lithostatic overburden both scale with r_c^2. This is the foundation for the scale-similarity of tidal disruption.) Since the closest one can get to Jupiter is $1R_J$, the split comets which imprinted themselves as chains of craters on the Galilean moons Ganymede and Callisto (Schenk *et al.* 1996) could have had strengths at most twice this value.

Shoemaker-Levy 9 provides a statistic of one. Whether all comets are strengthless is open to debate, for tidal break-up admits a bias, in that comets passing through the Roche zone which do *not* disrupt (strong bodies) bear no record of surviving this passage, other than being torqued into a new spin state and losing any loose surface material. For every split-comet crater chain (catena) on Ganymede and Callisto, there could be several impact craters by comets with had equally close Roche encounters with Jupiter but did not suffer disruption. But the number of catenae on the observed surfaces of Ganymede and Callisto is the same as one would expect (Schenk *et al.* 1996) if *every* comet disrupted in this fashion. While there is no guarantee that every comet in the Jupiter family is a rubble pile, catenae statistics provides strong evidence that most are. The break-up of Shoemaker-Levy 9, and by extention the cometary break-ups recorded at Ganymede and Callisto, constitute a direct record of cometary structure, and is thus worth contemplating further. The other direct record, that of cometary spin state, has further implications that are also consistent with the rubble pile structure of comets (Weissman *et al.* 2004).

5.1 Tidal disruption modeling

The first effort to physically model the tidal break-up of Shoemaker-Levy 9 resulted in a parent comet that was only 2 km across, five to ten times smaller than the more popular estimates. Scotti and Melosh (1993) assumed that Shoemaker-Levy 9 had consisted of 21 grains ("cometesimals") in a non-rotating aggregate spheroid, and its break-up was represented by evolving two non-gravitating test particles (extrema of the non-rotating aggregate) from identical but offset trajectories tangential to Jupiter at perijove. Their 2 km comet diameter was obtained by fitting the predicted end points to the extrema of the Shoemaker-Levy 9 fragment chain, neglecting self-gravity, rotation, and tidal torques.

On a different front, Asphaug and Benz (1994a) attempted modeling Shoemaker-Levy 9 as a solid elastic sphere undergoing brittle disruption. Using their fragmentation hydrocode (Benz and Asphaug 1994) they were able to reproduce a key theoretical result: if the comet's strength was lower than the small tidal stress, it would crack in two in the manner predicted by Dobrovolskis (1990). However, fracturing the comet into *many* pieces was impossible. Once the comet breaks in two, the tidal stress drops by a factor of four (by Eq. 4.2) and must build up to where it was (and then some, since the weakest flaws in the solid have been utilized) before further fragmentation can occur. In order to come apart into 20 pieces by the time of periapse, fracture would have to begin at a strength lower than ~ 30 dyn cm^{-2}, about a million times weaker than ice. Ironically, strength modeling led to the notion of the utterly strengthless comet.

Sensing a dead end to strength-based disruption scenarios, Asphaug and Benz (1994a) evaluated the Scotti and Melosh (1993) model by constructing an explicit cometesimal aggregate, using an N-body granular physics code with Jupiter as the central mass, and allowing the comet (a non-rotating aggregate) to approach Jupiter from far away. They observed that their cometesimal aggregate deformed into a "cigar" prior to periapse, and was strongly torqued by Keplerian shear to rotate retrograde. These factors evidently cancelled, for they obtained, in their fits to the Shoemaker-Levy 9 chain length, the same 2 km diameter parent comet derived by Scotti and Melosh (1993) with their simpler model.

But the model of ~ 21 discrete cometesimals failed, because self-gravitation was observed to form clumps or pairs among the grains unless density was decreased to very low values (0.05 g cm^{-3}). For reasonable densities, the number of observable fragments (clumps) was always significantly lower than the number of grains. To form ~ 20 fragments, ~ 100 or more grains were required – no longer a cometesimal model, but a rubble pile model instead. For $N > 200$ initial grains, self-similarity took over, and what was revealed was a tidal disruption process responding to self-gravitational instability (e.g., Chandrasekhar 1961). This mode of tidal disruption is scale-invariant (Asphaug and Benz 1996), and unlike the cometesimal model readily explains why large tidal catenae (crater chains) on Ganymede and Callisto do *not* consist of more craters than small ones (Schenk *et al.* 1996). The cometesimal model (e.g., Weidenschilling 1997) on the other hand implies that a comet twice as big will break up into 2^3 times as many pieces.

Sekanina *et al.* (1994) constructed the first detailed kinematical model for the break-up, evolving all the observed fragments backwards in time to the 1992 tidal break-up. To within error, the fragment trajectories emerge from a large object (10 km minimum diameter) at a time $\sim$2 h after perijove (1.5 h later in Sekanina *et al.* 1994; 2.5 h later in Sekanina 1996). This reconstruction is of great value, as it takes the fragments back to the point where more complex dynamics (tidal torques and self-gravitation) would apply. The interpretation they offered was that a very large (>10 km) parent comet suffered brittle catastrophic disruption, but waited until a few hours after the peak tidal stress to come apart. However, Asphaug and Benz (1996) found that these values were entirely consistent with a 1.5-km diameter rubble pile undergoing tidal distortion during periapse passage. The much smaller comet, in their rubble pile model, becomes a 10 km long "cigar" by $t = +2$ h, and is torqued to rotate at the same rate deduced by Sekanina *et al.* (1994). This also turns out to be the time that mutual gravity between the clumps can be ignored, so all aspects of the rubble pile description are entirely consistent with the kinematic requirements.

Asphaug and Benz (1994a), and shortly thereafter Solem (1994), estimated the size of the parent comet by fitting to the length of the fragment chain 1 year after the break-up, and obtained a best fit of 1.5 km diameter if the parent comet was not rotating, and 1.0 km if it was rotating prograde with a period of 6 h. A retrograde rotating progenitor was not possible, as this would have prevented the formation of the highly symmetric chain that was observed. A fast prograde rotation required a somewhat higher density for the parent comet, as large as $\sim$1.1 g cm^{-3}.

Unlike cometesimal or brittle fracture models, the rubble pile models predicted the symmetric appearance of the Shoemaker-Levy 9 "string of pearls," with the largest objects near the center. Moreover, the derived $\sim$1.5 km diameter is near the median for all disrupted comets striking the Jovian satellites Ganymede and Callisto (Schenk *et al.* 1996), making Shoemaker-Levy 9 a fairly typical cometary nucleus (despite its extraordinary fate).

Cometary density plays a key role in determining the final, post-break-up configuration of clumping in the tidally disrupted rubble pile. Asphaug and Benz (1994a) and Solem (1995) found that for densities less than $\sim$0.3 g cm^{-3}, the resulting chain of clumps was highly diffuse. For densities greater than 1.0 g cm^{-3}, the aftermath is one central clump containing >80% of the mass. These outcomes are, as expected from dimensional analysis (Asphaug and Benz 1996), self-similar for any density, when periapse distance is normalized to the density-dependent Roche limit. For Shoemaker-Levy 9, where periapse was known, morphology thus implicates nucleus density. The result is that the parent comet's density was between 0.5 and 0.9 g cm^{-3} (Asphaug and Benz 1996) or between 0.5 and 0.6 g cm^{-3} (Solem 1995).

Reassembly of the comet's material after tidal disruption, in the rubble pile model, will lead to post-disruption clumps of lower density. Crawford (1997) used a hydrocode with advanced thermodynamics (CTH) to obtain best matches to the Galileo near-infrared mapping spectrometer (NIMS) observations of the Shoemaker-Levy 9 impacts. His best fit was for impactor densities of 0.25 g cm^{-3}, and for a parent comet with the same mass as derived by Asphaug and Benz (1994a). Evidence thus indicates that the 1.5 km diameter, $\sim 0.6 \text{ g cm}^{-3}$ initial comet reassembled into nuclei about half the bulk density as the parent nucleus which then struck Jupiter.

The rubble pile hypothesis provides by far the most economical explanation for the ensemble data from Shoemaker-Levy 9's disruption and impact into Jupiter, and the record of catenae on Ganymede and Callisto. But it requires that comets are no more cohesive than an unpacked snowball. There is one caveat to this assumption. Asphaug and Benz (1994a) calculated that Shoemaker-Levy 9 might have encountered many kilograms of ring material at about 45 km s^{-1} upon ingress, 1 h prior to perijove. Interloping comets have a high probability of experiencing maximum tidal stress after crossing the ring plane near Jupiter, and in past epochs might have encountered more massive Jovian rings. A 1 m ring particle striking at 45 km s^{-1} delivers the equivalent energy of 1 kiloton of TNT, and this may be sufficient to break the structural bonds prior to tidal effects.

6 Size- and rate-dependent strength

Consider large rock or ice masses from a purely theoretical perspective. They are weak because strength is sensitive to target size. If the static failure stress σ_s of a rock decreases with flaw size L as $\sigma_s \propto 1/\sqrt{L}$ (Griffith 1920), and if the maximum flaw size in a target rock increases in proportion to its radius (as postulated by Fujiwara (1980) in his argument for size-dependent strength), then impact strength – if equated with static failure stress – depends on target size R as

$$Q^*_S \propto 1/\sqrt{R} \qquad (4.3)$$

When plotted on a graph of Q^*_D versus R this has a slope of $-1/2$ in the strength regime: a centimeter-size meteorite is ~ 1000 times stronger than a 10-km asteroid, according to this relation.[3] (See also Farinella *et al.* 1982.) This has tremendous implications for the survival of asteroids. If strength diminishes this rapidly, self-gravity becomes the dominant binding force at sizes considerably smaller than previously considered.

A problem crops up at this point: the larger asteroid is only weaker *somewhere*, otherwise every cubic centimeter of a large asteroid would be weaker than every cubic centimeter of a small asteroid of the same material. To avoid such nonsense,

---

[3] Neglecting strength-selection effects of meteorites surviving passage to Earth, which makes them stronger still. Observations of atmospheric disruption of bolides (Borovicka *et al.* 1998) shows meter-sized meteoroids to be quite weak, with early fragmentation strengths of order 10^6 dyn cm^{-2}.

concepts of heterogeneity are required. Instead of considering only the single weak-est flaw, consider an asteroid of volume $V \sim R^3$ which samples a theoretical contin-uum riddled with a power-law distribution of flaws. These can be grain boundaries, pores, inclusions, or cracks opened by previous collisions.

Suppose the number of flaws per unit volume that begin to grow at or below a stress σ is described by a power law

$$n(\sigma) = k\sigma^m \tag{4.4}$$

The probability (Weibull 1939) of finding a flaw in a random volume V that will begin to grow at or below σ is then $1 - \exp[-(\sigma/\sigma_{min})^m]$, where

$$\sigma_{min} = (kV)^{-(1/m)} \tag{4.5}$$

The threshold for failure σ_{min} thus goes with the $-3/m$ power of size. For reasons not yet understood, $m \sim 6$ may be favored by nature (Housen and Holsapple 1999), in which case one recovers Eq. 4.3 assuming $Q^*_S \propto \sigma_{min}$. But values of m as high as 57 and as low as 2 have been reported for rocks in the literature (Grady and Kipp 1980) (see Table 4.1) and no values of m or k have been reported for any meteorite. Hierarchical bolide fragmentation events in Earth's atmosphere (Artemieva and Shuvalov 2001) are consistent with nominal Weibull exponents ranging from 6 to 9.

In some instances m may be determined by direct measurement of the flaws in a geologic specimen. However, this introduces a bias, as flaw sizes approaching the sample size tend to be excluded, and flaws larger than the sample will not exist. More generally m is determined by fitting laboratory experiments to the strength–strain rate relations described below, although these experiments are limited to small specimens and to the small flaws activated by dynamic fragmentation. In most asteroid impact models, fracture parameters for basalt, granite, or ice are assumed; the sensitivity of these models upon m precludes one from having robust faith in blind forward modeling.

6.1 Rate dependence

The above analysis of σ_{min} provides a static explanation for why Q^*_S should dimin-ish with size. But Q^*_S is further diminished on account of dynamical fracture mechanics: rock strength is a function of the loading rate $\dot{\varepsilon} = d\varepsilon/dt$, where ε is the mechanical strain. It has long been established that the dynamic failure strength of rocks scales with the $\sim 1/4$ to $\sim 1/3$ power of $\dot{\varepsilon}$ (Rinehart 1965). Recognizing that $\dot{\varepsilon}$ decreases with the size of the collisional event, Holsapple and Housen (1986) explored the implication of low strain rate collisions in a revised set of scaling models.

Table 4.1 *Weibull dynamic fracture coefficients for various rocks. Although* k *ranges by ~150 orders of magnitude, this is because homogeneous materials (such as limestone reported below) have large* m. *Metals are approximated by* m→∞. *Because the static failure threshold* σ_{min} *is unchanged so long as* ln(k)/m *remains constant (see text), homogeneous materials of a given strength have arbitrarily large* k. *Equation 4.5 shows how rocks with large* m *exhibit little size dependence; metallic monoliths should be about equally resistant to fracture under a body strain (such as tides) regardless of their size, whereas rocky monoliths fracture more easily the larger they are. While dynamic fracture coefficients have yet to be measured for any meteorite, Housen and Holsapple (1999) propose that* m ~ 6 *might generally describe "well-cracked rocks at large scale," and hence might apply to asteroids. From time-resolved observations of meteoroid airbursts (Borovicka* et al. *1998) which exhibit hierarchical fragmentation as the ram pressure increases, there is a distinct possibility of constraining these parameters with genuine free-space meteoroids*

Material	Reference	m	k (cm^{-3})	ln(k)/m
Basalt[a]	Melosh *et al.* (1992)	9.5	1.0×10^{27}	6.54
Basalt[a]	Benz and Asphaug (1995)	9.0	4.0×10^{29}	7.17
Basalt[b]	Lindholm *et al.* (1974)	9.5	1.59×10^{30}	7.32
Granite[b]	Grady and Lipkin (1980)	6.2	4.14×10^{17}	6.54
Water ice[a,b,c]	Benz and Asphaug (1999)	9.6	1.4×10^{32}	7.71
30% Sand + water ice[b]	Stewart *et al.* (1999)	9.57	1.34×10^{30}	7.25
Concrete[b]	Grady and Lipkin (1980)	5.3	5.27×10^{12}	5.53
Oil shale[b]	Grady and Kipp (1980)	8.1	1.70×10^{21}	6.04
Limestone[b]	Grady and Lipkin (1980)	57.0	4.26×10^{167}	6.77

[a] Determined from simulation fits to laboratory data. The two-dimensional axisymmetric simulations of Melosh *et al.* (1992) require stronger fracture parameters than the non-symmetric three-dimensional simulations of Benz and Asphaug (1995) for the same impact experiment.

[b] Determined experimentally through measurements of tensile strength versus strain rate.

[c] Obtained by direct fits to the original experimental data of Lange and Ahrens (1983). Published values of $m = 8.7$, $k = 3.2 \times 10^{38}$ (Lange and Ahrens 1983) were incorrect fits to the data, and were corrected (Stewart *et al.* 1999) to values ($m = 9.57$, $k = 1.28 \times 10^{32}$), essentially the same as those derived (also from the original data) and applied by Benz and Asphaug (1999).

If $\dot{\varepsilon}$ is approximated as the impact speed divided by the impactor radius, then a 1-km diameter impactor striking at 5 km s^{-1} couples at a strain rate $\dot{\varepsilon} \sim 10$ s^{-1}. Though undeniably dynamic, this is a far cry from the $\dot{\varepsilon} \sim 10^6$ s^{-1} typical of laboratory experiments. Moreover, stress waves broaden and decay with distance, so $\dot{\varepsilon}$ drops with a steep ($\sim$4th) power of distance during hypervelocity collisions (Melosh *et al.* 1992). Strain rates responsible for shattering of $\sim$10 km asteroids can be as small as 10^{-3} to 10^{-5} s^{-1} (Asphaug and Melosh 1993).

Grady and Kipp (1980) applied Weibull statistics (Eq. 4.4) to derive a dynamic fragmentation model computing fracture stress and fragment size as a function of $\dot{\varepsilon}$. Except for dynamical loads of such intensity that brittle fragmentation is not the mode of failure, cracks grow at a rate c_g which is about half the sound speed (Lawn and Wilshaw 1975). Because cracks relieve stress over finite time, a dynamic equilibrium is established whereby strong flaws become active as needed, when weaker flaws cannot grow fast enough to accommodate the accumulating stress. Therefore the weakest flaws are sufficient to relieve the stress under low strain rates, resulting in large fragments (due to the low density of weak flaws) and a low measured strength. At high strain rates, on the other hand, stronger and more numerous flaws are called into play. This leads to the formation of small fragments (due to the high density of strong flaws) and at high measured strength. The resulting relationships for fracture strength and fragment size (Grady and Kipp 1980) are, respectively:

$$\sigma \propto \dot{\varepsilon}^{3/(m+3)} \tag{4.6}$$

where exponents of 1/3 and 1/4 (Rinehart 1965) corresponds to $m = 6$ and $m = 9$ respectively, and

$$L \propto \dot{\varepsilon}^{-m/(m+3)} \tag{4.7}$$

so that fragment size L is inversely proportional to strain rate for large m.

These same assumptions are the basis for numerical models of dynamic fracture of brittle solids (Melosh *et al.* 1992; Benz and Asphaug 1994, 1995), namely that (a) cracks initiate from a Weibull distribution in accordance with the applied tensile stress; (b) the crack tip accelerates instantaneously to half the sound speed; (c) planar cracks of radius a relieve deviatoric and tensile stress in a spherical circumscribed volume; and (d) a volume is damaged (stress-relieved) according to the ratio of cracked volume to total volume. Fully damaged rock, according to these models, is relieved of all stress except compressive pressure, and behaves as a fluid.[4] Differences in these models depend primarily upon implementation, with Grady

---

[4] A straightforward modification is to have fully damaged rock behave as a cohesionless Mohr–Coulomb material, where shear stress is still resisted according to the applied normal stress according to material's angle of internal friction (angle of repose).

and Kipp (1980) assuming $\dot{\varepsilon} =$ constant to derive closed integrals, Melosh *et al.* (1992) allowing integration over time-varying strain rate, and Benz and Asphaug (1994, 1995) introducing an explicit flaw distribution plus three-dimensional hydro-dynamics in addition to time-varying strain rate.[5] Without explicit flaws (assigned at random from Eq. 4.4 to all discretized volume elements) large rock volumes are modeled as homogeneously weak everywhere, when in fact rock strength is approximately fractal.

Experimental verification of rate dependence in catastrophic collisions was obtained by Housen and Holsapple (1999) in a series of experiments involving cylindrical granite targets ranging from 1.9 to 34.4 cm diameter. These targets, identical except for size, were impacted by proportional diameter aluminum cylin-ders at uniform speed ($\sim$0.6 km s^{-1}), with Q^* held constant at $\sim$1.0 $\times$ 10^7 erg g^{-1} for each event. The smallest target is merely chipped by the bullet, while the largest target (everything held constant but size) is catastrophically demolished. Unpublished hydrocode simulations by the author and L. S. Bruesch find that $m = 7.0$ and $k = 1 \times 10^{27}$ cm^{-3} reproduce the entire suite of laboratory collisions to within experimental error.

At low strain rates, not only is strength diminished, but so is ejection velocity. Fragment ejection energy ($1/2\ v_{ej}^2$) is proportional to the energy of fragmenta-tion (strength), which decreases with target size. Thus, the fragments from a large asteroid are ejected more slowly, not counting the effects of gravity. This reduced ejection velocity is a significant aspect of rate-dependent strength for asteroid colli-sional evolution and disruption, since as $v_{ej} \rightarrow v_{esc}$ self-gravity begins to dominate. If Eq. 4.3 applies, one obtains $v_{ej} \propto R^{-1/4}$ for a characteristic ejection speed during catastrophic disruption, whereas $v_{esc} \propto R$.

Rate-dependent and size-dependent strength, together with the diminished ejec-tion speeds for large/weak targets, combine to greatly increase gravity's influence over catastrophic disruption for small asteroids. Rate-dependent scaling theories (Holsapple and Housen 1986) pushed the threshold size for catastrophic disruption down to tens of kilometers diameter. This revised threshold was later pushed to hun-dreds of meters (Melosh and Ryan 1997; Benz and Asphaug 1999) in dynamical fragmentation models which included both rate and size dependence.

7 Rotation rates

Recent discussion regarding monolithic asteroids has centered around a simple but profound observational result by Pravec and Harris (2000). None of the $\sim$1000 aster-oids larger than 150 m with reliably measured spin periods rotates fast enough to

---

[5] In these numerical models, the total stress tensor is rotated into principal axis coordinates and the maximal value (most tensile) it taken as the stress σ which activates the Weibull flaws.

require global cohesion. Specifically, for reasonable density estimates none rotates faster than $\omega_o{}^2 = 4\pi G\rho/3$, where ω_o is the frequency at which material on a sphere's equator becomes orbital. The corresponding minimum rotation period is $P_{crit} = 3.3$ h/$\sqrt{\rho}$, with ρ in g cm^{-3}. The most dense common asteroids (S-type) appear to have $\rho \sim 2.7$ g cm^{-3} (Belton *et al.* 1996; Yeomans *et al.* 2000); thus an S-type gravitational aggregate can spin no faster than $P_{crit} \sim 2.0$ h (see Holsapple (2001) for a more detailed analysis of rotational break-up). From the Pravec–Harris asymptote at 2.2 h one might infer a common maximum density $\rho \sim 2.3$ g cm^{-3}, but because asteroids are not spherical the corresponding density is greater: equatorial speeds are faster on an elongated body and the mass distribution is non-central.

The easiest interpretation is that nearly all asteroids larger than $\sim$150 m lack cohesion. There are other plausible interpretations. Larger asteroids might be internally coherent but possess thick regolith, shedding mass into orbit whenever random spin-ups (or spin-up by the Yarkovsky effect) (Rubincam 2000) cause them to transgress the 2.2-h rotation period, and then transferring angular momentum to this orbiting material. In this case the $\sim$150 m transition may be the minimum size required for a body to retain regolith, and may have less to do with asteroid internal structure.

Another possibility is angular momentum drain (Dobrovolskis and Burns 1984). Angular momentum is lost whenever prograde ejecta (that launched in the direction of an asteroid's rotation) preferentially escapes while retrograde ejecta remains bound. This is a good explanation for the relatively slow rotations of $\sim$50–100 km diameter asteroids, but for this process to apply to $\sim$150 m asteroids the ejecta mass–velocity distribution must have a significant component slower than $v_{esc} \sim 5$–10 cm s^{-1}. Such slow ejection speeds are typical of gravity regime cratering; for craters on $\sim$150-m bodies to be governed by gravity, the target must be strengthless.

Only one asteroid smaller that 150 m diameter (out of $\sim$25 total) with measured rotation rate rotates *slower* than this limit. (These statistics change with monthly observation; see Pravec *et al.* 2002.) While no asteroid larger than $\sim$150 m shows evidence for global cohesion, almost all asteroids smaller than this must be cohesive. One might suppose that every larger asteroid is a gravitational aggregate of smaller pieces, whereas every smaller asteroid is a fast-rotating collisional shard. But the term "monolith" for these smallest asteroids is misleading. Consider a spherical object of uniform density ρ rotating with a frequency ω; the mean stress across its equator is $\sim R^2 \rho \omega^2$. For the well-studied fast rotator 1998 KY26 (Ostro *et al.* 1999), self-gravity is not capable of holding it together; however, its $\sim$11 min period and $\sim$30 m diameter requires only a tensile strength of $\sim$300 dyn cm^{-2} (presuming $\rho \sim 1.3$ g cm^{-3} for this C-type), orders of magnitude weaker than the tensile strength of snow.

8 Simulating the giant craters

Giant craters probe the brink of catastrophic disruption and thereby elucidate impacts at geologic scales – a process masked by gravity on Earth. With three-dimensional hydrocodes including rate-dependent strength running on modern computers (see Fig. 4.1), one can model the details of their formation. Impact models can now be tested against targets of asteroidal composition and asteroidal scale. Rather than relying on blind extrapolation from laboratory experiments, we can match the outcomes of detailed impact models against existing giant craters on asteroids. This begins with an asteroid shape model transformed into a simulation grid (Fig. 4.2) (see Asphaug *et al.* 1996). For a giant crater, one must then reconstitute the shape so as to fill in the volume deficit, hopefully leaving a good approximation of the pre-impact target. One then makes initial guesses as to interior structure and composition, as these are to be tested. For the impactor, one can assume nominal incidence angle and impact velocity and adopt some value (that predicted by gravity scaling perhaps[6]) for the mass. One can then iterate upon interior geology, and if required, upon the mass and velocity of the impactor, until one arrives at a satisfactory fit to observables: distribution or absence of crater ejecta, global or regional fractures associated with the event, boulder distributions, etc.

Because high-fidelity modeling remains computationally expensive, and data analysis is not yet automated for this kind of inversion, only rough iterations have been performed to date for a handful of asteroids.

8.1 Ida

Ida is an irregular-shaped S-type main belt asteroid with mean radius 15.7 km (Thomas *et al.* 1996). Figure 4.2 shows a hydrocode representation of this target. As with Gaspra (Carr *et al.* 1994) and Eros (Prockter *et al.* 2002), this S-type asteroid exhibits expressions of internal shattering in the form of linear furrows almost certainly related to large collisions. Asphaug *et al.* (1996) attempted to recreate major impacts at Ida including the ~12 km Vienna Regio (Fig. 4.2 upper panel). Figure 4.2 (lower panels) plots particle velocity through a slice of the target 9, 10, and 12 seconds after a gravity-scaled 330 m diameter projectile strikes at 3.55 km s^{-1} (the average v_{impact} at Ida) (Bottke *et al.* 1994). Gravity scaling assumes constant g (adopting the local effective gravity at the impact site) when in fact g varies by a factor of ~4 across the asteroid (Thomas *et al.* 1996), and by a factor of ~2 over the region of this particular collision. Wave interference is complex in this

---

[6] Craters larger than about ~1 km diameter appear to form in the gravity regime for ~20 km diameter monolithic asteroids (Asphaug *et al.* 1996), so this may be a reasonable starting assumption.

E. Asphaug

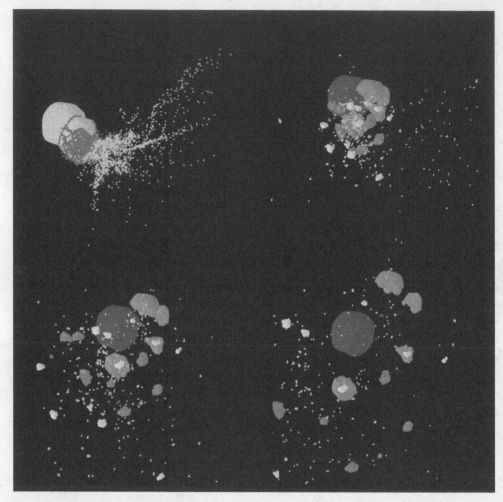

Figure 4.1 A numerical simulation of the catastrophic disruption of near-Earth asteroid Castalia, upon impact by a 16-m diameter rocky meteoroid traveling at $5\,\mathrm{km\,s^{-1}}$. The target asteroid is modeled as a basalt contact binary ($2.7\,\mathrm{g\,cm^{-3}}$), that is, two competent rocky spheroids, each about 800 m diameter, in close contact (from Asphaug and Scheeres 1999). In the first panel light color corresponds to unfractured rock (for computational efficiency the impactor fails by plastic deformation not fracture, and is hence colored white). In the remaining panels fragments are color coded logarithmically by size. Note the intact survival of the distal half of the binary. The structural parameters of the target body can only be guessed; its shape is rendered from radar observations (Ostro *et al.* 1990). This simulation gives some idea of the detail in which catastrophic outcomes can be modeled, but *only if target properties are known*. Asphaug *et al.* (1998) and Michel *et al.* (2003) find that varying target structures lead to radically different outcomes.

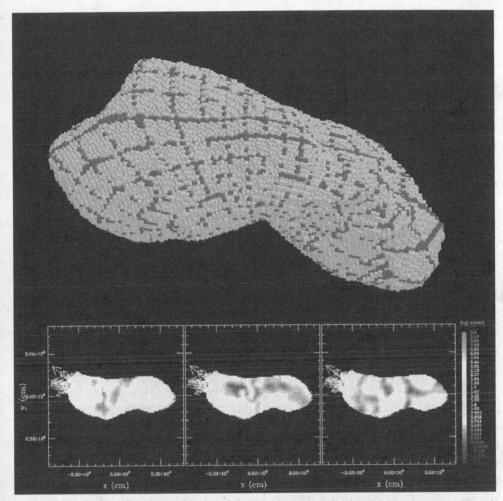

Figure 4.2 Asteroid 243 Ida, rendered in a smooth particle hydrocode (SPH) simulation (from Asphaug *et al*. 1996). The upper panel shows the asteroid with longitudinal and latitudinal grid lines overprinted. The large (∼14 km diameter) cavity at the upper left is filled in with particles to create the pre-impact asteroid. The lower three panels show an impact in progress to re-create this Vienna Regio impact structure. Vectors show ejecta leaving the crater; dark lines show zero-velocity nodes as the complex stress wave propagates through the complex shape. Models such as these can be used, in combination with detailed surface images, to determine structural properties of a small body's interior (see also Bruesch and Asphaug 2004).

irregular target, and potentially important focusing might occur if the asteroid is a homogeneous transmitter of compressive stress.

The simulation produced opening of fractures in the narrow opposite end of Ida (Pola Regio) where fracture grooves are observed in spacecraft images. A heterogeneous or porous interior would have dissipated or scattered these stresses (Asphaug and Benz 1994b) so this simulation lends support to the hypothesis that the deep interior of Ida is competent and homogeneous, in agreement with its presumably low porosity (from its satellite-derived density of 2.6 ± 0.5 g cm^{-3}) (Belton et al. 1996). As with fractured Phobos (the small Martian moon similarly modeled by Asphaug and Melosh (1993)), mechanical competence is needed only under compression. A well-connected interior is required, but tensile strength is not.

8.2 Eros

In composition and shape, and roughly in size, Eros is akin to Ida, also an S-type with bulk density $\sim$2.7 g cm^{-3}. This is somewhat lower than the mean density of ordinary chondrites (the closest compositional analog according to Trombka et al. (2000)), so Eros is probably as nominally porous as any heavily fractured large rock mass. Given its battered appearance (Prockter et al. 2002) Eros' density is not a mystery.

Despite this evidence for disruption, a number of observations exhibit structural competence: the twisted planform of Eros, its clustered regions of high slopes (Zuber et al. 2000), its long, continuous grooves (Prockter et al. 2002), its polygonal craters (indicative of fault structure, just as at Meteor Crater), and its subdued or missing crater rims (particularly at the low-gravity ends of the asteroid). The latter suggest crater formation in the strength-regime, where ejecta velocities are higher than v_{esc}. But geological competence is different from dynamical competence, and this has led to a healthy debate about structural nomenclature (perhaps now resolved; see Richardson et al. (2002)) and asteroid mechanics.

The saddle-shaped depression Himeros is probably an impact crater. Crater fracture models (Asphaug et al. 1996) and seismic profiles at Meteor Crater (Ackerman et al. 1975) show that impact craters have major fractures extending about one crater radius beyond the crater in all directions. For Himeros, fractures would extend through the asteroid's narrow waist. If so, the asteroid is disconnected, possibly evidenced by the complex fault structure Rahe Dorsum which strikes through Himeros. Despite its irregular shape, actual slopes on Eros, as on Ida, are moderate. At 100 m baseline, only $\sim$2% of Eros is steeper than can be maintained by talus (see Asphaug et al. 2002), and $\sim$3–4% of the slopes are steeper than can be maintained by sand. In order to have some slopes steeper than the presumed angle of repose, Eros must possess some cohesion at $\sim$100 m scales. Conversely, the total area steeper than repose is less than a few percent, and highly localized (Zuber *et al.* 2000). Almost all slopes exceeding ~30° lie inside the rims of craters, which on Eros (as on the Moon and Earth and other bodies) have slopes consistent with the angle of repose of unconsolidated rock.

The only interior probe of Eros was the radio science experiment (Doppler tracking of the spacecraft position and velocity) (Yeomans *et al.* 2000) and yielded a gravity map of the asteroid's interior. Because orbital speeds were only a few meters per second, intrinsic limits to Doppler accuracy allowed only about kilometer-scale determination of interior density distribution, which proved homogeneous. Most of the power in gravity variation is found in the asteroid's exterior, so that interior structure is not as well constrained.

The highly fractured state of Eros (Prockter *et al.* 2002; Sullivan *et al.* 2002) has been used to support the notion of geologic competence. On the one hand the asteroid can support global fault structures. But those global fault structures are probably what disconnect the asteroid, in a dynamical sense. The distinction between a fractured and unfractured monolith is critical, for an asteroid which is broken behaves very differently during impact than one which is not (Asphaug *et al.* 1998; Richardson *et al.* 2002).

8.3 Mathilde

En route to Eros, NEAR encountered the primitive C-type asteroid 253 Mathilde (Veverka *et al.* 1997). Mathilde is the first visited of those dark, primitive asteroids which dominate the main belt; it may be typical of unequilibrated or carbonaceous asteroids and perhaps representative of early planetesimals. Orbiting the sun at $a = 2.6$ AU, Mathilde measures $66 \times 48 \times 46$ km (best-fit ellipsoid). Its mass, determined by deflection of the spacecraft, is $\sim 1.0 \times 10^{20}$ g, yielding a bulk density of $\rho \sim 1.3$ g cm^{-3}. Ground-based astronomy detects no water on the surface (Rivkin *et al.* 1997). Mathilde is a typically red C-type asteroid whose spectrum is best matched by primitive carbonaceous meteorites with densities over twice as great, implying >50% porosity. Mathilde's rotation period is among the longest measured, $P = 17.4$ days, which is difficult to reconcile with its incredible cratering history: five to seven giant craters, at least four larger in diameter than the asteroid's mean radius, none exhibiting clear signs of structural degradation or overprinting by subsequent collisions, and no evidence for the impact fragmentation that is widespread on Phobos and Eros (Thomas *et al.* 1999).

Following NEAR's determination of Mathilde's low density, a half dozen primitive asteroids have been found to have similar densities, computed from orbital periods of their newly discovered companion satellites (see Merline *et al.* 2002). Once density is known, structure and composition are inseparable, since molecular weight then depends on the distribution of voids. Along with previous determinations of low density for the martian moons and for comets Halley and Shoemaker-Levy 9 (Sagdeev *et al.* 1988; Asphaug and Benz 1994a), these low C-type asteroid densities support a consensus that primitive small bodies are highly porous – rubble piles, primordial aggregates, or volatile-depleted residues. Whatever the origin of a body's porosity, it is significant to impact mechanics (Trucano and Grady 1995).

The absence of ejecta deposits around a crater can signify formation in the strength regime. Despite earlier modeling to the contrary for a much smaller low-density body (Phobos) (Asphaug and Melosh 1993), a strength-regime crater model was attempted for Mathilde by Asphaug and Thomas (1999) in which the largest crater Karoo was filled in, and a strength-scaled impactor (1.2 km diameter) was introduced at 5 km s^{-1} into the 1.3 g cm^{-3} continuum. The result was a globally fragmented asteroid covered in gravity-regime ejecta deposits – entirely opposite to observation. But Mathilde cannot conceivably be any stronger than basalt, because primitive meteorites are notoriously fragile.

9 Survival of the weakest

An alternative has emerged from these failed attempts: perhaps Mathilde is too *weak* to be disrupted by collisions. This scenario (Asphaug 1999) is adopted from armament lore and experimental literature (Love *et al.* 1993). Commenting upon Mathilde, Davis (1999) recounts the use of soft palmetto logs to dissipate the energy of cannon balls colliding into the walls of Revolutionary War forts. A similar heterogeneity, porosity, and/or compactibility has almost certainly enabled Mathilde to survive six giant cratering events (Chapman *et al.* 1999), each without apparent disturbance to pre-existing morphology.

The effects of structure and pre-fragmentation on collisional evolution was examined by Asphaug *et al.* (1998) in which the approximately kilometer-sized near-Earth asteroid Castalia was rendered as monolith, a contact binary, and a rubble pile. Each target was impacted by the same projectile (16 m diameter, 2.7 g cm^{-3}, 5 km s^{-1}). This modeling showed that monolithic asteroids are actually *easier* to disperse than rubble piles or gravitational aggregates.

Using the same technique, Mathilde was assembled out of approximately kilometer-sized basalt spheres just touching, with bulk porosity ~0.5 and bulk density ~1.3 g cm^{-3}, using the Tillotson equation of state for basalt (2.7 g cm^3). Resolving shock waves in each sphere requires a resolution of ~1000 particles per component,[7] or ~500 000 particles total. The contact portion of touching spheres is replaced with damaged rock of slightly lower material density (1.7 g cm^{-3}). As

---

[7] This places about six particles across each component's radius. Because numerical shock waves cannot be treated with fewer than three zones (von Neumann and Richtmyer 1950), that is probably a safe minimum for such simulations.

damaged rock cannot support tensile or shear stress, the model is just like the modeled contact binary, only with thousands of components. The size of the component spheres is established by the resolution requirement of modeling each individual sphere by ∼1000 particles, not by any guiding philosophy of rubble pile structure; simulations at much finer grain fidelity may be required, but are not at present feasible.

A shock wave can propagate freely through a rubble pile so long as it melts, vaporizes, or collapses pores in the rock it encounters. But shocks attenuate rapidly even in competent rock (Rodionov *et al.* 1972), and powerful elastic waves have particular difficulty propagating in an aggregate. Stress energy becomes trapped, and impact energy that would otherwise be transmitted to distant regions is trapped, with significant effects for asteroids with escape velocities of tens of meters per second.

Because of this energy confinement, ejection velocities in the damaged volume of the asteroid are greater. For simulations of Karoo forming in a rubble pile Mathilde (Asphaug *et al.* 2002) the escaping volume from the giant crater equals the damaged volume, leaving a clean crater with no outside damage and no ejecta. Because the post-impact flow field does not depend on asteroid size, these model results predict that rubble pile asteroids larger than ∼100 km diameter, with correspondingly high escape velocity, should show crater ejecta deposits, while smaller asteroids should have their large craters blasted clean of ejecta.

As a caveat to this model, note that Michel *et al.* (2003) use the same hydrocode (Benz and Asphaug 1995) to simulate collisions into shattered masses without voids or pores. These shattered monoliths (different from rubble piles; see above) appear to be *easier* to disrupt than unfractured monoliths. This is simply because compressive impact stresses propagate unimpeded across well-connected fracture planes: momentum is deposited throughout the body, which then disassembles readily since tensile stress is unsupported. These fundamentally different outcomes for shattered monoliths versus rubble piles attest profoundly to the fact that any NEO mitigation strategy involving shocks will be sensitive to target structure. Are common S-type asteroids (the most hazardous NEO class) the shattered-in-place masses that Michel *et al.* (2003) have modeled, or are they less-well-connected rubble piles, or even well-connected monoliths? The final answer may await direct probes of asteroid and comet interiors (Huebner and Greenberg 2000; Safaeinili *et al.* 2002).

9.1 Compaction cratering

Mathilde's absence of impact ejecta deposits, and of distal damage, has been otherwise interpreted as evidence for a more finely comminuted, highly compactible rubble pile.[8] Housen et al. (1999) propose that Mathilde is so underdense that craters form by crushing rather than ejection. In their model no ejecta leave the crater – just opposite the previous scenario in which all ejecta leave the crater. Under compaction cratering, an asteroid crushes up to higher density over time, leaving mass concentrations at each crater floor. Compaction cratering, if it occurs, might have facilitated planetary growth, as it allows primitive bodies to accrete material with ease.

That is also the model's problem, given Mathilde's near-absence of spin. In a regime of perfectly inelastic collisions (see Agnor et al. 1999) accreting bodies almost always begin to spin with periods of hours, not tens of days. One can compute the probability of seven (or so) major impacts knocking Mathilde around by impact momentum conservation, and ending up at its present slow spin. This is a drunkard's walk home in three dimensions, and a simple Monte Carlo analysis shows that the probability of Mathilde spinning as slowly as it does is only ~1.5% if its major craters all formed from momentum-conserving compaction.

10 Tomographic and seismic imaging of asteroid and comet interiors

Exploration tools commonly deployed on Earth to image kilometer-scale geology can be flown on spacecraft to image asteroid interiors directly. Many are designed for field deployment in remote areas and are adaptable to spacecraft designs. Ground-penetrating radar is commonly used to image construction and archeological sites; a full inversion of radar reflection data can, in principle, produce volumetric images of asteroid interiors from orbit (Safaeinili et al. 2002) directly comparable to ultra-sound medical imaging. Transmission radio science (orbiter to lander) (Kofman et al. 1998) is also a central investigation of the upcoming Rosetta rendezvous. Another popular field-deployable tool, magnetotelluric imaging, measures the magnetic to electric field ratio (impedance) which is constant at given frequency for constant resistivity. On Earth this technique probes the crust by taking advantage of terrestrial fluctuations in the background magnetic field; on an asteroid, this field generation would be artificial or solar-wind induced.

Some candidate asteroid and candidate comet materials (clays; frozen brines; metal-rich bodies) may be opaque to electromagnetic energy and better explored by other means. Deep Impact, a NASA Discovery mission, will pioneer the technique of kinetically blasting holes in small bodies (Belton and A'Hearn 1999), slamming ~350 kg of copper into comet Tempel 1 at ~10 km s^{-1} to investigate outer crust and mantle physical and compositional properties.

---

[8] Note that the fluffiest puffball can still be a serious impact hazard; the fragments of comet Shoemaker-Levy 9 that struck Jupiter were probably rubble piles (Asphaug and Benz 1996). For an unmitigated comet or asteroid, what matters most is impact energy and momentum, not target structure.

A third technique is seismic imaging, which teaches us the internal structure of the Earth and has recently been refined for detailed imaging of kilometer-scale geologic masses (Wu and Yang 1997). Seismic imaging may prove to be an ideal complement to electromagnetic imaging of asteroid interiors, since its spatial resolution can be comparable but its means of data acquisition and inversion are entirely distinct. But if NEO surfaces are characterized by loose regolith in low gravity, as is the case for Eros (Robinson *et al.* 2002), seismometers will be complex to deploy, as they require good mechanical coupling. In cases where seismic imaging may be challenging (highly attenuative porous bodies such as comets; regolith-shrouded asteroids) radio imaging may be optimal, and perhaps vice versa.

Seismic experiments require surface probes. In one scenario instrumented penetrators detect signals broadcast from explosions. Blasts produce white noise and are not optimal for tomographic data inversion (compared with drills or thumpers), but they are convenient, reliable, and cheap. They also produce small-scale cratering experiments as a bonus, enabling the imaging and spectroscopy of the buried upper layers (Asphaug *et al.* 2003a). Alternatively, without requiring a rendezvous, an armada of ballistic penetrators could be deployed to strike an NEO in the manner that the pieces of comet Shoemaker-Levy 9 struck Jupiter, striking one after another, with each embedded penetrator acquiring seismic reverberations from its own impact and from successive impacts at diverse locations. The main craft would fly by a few minutes later to acquire the ballistic penetrometry seismic data.

At the NASA Discovery cost level, one can deploy a dual-wavelength radar tomography mission pursuing multiple rendezvous with a variety of near-Earth objects (Asphaug *et al.* 2003b). This population is fairly representative of the asteroid population at large, and may include dormant comet nuclei (Morbidelli *et al.* 2002; Weissman *et al.* 2002). Spring-fired grenades (a precursor to seismic studies) would blast several small (~10 m) craters at selected sites on each asteroid or comet nucleus. Ejecta ballistics and crater formation filmed from orbit, at high time and spatial resolution, would facilitate the development of comet- and asteroid-analog simulation chambers at impact research laboratories. Together with long-term imaging of ejecta orbital evolution, this would greatly reduce uncertainties and design parameters for future landed spacecraft, including penetrators and seismic stations.

10.1 Recommendations

It should be clear from this chapter that until we proceed with direct geophysical exploration of asteroids and comets, our understanding of their interiors will be a matter of educated speculation. To divert or disrupt a potentially hazardous

near-Earth asteroid, one must understand its internal structure, as porous or dis-
continuous asteroids can absorb or divert disruptive energy. Composite bodies can
sacrifice small regions near an impact or explosion without perturbing disconnected
regions. Monoliths, rubble piles, and porous ice–dust mixtures might each require
a different mode of diversion, disruption, or resource exploitation (Huebner and
Greenberg 2000). Except for the mean densities of a dozen asteroids and the coarsely
resolved homogeneous mass distribution of Eros (Yeomans *et al.* 2000), we know
none of the basic bulk constitutive properties of any asteroid. The modeling presented above is fraught with guesswork, and must be considered a snapshot in time
as the science evolves.

The pragmatic implications are several. While the efficacy of a stand-off nuclear
explosion was demonstrated by Ahrens and Harris (1992) for coherent bodies, it is
now believed that most asteroids down to a few 100 m are rubble piles or multi-
component bodies, for which a stand-off blast could impart orders of magnitude less
momentum (see Chapter 6). Antipodal spallation may or may not occur, in which
case even the *direction* of Δv might be indeterminate. Or a blast might disrupt a
body without diverting it, sending a cluster of fragments toward Earth.

Because shock-based methods of NEO diversion depend sensitively upon interior structure, and because NEO interiors are probably diverse and are perhaps
unknowable by visual remote sensing, shock-based technologies may require precursor reconnaissance missions to characterize adequately a given interloper. This
makes them inefficient and perhaps impossible for mitigation scenarios with little
lead time, contrary to popular assumption, unless they are fully redundant (several
impulses or blasts applied in sequence) or overwhelmingly powerful, and far more
massive than current rockets can launch to the required intercept velocities.

If the search for NEOs continues on track (Jedicke *et al.* 2003) then the most
likely scenario, in a few decades, will be lead times of decades to centuries for
nearly all the major hazards, in which case non-impulsive technologies are more
viable (Melosh *et al.* 1994), more cost effective, and far better aligned with the
broader goals of solar system exploration. Soon enough, if not already, the nuclear
blast scenario will become more hazardous than the threat it seeks to mitigate. But
in the meantime no practical solution exists to take its place. We should therefore
keep blast mitigation concepts alive, in models, while we focus upon gaining basic
knowledge of asteroids, comets, and their populations and orbits.

While some non-nuclear mitigation strategies do require detailed awareness of
NEO geology and composition for the purposes of anchoring, momentum loading,
and resource utilization, others do not. A current favorite, given adequate lead
time, is the station-keeping solar reflector (see Chapter 9) which requires almost
no knowledge regarding composition, spin state, or internal structure, and which
works best for materials (such as regolith) with low thermal conductivity.

11 Conclusions

The near-Earth asteroids include about 1000 objects larger than 1 km across, and millions of objects smaller than 100 m. Between these sizes there is evidence for a transition from strength to self-gravity, and from slow to rapid spin rate, with objects larger than a few hundred meters diameter appearing to be rubble piles. About 15% of the larger near-Earth asteroids are binary systems orbiting about a mutual center of mass, either the result of tidal disruption by Earth or catastrophic collision, or both.

A fraction of the near-Earth population, notably the most primitive C- and D-type asteroids, may include extinct comets. Active short- and long-period comets are also potential Earth impactors, and although they represent a smaller impact flux than asteroids, they are much more challenging (sometimes impossible) to detect with long lead time and have, on average, greater impact speeds. We may never be entirely safe from large, long-period comets. However, the probability of global calamity from such a comet approaches the infinitesimal, recommending us to wait a few thousand years for technological advances in detection and mitigation.

Structures proposed for rubble piles range from shattered monoliths sufficiently competent to transmit compressive stress, to highly porous bodies with attenuative interiors. This porosity may be at the scale of giant boulders, or may be microstructural; in the latter case asteroids may crater by compaction rather than by excavation. For impulsive modes of mitigation (kinetic impactors and nuclear blasts) the detailed characterization of rubble piles is of critical concern, while for more gradual modes of mitigation, such as application of reflected solar energy, it is not. These considerations indicate that science must advance along two fronts, one aimed towards a better understanding of asteroids, comets, and their interiors, and another aimed towards securing long lead times of detection. Both of these happen to be fundamental solar system exploration goals.

Acknowledgments

I thank Paul Weissman for insight into cometary nuclei, and Jay Melosh and Don Davis for helpful reviews. My research and writing was made possible by the support from National Aeronautics and Space Administration's Planetary Geology and Geophysics Program and by the Discovery Data Analysis Program.

References

Ackermann, H. D., Godson, R. H., and Watkins, J. S. 1975. A seismic refraction technique used for subsurface investigations at Meteor Crater, Arizona. *J. Geophys. Res.* **80**, 765–775.

Agnor, C. B., Canup, R. M., and Levison, H. F. 1999. On the character and consequences of large impacts in the late stage of terrestrial planet formation. *Icarus* **142**, 219–237.

Ahrens, T. J. and Harris, A. W., 1992. Deflection and fragmentation of near-earth asteroids. *Nature* **360**, 429–433.

Artemieva, N. A. and Shuvalov, V. V. 2001. Motion of a fragmented meteoroid through the planetary atmosphere. *J. Geophys. Res.* **106**, 3297–3310.

Asphaug, E. 1999. Survival of the weakest. *Nature* **402**, 127–128.

Asphaug, E. and Benz, W. 1994a. Density of comet Shoemaker-Levy 9 deduced by modelling breakup of the parent rubble pile. *Nature* **370**, 120–124.

1994b. The surface and interior of Phobos. *Lunar Planet. Sci. Conf.* **25**, 43–44.

1996. Size, density, and structure of comet Shoemaker-Levy 9 inferred from the physics of tidal breakup. *Icarus* **121**, 225–248.

Asphaug, E., and Melosh, H. J. 1993. The Stickney impact of Phobos: a dynamical model. *Icarus* **101**, 144–164.

Asphaug, E., and Scheeres, D. J. 1999. Deconstructing Castalia: evaluating a postimpact state. *Icarus* **139**, 383–386.

Asphaug, E. and Thomas, P. C. 1999. Modeling mysterious Mathilde. *Lunar Planet. Sci. Conf.* **30**, Abstract no. 2028.

Asphaug, E., Moore, J. M., Morrison, D., *et al.* 1996. Mechanical and geological effects of impact cratering on Ida. *Icarus* **120**, 158–184.

Asphaug, E., Ostro, S. J., Hudson, R. S., *et al.* 1998. Disruption of kilometre-sized asteroids by energetic collisions. *Nature* **393**, 437–440.

Asphaug, E., Petit, J.-M., and Rivkin, A. S. 2002. Removing mantles from cores: tidal disruption of ancient asteroids. *Lunar Planet. Sci. Conf.* **33** Abstract no. 2066.

Asphaug, E., Colwell, J., Dissly, R., *et al.* 2003a. Meteoroid bombardment and blast experiments on asteroids. *Lunar Planet. Sci. Conf.* **34**, Abstract no. 1537.

Asphaug, E., Belton, M. J. S., Cangahuala, A., *et al.* 2003b. Exploring asteroid interiors: the Deep Interior mission concept. *Lunar Planet. Sci. Conf.* **34**, Abstract no. 1906.

Bell, J. F. 2001. Eros: a comprehensive model. *Lunar Planet. Sci. Conf.* **32**, Abstract no. 1964.

Belton, M. J. S. and A'Hearn, M. F. 1999. Deep sub-surface exploration of cometary nuclei. *Adv. Space Res.* **24**, 1167–1173.

Belton, M. J. S. and 20 colleagues 1996. The discovery and orbit of 1993 (243)1 Dactyl. *Icarus* **120**, 185–199.

Benz, W., and Asphaug, E. 1994. Impact simulations with fracture. I. Method and tests. *Icarus* **107**, 98–116.

1995. Simulations of brittle solids using smooth particle hydrodynamics, *Comput. Phys. Commun.* **87**, 253–265.

1999. Catastrophic disruptions revisited. *Icarus* **142**, 5–20.

Borovicka, J., Popova, O. P., Nemtchinov, I. V., *et al.* 1998. Bolides produced by impacts of large meteoroids into the Earth's atmoshpere: comparison of theory with observations. I. Benesov bolide dynamics and fragmentation. *Astron. Astrophys.* **334**, 713–728.

Bottke, W. F. and Melosh, H. J. 1996. The formation of asteroid satellites and doublet craters by planetary tidal forces. *Nature* **381**, 51–53.

Bottke, W. F., Nolan, M. C., Greenberg, R., *et al.* 1994. Collisional lifetimes and impact statistics of near-Earth asteroids. In *Hazards due to Comets and Asteroids*, eds. T. Gehrels, M. S. Matthews, and A. Schumann, pp. 337–357. Tucson, AZ: University of Arizona Press.

Bottke, W. F., Jr., Cellino, A., Paolicchi, P., *et al.* (eds.) 2002. *Asteroids III*. Tucson, AZ: University of Arizona Press.

Bridges, F. G., Supulver, K. D., Lin, D. N. C., *et al.* 1996. Energy loss and sticking mechanisms in particle aggregation in planetesimal formation. *Icarus* **123**, 422–435.

Bruesch, L. S. and Asphaug, E. 2004. Modeling global impact effects on middle-sized icy bodies: application to Saturn's Moons. *Icarus* **168**, 457–466.

Canup, R. 2004. Simulations of a late lunar-forming impact. *Icarus*, **168**, 433–456.

Carr, M. H., Kirk, R. L., McEwen, A., *et al.* 1994. The geology of Gaspra. *Icarus* **107**, 61–71.

Chandrasekhar, S. (1961) *Hydrodynamic and Hydromagnetic Stability*. London, UK: Oxford University Press.

Chapman, C. R. and Davis, D. R. 1975. Asteroid collisional evolution: evidence for a much larger early population. *Science* **190**, 553.

Chapman, C. R., Paolicchi, P., Zappala, V., *et al.* 1989. Asteroid families: physical properties and evolution. In *Asteroids II*, eds. R. P. Binzel, T. Gehrels, and M. S. Matthews, pp. 386–415. Tucson, AZ: University of Arizona Press.

Chapman, C. R., Merline, W. J., and Thomas, P. 1999. Cratering on Mathilde. *Icarus* **140**, 28–33.

Chodas, P. W., and Yeomans, D. K. 1996. The orbital motion and impact circumstances of comet Shoemaker-Levy 9. In *The Collision of Comet Shoemaker-Levy 9 and Jupiter*, eds. K. S. Noll, H. A. Weaver, and P. D. Feldman, pp. 1–30. Cambridge, UK: Cambridge University Press.

Clark, B. E., Lucey, P., Helfenstein, P., 2001. Space weathering on Eros: constraints from albedo and spectral measurements of Psyche crater. *Meteor. and Planet. Sci.* **36**, 1617–1637.

Consolmagno, G. J. and Britt, D. T. 1998. The density and porosity of meteorites from the Vatican collection. *Meteor. and Planet. Sci.* **33**, 1231–1241.

Crawford, D. A. (1997). Comet Shoemaker-Levy 9 fragment size and mass estimates from light flux observations. *Lunar Planet. Sci. Conf.* **28**, Abstract no. 1351, pp. 267–268.

Davis, D. R. 1999. The collisional history of asteroid 253 Mathilde. *Icarus* **140**, 49–52.

Davis, D. R., Chapman, C. R., Greenberg, R., *et al.* 1979. Collisional evolution of asteroids: populations, rotations and velocities. In *Asteroids*, ed. T. Gehrels, pp. 528–557. Tucson, AZ: University of Arizona Press.

Dobrovolskis, A. R. 1990. Tidal disruption of solid bodies. *Icarus* **88**, 24–38.

Dobrovolskis, A. R. and Burns, J. A. 1984. Angular momentum drain: a mechanism for despinning asteroids. *Icarus* **57**, 464–476.

Durda, D. D. 1996. The formation of asteroidal satellites in catastrophic collisions. *Icarus* **120**, 212–219.

Durda, D. D., Flynn, G. J., and van Veghten, T. W. 2003. Impacts into porous foam targets: possible implications for the disruption of cometary nuclei. *Icarus* **163**, 504–507.

Farinella, P., Paolicchi, P., and Zappala, V. 1982. The asteroids as outcomes of catastrophic collisions. *Icarus* **52**, 409–433.

Festou, M., Keller, H. U., and Weaver, H. A. (eds.) 2004. *Comets II*, Tucson, AZ: University of Arizona Press.

Fujiwara, A. 1980. On the mechanism of catastrophic destruction of minor planets by high-velocity impact. *Icarus* **41**, 356–364.

Fujiwara, A., Cerroni, P., Davis, D. R., *et al.* 1989. Experiments and scaling laws for catastrophic collisions. In *Asteroids II*, eds. R. P. Binzel, T. Gehrels, and M. S. Matthews, pp. 240–265. Tucson, AZ: University of Arizona Press.

Goldreich, P., Lithwick, Y., and Sari, R. 2002. Formation of Kuiper-belt binaries by dynamical friction and three-body encounters. *Nature* **420**, 643–646.

Grady, D. E. and Kipp, M. E. 1980. Continuum modeling of explosive fracture in oil shale. *Int. J. Rock Mech. Min. Sci. Geomech. Abstr.* **17**, 147–157.

Grady, D. E. and Lipkin, J. 1980. Criteria for impulsive rock fracture. *Geo. Phys. Lett.* **7**, 255–258.

Griffith, A. A. 1920. The phenomena of rupture and flow in solids. *Phil. Trans. R. Soc. London Ser. A* **221**, 163–198.

Harris, A. W. 1996. The rotation rates of very small asteroids: evidence for "rubble pile" structure. *Lunar Planet. Sci. Conf.* **27**, Abstract no. 493.

Hayward, T. L., Hanner, M. S., and Sekanina, Z. 2000. Thermal infrared imaging and spectroscopy of comet Hale-Bopp (C/1995 O1). *Astrophys. J.* **538**, 428–455.

Holsapple, K. A. 2001. Equilibrium configurations of solid cohesionless bodies. *Icarus* **154**, 432–448.

Holsapple, K. A. and Housen, K. R. 1986. Scaling laws for the catastrophic collisions of asteroids, *Mem. Soc. Astron. Italia* **57**, 65–85.

Housen, K. R. and Holsapple, K. A. 1999. Scale effects in strength-dominated collisions of rocky asteroids. *Icarus* **142**, 21–33.

Housen, K. R., Wilkening, L. L., Chapman, C. R. 1979. Asteroidal regoliths. *Icarus* **39**, 317–351.

Housen, K. R., Holsapple, K. A., and Voss, M. E. 1999. Compaction as the origin of the unusual craters on the asteroid Mathilde. *Nature* **402**, 155–157.

Huebner, W. F. and Greenberg, J. M. 2000. Needs for determining material strengths and bulk properties of NEOs. *Planet. Space Sci.* **48**, 797–799.

Jedicke, R., Morbidelli, A., Spahr, T., *et al.* 2003. Earth and space-based NEO survey simulations: prospects for achieving the Spaceguard goal. *Icarus* **161**, 17–33.

Jewitt, D. C. 1999. Kuiper belt objects. *Ann. Rev. Earth and Planet. Sci.* **27**, 287–312.

Julian, W. H., Samarasinha, N. H., and Belton, M. J. S. 2000. Thermal structure of cometary active regions: Comet 1P/Halley. *Icarus* **144**, 160–171.

Keil, K., Haack, H., and Scott, E. R. D. 1994. Catastrophic fragmentation of asteroids: evidence from meteorites. *Planet. Space Sci.* **42**, 1109–1122.

Keller, H. U., Arpigny, C., Barbieri, C., *et al.* 1986. First Halley multicolour camera imaging results from Giotto. *Nature* **321**, 320–326.

Kofman, W., Barbin, Y., Klinger, J., *et al.* 1998. Comet nucleus sounding experiment by radiowave transmission. *Adv. Space Res.* **21**, 1589–1598.

Korycansky, D. G. and Asphaug, E. 2003. Impact evolution of asteroid shapes. I. Random mass redistribution. *Icarus* **163**, 374–388.

Lange, M. A. and Ahrens, T. J. (1983). The dynamic tensile strength of ice and ice–silicate mixtures. *J. Geophys. Res.* **88**, 1197–1208.

Lawn, B. R. and Wilshaw, T. R. 1975. *Fracture of Brittle Solids.* Cambridge, UK: Cambridge University Press.

Lindholm, U. S., Yeakley, L. M., and Nagy, A. 1974. The dynamic strength and fracture properties of Dresser basalt. *Int. J. Rock Mech. Min. Sci. Geomech. Abstr.* **11**, 181–191.

Love, S. G., Hörz, F., and Brownlee, D. E. 1993. Target porosity effects in impact cratering and collisional disruption. *Icarus* **105**, 216–224.

Mäkinen, J. T. T., Bertaux, J.-L., Combi, M. R., *et al.* 2001. Water production of comet C/1999 S4 (LINEAR) observed with the SWAN instrument. *Science* **292**, 1326–1329.

Margot, J. L., Nolan, M. C., Benner, L. A. M., *et al.* 2002. Binary asteroids in the near-Earth object population. *Science* **296**, 1445–1448.

McSween, H. Y. 1999. *Meteorites and their Parent Planets*. New York: Cambridge University Press.

Meech, K., Svoren, J., and Benkhoff, J. 2004. Physical and chemical evolution of cometary nuclei. In *Comets II*, eds. Festour, M., Keller, H. U., and Weaver, H. A. Tucson, AZ: University of Arizona Press.

Melosh, H. J. and Ryan, E. V. 1997. Asteroids: shattered but not dispersed. *Icarus* **129**, 562–564.

Melosh, H. J. and Stansberry, J. A. 1991. Doublet craters and the tidal disruption of binary asteroids. *Icarus* **94**, 171–179.

Melosh, H. J., Ryan, E. V., and Asphaug, E. 1992. Dynamic fragmentation in impacts: hydrocode simulation of laboratory impacts. *J. Geophys. Res.* **97**, 14735–14759.

Melosh, H. J., Nemchinov, I. V., and Zetzer, Y. I. 1994. Non-nuclear strategies for deflecting comets and asteroids. In *Hazards due to Comets and Asteroids*, ed. T. Gehrels, pp. 1111–1132. Tucson, AZ: University of Arizona Press.

Merline, W. J., Weidenschilling, S. J., Durda, D. D., *et al.* 2002. Asteroids do have satellites. In *Asteroids III*, eds. W. F. Bottke Jr., A. Cellino, P. Paolicchi, and R. P. Binzel, pp. 289–312. Tucson, AZ: University of Arizona Press.

Michel, P., Benz, W., and Richardson, D. C. 2003. Disruption of fragmented parent bodies as the origin of asteroid families. *Nature* **421**, 608–611.

Miller, J. K., Konopliv, A. S., Antreasian, P. G., *et al.* 2002. Determination of shape, gravity and rotational state of asteroid 433 Eros. *Icarus* **155**, 3–17.

Morbidelli, A., Bottke, W. F., Jr., Froeschle, C., *et al.* 2002. Origin and evolution of near-Earth objects. In *Asteroids III*, eds. W. F. Bottke, Jr., A. Cellino, P. Paolicchi, and R. P. Binzel, pp. 409–422. Tucson, AZ: University of Arizona Press.

Noll, K. S., Stephens, D. C., Grundy, W. M., *et al.* 2002. Detection of two binary trans-Neptunian objects, 1997 CQ29 and 2000 CF105, with the Hubble Space Telescope. *Astron. J.* **124**, 3424–3429.

Ostro, S. J., Chandler, J. F., Hine, A. A., *et al.* 1990. Radar images of asteroid 1989 PB. *Science* **248**, 1523–1528.

Ostro, S. J., Pravec, P., Benner, L. A. M., *et al.* 1999. Radar and optical observations of asteroid 1998 KY26. *Science* **285**, 557–559.

Peale, S. J. 1989. On the density of Halley's comet. *Icarus* **82**, 36–49.

Pravec, P. and Harris, A. W. 2000. Fast and slow rotation of asteroids. *Icarus* **148**, 12–20.

Pravec, P., Harris, A. W., and Michalowski, T. 2002. Asteroid rotations. In *Asteroids III*, eds. W. F. Bottke, Jr., A. Cellino, P. Paolicchi, and R. P. Binzel, pp. 113–122. Tucson, AZ: University of Arizona Press.

Prialnik, D. and Podolak, M. 1999. Changes in the structure of comet nuclei due to radioactive heating. *Space Sci. Rev.* **90**, 169–178.

Prockter, L., Thomas, P., Robinson, M., *et al.* 2002. Surface expressions of structural features on Eros. *Icarus* **155**, 75–93.

Richardson, D., Leinhardt, Z. M., Melosh, H. J., *et al.* 2002. Rubble pile asteroids: provenance, evolution, and properties. In *Asteroids III*, eds. W. F. Bottke, Jr., A. Cellino, P. Paolicchi, and R. P. Binzel, pp. 00–00. Tucson, AZ: University of Arizona Press.

Rinehart, J. S. 1965. Dynamic fracture strength of rocks. *7th Symp. Rock Mechanics* **1**, 205–208.

Rivkin, A. S., Clark, B. E., Britt, D. T., *et al.* 1997. Infrared spectrophotometry of the NEAR flyby target 253 Mathilde. *Icarus* **127**, 255–257.

Robinson, M. S., Thomas, P. C., Veverka, J., *et al.* 2002. The geology of 433 Eros. *Meteor. and Planet. Sci.* **37**, 1651–1684.

Rodionov, V. N., Adushkin, V. V., Kostyuchenko, V. N. 1972. *Mechanical Effect of an Underground Explosion*. Los Alamos, NM: US Atomic Energy Commission.

Rubincam, D. P. 2000. Radiative spin-up and spin-down of small asteroids. *Icarus* **148**, 2–11.

Ryan, E. V., Hartmann, W. K., and Davis, D. R. 1991. Impact experiments. III. Catastrophic fragmentation of aggregate targets and relation to asteroids. *Icarus* **94**, 283–298.

Safaeinili, A., Gulkis, S., Hofstadter, M. D., *et al.* 2002. Probing the interior of asteroids and comets using radio reflection tomography. *Meteor. and Planet. Sci.* **37**, 1953–1963.

Sagdeev, R. Z., Elyasberg, P. E., and Moroz, V. I. 1988. Is the nucleus of Comet Halley a low density body? *Nature* **331**, 240–242.

Samarasinha, N. H. 2001. A model for the breakup of comet LINEAR (C/1999 S4). *Icarus* **154**, 540–544.

Samarasinha, N. H. and Belton, M. J. S. 1995. Long-term evolution of rotational stress and nongravitational effects for Halley-like cometary nuclei. *Icarus* **116**, 340–358.

Schenk, P., Asphaug, E., McKinnon, W. B., *et al.* 1996. Cometary nuclei and tidal disruption: the geologic record of crater chains on Callisto and Ganymede. *Icarus* **121**, 249–274.

Scotti, J. V. and Melosh, H. J. 1993. Estimate of the size of comet Shoemaker-Levy 9 from a tidal breakup model. *Nature* **365**, 733–735.

Sekanina, Z. 1996. Tidal breakup of the nucleus of comet Shoemaker-Levy 9. In *The Collision of Comet Shoemaker-Levy 9 and Jupiter*, eds. K. S. Noll, H. A. Weaver, and P. D. Feldman, pp. 55–80. Cambridge, UK: Cambridge University Press.

Sekanina, Z., Chodas, P. W., and Yeomans, D. K. 1994. Tidal disruption and the appearance of periodic comet Shoemaker-Levy 9. *Astron. Astrophys.* **289**, 607–636.

Shoemaker, E. M. and Wolfe, R. A. 1982. Cratering timescales for the Galilean satellites. In *Satellites of Jupiter*, ed. D. Morrison, pp. 277–339. Tucson, AZ: University of Arizona Press.

Soderblom, L. A., Becker, T. L., Bennett, G., *et al.* 2002. Observations of comet 19P/Borrelly by the miniature integrated camera and spectrometer aboard Deep Space 1. *Science* **296**, 1087–1091.

Solem, J. C. 1994. Density and size of comet Shoemaker-Levy 9 deduced from a tidal breakup model. *Nature* **370**, 349–351.

   1995. Cometary breakup calculations based on a gravitationally-bound agglomerationmodel: the density and size of Shoemaker-Levy 9. *Astron. Astrophys.* **302**, 596–608.

Stern, S. A. 2002. Implications regarding the energetics of the collisional formation of Kuiper Belt satellites. *Astron. J.* **124**, 2300–2304.

Stern, S. A. and Weissman, P. R. 2001. Rapid collisional evolution of comets during the formation of the Oort cloud. *Nature* **409**, 589–591.

Stewart, S. T., Ahrens, T. J., and Lange, M. A. 1999. Correction to the dynamic tensile strength of ice and ice–silicate mixtures (Lange and Ahrens 1983). *Lunar Planet. Sci. Conf.* **30**.

Sullivan, R. Thomas, P. C., Murchie, S. L., *et al.* 2002. Asteroid geology from Galileo and NEAR data. In *Asteroids III*, eds. W. F. Bottke, Jr., A. Cellino, P. Paolicchi, and R. P. Binzel, pp. 000–000. Tucson, AZ: University of Arizona Press.

Thomas, P. C. 1998. Ejecta Emplacement on the Martian Satellites. *Icarus* **131**, 78–106.

Thomas, P. C., Belton, M. J. S., Carcich, B., *et al.* 1996. The Shape of Ida. *Icarus* **120**, 20–32.

Thomas, P. C., Veverka, J., Bell, J. F., *et al.* 1999. Mathilde: size, shape, and geology. *Icarus* **140**, 17–27.

Trombka, J. I. Squyres, S. W., Brückner, J., *et al.* 2000. The elemental composition of asteroid 433 Eros: results of the NEAR–Shoemaker X-ray spectrometer. *Science* **289**, 2101–2105.

Trucano, G. T. and Grady, D. E. 1995. Impact shock and penetration fragmentation in porous media. *J. Impact Eng.* **17**, 861–872.

Veverka, J., Thomas, P., Johnson, T. V., *et al.* 1986. The physical characteristics of satellite surfaces. In *Satellites*, eds. J. A. Burns and M. S. Matthews, pp. 342–402. Tucson, AZ: University of Arizona Press.

Veverka, J., Thomas, P., Harch, A., *et al.* 1997. NEAR's flyby of 253 Mathilde: images of a C asteroid. *Science* **278**, 2109–2112.

von Neumann, J. and Richtmyer, R. D. 1950. A method for the numerical calculation of hydrodynamic shocks. *J. Appl. Phys.* **21**, 232–237.

Weibull, W. A. 1939. A statistical theory of the strength of materials. (English translation) *Ingvetensk. Akad. Handl.* **151**, 5–45.

Weidenschilling, S. J. 1997. The origin of comets in the solar nebula: a unified model. *Icarus* **127**, 290–306.

2002. On the origin of binary transneptunian objects. *Icarus* **160**, 212–215.

Weissman, P. R. 1980. Physical loss of long-period comets. *Astron. Astrophys.* **85**, 191–196.

1987. Post-perihelion brightening of Halley's comet: spring time for Halley. *Astron. Astrophys.* **187**, 873–878.

Weissman, P. R. and Lowry, S. C. 2004. The size distribution of cometary nuclei. *Science*, in press.

Weissman, P. R., Bottke, W. F., Jr., and Levison, H. F. 2002. Evolution of comets into asteroids. In *Asteroids III*, eds. W. F. Bottke, Jr., A. Cellino, P. Paolicchi, and R. P. Binzel, pp. 669–686. Tucson, AZ: University of Arizona Press.

Weissman, P. R., Asphaug, E., and Lowry, S. 2004. Structure and density of cometary nuclei. In *Comets II*, eds. M. Festou, H. U. Keller, and H. A. Weaver, pp. 000–000. Tucson, AZ: University of Arizona Press.

Wu, R. and Yang, F. 1997. Seismic imaging in wavelet domain: decomposition and compression of imaging operator. *Proc. Int. Soc. Optical Eng.* **3169**, 148–162.

Yeomans, D. K., Antreasian, P. G., Bariiot, J.-P., *et al.* 2000. Radio science results during the NEAR-Shoemaker spacecraft rendezvous with Eros. *Science* **289**, 2085–2088.

Zahnle, K., Schenk, P., Levison, H., *et al.* 2003. Cratering rates in the outer solar system. *Icarus* **162**, 263–289.

Zuber, M. T., Smith, D. E., Cheng, A. F., *et al.* 2000. The shape of 433 Eros from the NEAR–Shoemaker laser rangefinder. *Science* **289**, 2097–2101.

5

What we know and don't know about surfaces of potentially hazardous small bodies

Clark R. Chapman

Southwest Research Institute

1 Introduction

One of the most fundamental aspects of mitigating an impact threat by moving an asteroid or comet involves physical interaction with the body. Whether one is bathing the body's surface with neutrons, zapping it with a laser or solar-reflected beam, bolting an ion thruster or mass driver onto the surface, or trying to penetrate the surface in order to implant a device below the surface, we need to understand the physical attributes of the surface and sub-surface. Of course, we would critically wish to understand the surface of the particular body that is, most unluckily, found to be headed for Earth impact – should that eventuality come to pass. But, in the event that we have relatively little warning time, it might behoove us to examine well in advance the potential range of small-body surface environments that we might have to deal with. It will improve our ability to design experiments and understand data concerning the particular body if we have evaluated, beforehand, the range of surface properties we might encounter and have specified the kinds of measurement techniques that will robustly determine the important parameters that we would want to know.

We already know, from meteorite falls, that asteroidal materials can range from strong nickel–iron alloy (of which most smaller crater-forming meteorites, like Canyon Diablo, are made) to mud-like materials (like the remnants of the Tagish Lake fireball event). But the diversity could be even greater, especially on the softer/weaker end of the spectrum, because the Earth's atmosphere filters out such materials. That is why many meteoriticists doubt that we have any macroscopic meteorites from a comet. We could readily expect some icy, snowy, frothy, and dusty materials in the surface layers of asteroids and comets, and perhaps still stranger materials (e.g., with the honeycomb structure of styrofoam).

Mitigation of Hazardous Comets and Asteroids, ed. M. J. S. Belton, T. H. Morgan, N. H. Samarasinha, and D. K. Yeomans. Published by Cambridge University Press. © Cambridge University Press 2004.

A common framework for thinking about asteroid surfaces is to extrapolate from our very extensive knowledge of the lunar regolith. Indeed, a considerable literature concerning asteroid regoliths was published, mostly in the 1970s and 1980s, based on theoretical extrapolation from lunar regolith models and based on inferences from what are termed "regolith breccia" meteorites. These studies suggested that we should expect both general similarities (e.g., in depths and spectrum of morphologies of degraded craters) and relatively minor differences (e.g., coarser, less mature regolith) from our lunar experience, for asteroids several kilometers in diameter and larger. Less thought was given to smaller asteroids, except that – at small sizes – properties must eventually transition to those of a "bare rock in space."

Beginning in the early 1990s, it finally became possible to study asteroids as geological bodies, using imaging and other measurements from flybys of two main-belt asteroids by the Galileo spacecraft. Ground-based delay-Doppler radar mapping of some Earth-approaching asteroids also began to provide spatial resolution on geologically interesting spatial scales. While such data have been very fascinating from the perspective of planetary science, the best resolution achieved (several tens of meters) is very coarse in the context of addressing questions relating to human-scale interaction with the surface of an asteroid. That all changed when the NEAR–Shoemaker spacecraft orbited the large, Earth-approaching asteroid Eros for a year, including some low-altitude passes that obtained several-meter imaging resolution, and finally landed on the surface in February 2001, taking a final image that resolved features just a few centimeters across.

2 Eros

The Earth-approaching asteroid Eros is large enough that it was expected to have a roughly lunar-like regolith, although perhaps somewhat coarser and less well mixed. A major surprise from the NEAR–Shoemaker mission to Eros is that its surface is totally unlike the Moon's, particularly at spatial scales of centimeters to tens of meters – just the scales relevant for human interaction with an asteroid. The Moon is covered with a well-churned regolith (basically a dusty and sandy soil, with occasional larger rocks and boulders, especially near recent craters large enough to have penetrated the several-meter-deep regolith down to bedrock), and its surface is characterized by innumerable small craters. Eros, on the other hand and despite its lunar-like appearance at spatial scales larger than ~100 meters, has been found to have relatively few craters tens of meters in size, and almost no craters centimeters to meters in size. Instead, the surface of Eros is dominated by countless rocks and boulders . . . except in localized flat areas called "ponds," which are nearly devoid of both craters and rocks. (Examples of high-resolution views of Eros are shown in

Figure 5.1 Fifth from last image taken by the NEAR–Shoemaker camera during its February 12, 2001 landing on the surface of Eros. The patch is roughly 18 m across and illumination is from the right.

Figs. 5.1 and 5.2, with solar illumination from the right. Fig. 5.1 is the fifth from last picture of Eros's surface, showing a terrain strewn with rocks and boulders, the largest about 3 m across. The final, partial frame taken by NEAR–Shoemaker, Fig. 5.2, shows the transition to a flat terrain, probably a "pond"-like surface within a small crater in which the spacecraft apparently landed; a basketball has been inserted for scale.)

Clearly, simple imaging of a surface cannot provide an understanding of the detailed physical properties, nor properties of the sub-surface, unless the nature of the visible surface features can be understood (i.e., interpreted in terms of geological or physical processes). Preliminary attempts to understand the small-scale attributes of Eros have not yet yielded definitive results. Chapman *et al.* (2002) examined several hypotheses that might individually or together explain the unexpected differences between Eros and the Moon. These include: (a) inhibition of crater formation by the abundant boulders on the surface (this requires, then, an explanation for the superabundance of small boulders on Eros compared with the Moon, but there are ideas about that including distribution of ejecta from the Shoemaker crater (Thomas *et al.* 2001) and the "Brazil nut effect" (Asphaug *et al.* 2001)

Figure 5.2 Final partial frame taken before NEAR–Shoemaker landed on the surface of Eros. Basketball inserted for scale. Illumination is from the right. The bouldery region shown in Fig. 5.1 grades from the right into the smooth area (probably a "pond") shown in the lower-left portion of this frame. Note that virtually all equant features are positive relief features (i.e., pebbles and rocks); there are several curving depressions but no circular craters visible.

due to impact shaking of the surface), (b) impact-induced seismic shaking and consequent degradation of smaller craters on asteroids, (c) traditional erasure of small craters by regolith processes that might be more effective on small bodies, and (d) relative lack of smaller projectiles impacting Eros (which spent virtually all of its life in the asteroid belt) compared with the impactor population that craters the Moon (which can be explained qualitatively by the Yarkovsky effect (Bell 2001), but which has yet to be explained quantitatively.)

Chapman *et al.* (2002) concluded that several processes could contribute to the relative lack of small craters and abundance of boulders on Eros compared with the Moon (e.g., seismic shaking and consequent "Brazil nut effect," which brings larger boulders to the surface, augmented by armoring of the surface by the boulders, inhibit crater formation). However, such processes can probably affect the crater numbers by only factors of several, falling far short of the orders of magnitude required. Veverka *et al.* (2001) and Robinson *et al.* (2002) have argued that traditional regolith processes operating on the variably sloping surfaces of Eros resemble effects that reduce small-crater densities on the lunar highlands: Lunar Orbiter imaging long ago showed that the flat lunar maria have higher small-crater densities (craters tens to hundreds of meters across) than adjacent highlands, and the lower densities on highlands were attributed by some to downslope creep and degradation of craters on slopes. However, serious questions were raised about this interpretation at the time; the hypothesis can account for diminishing crater

Figure 5.3 Apollo 16 photograph of small craters on slopes in the lunar highlands (astronaut and lunar rover show approximate scale).

densities by only a factor of a few, and it is evident that very small craters exist in profusion on appreciable slopes in the lunar highlands (Fig. 5.3). Robinson *et al.* (2002) present crater statistics demonstrating that spatial densities of craters ~20 m in size on both Eros and the lunar highlands are depressed by a factor of a few. However, their analysis does not address at all the orders-of-magnitude lower crater densities at much smaller sizes and obviously completely fails to account for the virtual absence of small craters on the demonstrably flat surfaces of the "ponds."

Unless the Yarkovsky effect can come to the rescue and augment the other effects by orders of magnitude, we are left with an unsolved mystery concerning the small-scale structure of Eros. Therefore, we can infer other properties of the surface of Eros only with great uncertainty. For example, certain features on Eros have been interpreted in terms of debris flows (Cheng *et al.* 2002); ponds may be collections of electrostatically levitated dust; many boulders on Eros may be ejecta blocks from the relatively recent, large crater named Shoemaker (Thomas *et al.* 2001); and so on. Each idea has consequent implications for the strength and other physical properties of the relevant materials, but the inferences are not robust. For example, porous dust seems like a reasonable model for the "ponds" but are we really sure from the images that these surfaces aren't solid, like concrete? The numerous positive relief features called "boulders" may be presumed to be strong, solid rocks, but some of these features have a morphology more resembling a dirt clod that has broken apart when landing (at presumably very low velocities) on the surface of Eros. There remains the possibility, for instance, that many of the boulders are fractured bedrock being exhumed from below; implications for sub-surface structure would be very different, however, if one accepts other published ideas that Eros has a regolith generally tens of meters deep, or more. A comparison of the bulk density of Eros compared with the mineralogy inferred from NEAR–Shoemaker remote-sensing instruments suggests that there is considerable void space within Eros, but the spatial scale of the voids (e.g., millimeters vs. hundreds of meters) is unknown, nor is it well known if the voids pervade the interior of Eros or are preferentially concentrated in a surface megaregolith. Unfortunately, these uncertainties cannot be elucidated much by studies of the surface. Under the best of conditions, the characteristics of planetary surfaces provide only weak constraints about circumstances at depth; given the uncertainties about surface processes on Eros, even less can be said with any confidence about implications of surface features for the asteroid's interior. Most NEAR Science Team researchers have considered Eros to be a "fractured shard" rather than a cohesionless rubble pile, and certainly it has retained large craters like Psyche and Himeros without being disrupted, but the nature of its interior (and of the interiors of other asteroids) awaits geophysical exploration.

The lesson is that extrapolations from meteoritical and lunar studies to asteroids proved to be wrong, at least in the case of Eros. Evidently, our generalized under-standing of the processes that shape asteroid surfaces has been wrong in one or more fundamental ways. The way that we can really tell what an asteroid surface is like is to measure it directly rather than to theorize about it.

3 Other asteroids

It is tempting to draw inferences from the NEAR–Shoemaker data about what the surfaces of other asteroids, or at least of S-type asteroids, might be like. This would be unwise, however. S-types encompass a very broad range of mineralogies, ranging from assemblages that might have the strength of metal (Chapman (1997, 2002) has suggested that the S-type asteroid Gaspra might be such a stony-iron) all the way to weak, primitive materials. C-type asteroids, by far the most common type in the main asteroid belt and still a major component of the Earth-approaching asteroids, remain mysterious despite the tendency to associate them with carbonaceous meteorites. The NEAR–Shoemaker spacecraft flew past one large, main-belt C-type, Mathilde, but the best images have coarse resolution. The chief result from the flyby was that Mathilde has a very low bulk density, only 30% greater than water. Some have interpreted Mathilde to be a "rubble pile" (a cohesionless assemblage of components held together by their own gravity), but the voids within Mathilde could even be at microscopic scales, an inherent property of Mathilde's constituent material. Such material might never successfully penetrate the Earth's atmosphere to be collected as meteorites for study, in which case we have no known analog for material property studies.

Another major factor affecting surface properties involves asteroid size. Eros has roughly one-ten-millionth the mass of the Moon, and I have described above how radically different the surface properties of those two bodies are. A 100-m NEA in an Earth-threatening orbit has about one-ten-millionth the mass of Eros, hence it is doubtful that NEAR–Shoemaker's studies of Eros provide much insight concerning the properties of such a small body, even if its mineralogy were identical to that of Eros. Gravity should dominate surface processes on Eros but not on sub-kilometer bodies. The kinds of bodies that we might need to deflect will likely have essentially no steady-state regolith on them at all. But there may be legacy regoliths, evolved on the larger bodies from which the small bodies were formed, or almost any kind of unexpected structure. Evidence from several kinds of observations, especially radar but also interpretation of photometric lightcurves, suggests that most NEAs several hundred meters in size have the structure of rubble piles (i.e., composed of multiple components, such as a contact binary, held together solely by mutual gravity rather than by cohesive strength). Clearly, any attempt to move such an object might risk disassembling it into its separate components, although disassembly would be resisted by sliding friction. NEAs smaller than about 100 m, in sharp contrast, all seem to be monolithic, cohesive bodies (i.e., "rocks"). This is deduced from the statistics of their spin periods inasmuch as most of them spin very rapidly on their axes (with spin periods as short as minutes compared with the more common periods of hours for larger bodies); however, rapid spins present a different kind of challenge in trying to interact with them.

4 Comets

As uncertain as our knowledge of asteroid surface properties is, cometary properties are even more uncertain. While Whipple's dirty-snowball model of a comet nucleus has provided a framework for ever more detailed models of cometary properties, properties pertinent to potential mitigation requirements remain almost wholly unconstrained by observations. The best spatial resolution yet achieved on a comet nucleus (Deep Space 1's imaging of Comet Borrelly (Soderblom *et al.* (2002), showing features as small as 100 m across on the 8 km long comet) has been inadequate to infer more than the most basic geological information and is wholly inadequate to inform speculations about bearing strength or other pertinent properties of a comet's surface. Deep Space 1 images can be interpreted stereoscopically, permitting inferences that certain features are "plateaus" or "depressions," but connections between even these apparent features and the production of the observed jets remain elusive. Few impact craters are visible, and there may be none at all. Such absence of craters is unsurprising, given the rates at which comets lose volatiles during their periodic approaches to the Sun.

Of course, the kinds of comets that might threaten impact with Earth are not those familiar comets already discovered and named, which are undergoing rapid disintegration during their frequent visits to the inner solar system, but rather long-period or new comets that have lived the vast majority of their lives very far from the Sun. Thus whatever might be inferred about Halley or Borrelly from the crude imaging available probably is of marginal relevance to any comet that may suddenly threaten us. Of course, the frequency with which comets break up and disintegrate, sometimes for understandable reasons (e.g., Shoemaker-Levy 9's tidal break-up during close passage to Jupiter a year before the fragments impacted that planet) but usually just spontaneously for no obvious reason, suggests that comets are inherently fragile and composed of weak materials that might be very difficult to grab onto for mitigation purposes. As with rubble-pile asteroids, it might be necessary to apply very gentle forces across a broad portion of a comet's surface in order to move the body as a coherent whole. Fortunately, there are missions in various stages of development – even flight – that may improve on our nearly total absence of knowledge about the physical properties of cometary nuclei.

5 Conclusions

Eventually, scientifically motivated spacecraft missions will improve on our very poor knowledge of surface properties of asteroids and comets, although there is not always good correspondence between the first-order scientific questions that are the goals of such missions and the knowledge required for developing engineering approaches to mitigation. Missions have been advocated to study the "geophysics" of small bodies in a context that would address both scientific and mitigation-related issues. Of course, the best way to grapple with the practical realities of mitigation is to actually attempt to move a small body, as advocated by the B612 Foundation (http://www.b612foundation.org); indeed, initial work by the Foundation to develop a mitigation mission has already revealed some important practical and policy-oriented issues not previously considered and which would be unlikely to emerge from endeavors motivated solely by science. For example, it has recently been understood that it is better to change the spin axis of the object to be moved rather than to de-rotate it. An inadequacy, however, associated with concentrating on just a few mission targets is the wide diversity of small bodies. Until dozens of bodies of different mineralogies, geophysical configurations, sizes, spin periods, etc. are visited, we will always have to interpolate among them using ground-based telescopic data and theoretical models, in order to try to understand the full range of conditions we might have to deal with.

Although it would seem to be imperative to understand the nature of the surface of any body we want to deflect, our knowledge will always be imperfect. Thus the best deflection technology would be one that is least sensitive to differences in the nature of the target. Those technologies that are inherently low thrust and that distribute force across the broadest cross-section of the body would seem best able to meet such an objective.

References

Asphaug, E., P. J. King, M. R. Swift, *et al.* 2001. Brazil nuts on Eros: size-sorting of asteroid regolith. *Lunar Planet. Sci. Conf.* **32**, Abstract no. 1708.

Bell, J. F. 2001. Eros: a comprehensive model. *Lunar Planet. Sci. Conf.* **32**, Abstract no. 1964.

Chapman, C. R. 1997. Gaspra and Ida: implications of spacecraft reconnaissance for NEO issues. *Ann. New York Acad. Sci.* **822**, 227–235.

    2002. Cratering on asteroids from Galileo and NEAR Shoemaker. In *Asteroids III*, eds. W. F. Bottke, Jr., A. Cellino, P. Paolicchi, and R. P. Binzel, pp. 315–330. Tucson, AZ: University of Arizona Press.

Chapman, C. R., Merline, W. J., Thomas, P. C., *et al.* 2002. Impact history of Eros: craters and boulders. *Icarus* **155**, 104–118.

Cheng, A. F., Izenberg, N., Chapman, C. R., *et al.* 2002. Ponded deposits on asteroid 433 Eros. *Meteor. and Planet. Sci.* **37**, 1095–1105.

Robinson, M. S., Thomas, P. C., Veverka, J., *et al.* 2002. The geology of 433 Eros. *Meteor. and Planet. Sci.* **37**, 1651–1684.

Soderblom, L. A., Becker, T. L., Bennett, G., *et al.* 2002. Observations of Comet 19P/ Borrelly by the Miniature Integrated Camera and Spectrometer aboard Deep Space 1. *Science* **296**, 1087–1091.

Thomas, P. C., Robinson, M. S., Veverka, J., *et al.* 2001. Shoemaker crater as the source of most ejecta blocks on the asteroid 433 Eros. *Nature* **413**, 394–396.

Veverka, J., Farquhar, B., Robinson, M., *et al.* 2001. Imaging of small-scale features on 433 Eros from NEAR: evidence for a complex regolith. *Science* **292**, 484–488.

6

About deflecting asteroids and comets

Keith A. Holsapple

University of Washington

1 Introduction

In the 1994 book edited by Gehrels, *Hazards due to Comets and Asteroids*, chapters by Ahrens and Harris (1994), Shafer *et al*. (1994), Simonenko *et al*. (1994), Solem and Snell (1994), and Melosh *et al*. (1994) present and study a number of ways of preventing an oncoming asteroid from colliding with the Earth. Most methods considered nudging it sufficiently at 10 or so years before the impending collision to change its course so it would miss the Earth.

The methods studied include the use of conventional or nuclear explosives on or below the surface, the impact by large masses at high velocities, the blowing off of material by standoff nuclear weapons or by the concentration of solar energy using giant mirrors or by zapping it with lasers, and more gentle methods such as simply attaching a propulsion rocket, a solar sail or launching surface material off at sufficient velocity to escape the asteroid.

The analyses of the different methods rely primarily on data and estimates accumulated for cratering and disruption using the material properties of terrestrial materials. In most cases those were silicate materials with mass densities of ~ 3 g cm^{-3} or iron asteroids of density ~ 8 g cm^{-3}. However, it is becoming generally accepted that many of the asteroids are re-accumulated rubble pile bodies of very low density and strength, and comets have been thought for some time to have that structure. Low porosity may have little consequence for the pushing methods (except anchoring may be problematic), but the zapping and whacking methods depend crucially on material behavior.

Here I revisit some of the methods and put them into a common framework for comparisons. I also redo and improve some of the analyses and obtain different answers. I also report some studies of the effects of a low-strength porous structure

Mitigation of Hazardous Comets and Asteroids, ed. M. J. S. Belton, T. H. Morgan, N. H. Samarasinha, and D. K. Yeomans. Published by Cambridge University Press. © Cambridge University Press 2004.

on the various mitigation methods. I show that the presence of porosity can have a marked effect on some of the methods.

2 Previous results

Given about a decade lead time,[1] a change in velocity of a threatening asteroid of about 1 cm s^{-1} is sufficient to make it miss the Earth (Ahrens and Harris 1994). However, the estimates given for the energy or mass of a system to achieve that increment vary considerably. I have taken the material from the five references listed above and put them all into a common form for comparison, scaling them where necessary. I have assumed a velocity increment requirement of 1 cm s^{-1} for a 1 km diameter asteroid of mass density 3 g cm^{-3}.

The energy and mass requirements increase by at least a factor of 10^3 for a 10 km diameter asteroid. The exact increase is a scaling issue which has not been worked out, and will differ among the different methods.

Table 6.1 summarizes the methods and the estimates of the energy required. A discussion follows.

2.1 Impacts

The first three cases in Table 6.1 are for a deflection by a large mass impactor, at impact velocities of 12, 20, and 40 km s^{-1}. Ahrens and Harris (1994) use the impact scaling results of Holsapple and Schmidt (1982) for crater mass and ejecta velocity in each of the strength and gravity regimes to obtain their estimates. Their results range over a mass of 50 to 450 tons, having a kinetic energy[2] of 4 to 24 kt. They conclude that this approach might be feasible for smaller asteroids, but not competitive with nuclear devices for kilometer diameter and larger asteroids.

Shafer et al. (1994) rely on some very old data and calculations for impacts into metals to obtain their estimate. They get an impulse per unit energy dimensionless measure $I^* = (I\,v_{ref}/E)$, with the reference velocity $v_{ref} = 10^5$ m s^{-1}, of 10–30. The 1 cm s^{-1} deflection of a 1 km asteroid of mass density 3 g cm^{-3} requires an impulse of 1.6 10^{10} N s, so the required energies are from 12 to 36 kt, with a mass from 300 to 800 tons.

Melosh et al. (1994) also use the crater scaling laws of Holsapple and Schmidt (1982) and Housen et al. (1983) and derive the formula

$$d(\text{km}) = 0.14L(\text{m})^{0.81}t(\text{yr})^{0.27}v(\text{km s}^{-1})^{0.46} \tag{6.1}$$

[1] For short timescales, the needed velocity increment is $(75 \text{ m s}^{-1})/t(\text{days})$.
[2] A kiloton of energy is the energy of 1 kiloton of TNT explosive, or 4.2×10^6 MJ.

Table 6.1 *Estimates of the mass and energy required for various ways of diverting a silicate 1 km diameter asteroid by 1 cm s^{-1}, from five chapters of the book Hazards due to Comets and Asteroids (Gehrels 1994)*

Method	Reference	Energy (equivalent TNT)	Notes
Impact at 12 km s^{-1}	Ahrens and Harris (1994)	4–8 kt (250–450 tons mass)	Tables I, II in Reference
	Melosh *et al.* (1994)	0.1 kt (4.7 tons mass)	Equation (8) in Reference
Impact at 20 km s^{-1}	Shafer *et al.* (1994)	12–36 kt (300–800 tons mass)	
	Melosh *et al.* (1994)	0.1 kt (2 tons mass)	Equation (8) in Reference
Impact at 40 km s^{-1}	Ahrens and Harris (1994)	10–20 kt (50–100 tons mass)	Tables I, II in Reference
	Melosh *et al.* (1994)	0.11 kt (0.6 tons mass)	Equation (8) in Reference
Nuclear, 200 m standoff, neutrons (solid spall)	Ahrens and Harris (1994)	300 kt	Assumes no melt or vapor, 10% neutron yield, maximum heat is ~3 MJ kg^{-1}
Nuclear, 125 m standoff, neutrons (vapor expansion)	Shafer *et al.* (1994)	700 kt	Vaporizes surface. Maximum heat ~15 MJ kg^{-1}. I assume 10% neutron efficiency. Mass-scaled from 2.4 km extrapolated result
Nuclear, 90 m standoff, X-rays (vapor expansion)	Shafer *et al.* (1994)	3.5 Mt	Assumes vaporizes surface. Scaled up from above result
Surface nuclear explosion, cratering	Ahrens and Harris (1994)	40–90 kt	From crater scaling results of Schmidt *et al.* (1996)
	Simonenko *et al.* (1994)	100 kt	Simply assumes 1 ton of ejected mass per ton of nuclear explosive, at 100 m s^{-1}
Buried explosive nuclear or high-energy	Solem and Snell (1994)	0.3–6 kt	20–60 m burial. Based on scaling from their code cratering calculations
Fragmentation and dispersal	Ahrens and Harris (1994)	1–3 Mt	
	Simonenko *et al.* (1994)	1–100 Mt	

for the asteroid diameter d in km that can be deflected 1 cm s^{-1} if struck 7 years prior at a velocity v and an impactor diameter L in meters. Using the time of 7 years gives the impactor diameter required to divert a 1 km asteroid as

$$L = 5.92v^{-0.57} \tag{6.2}$$

which, for the three impact velocities shown give masses from less than 1 ton to 5 tons. Those values are well below the others, by about a factor of 100. That large discrepancy is due to a different assumption about the low velocity cutoff to the ejecta, and the use of gravity scaling at very low gravity levels, as addressed below. Note that the gravity result gives the impactor mass required for larger asteroids increasing with the 1.23 power of the asteroid mass.

2.2 Standoff nuclear weapons

A nuclear bomb detonated at some distance from the asteroid surface will deposit some of its energy in the form of X-rays or neutrons into a thin surface layer of material. That heated material will then rapidly expand, or "blow off" the surface imparting an impulse to the asteroid. The distance from the surface determines the resulting specific energy, which determines whether the heated material will remain in a solid state or will be melted or vaporized.

Ahrens and Harris (1994) assume conditions such that the resulting energy deposition does not melt nor vaporize the surface material, and that the energy is from neutrons that penetrate a uniform depth l. If y is the neutron energy, h the standoff height, and ρ the mass density, the resulting specific energy just under the bomb has the value

$$Q = \frac{y}{4\pi h^2 \rho l} \tag{6.3}$$

which must be less than the 2–3 MJ kg^{-1} melt energy of a silicate material. Therefore with 10% neutron energy from a 1 Mt device, the detonation height must be approximately 150 m or more.

As a consequence of the spherical geometry, as the detonation height becomes smaller, the percent of the device energy deposited on the surface increases, but the area of deposition decreases. By minimizing the product of the percentage area times the percentage energy, they find that the optimum standoff distance is about 40% of the asteroid radius: 200 m for a 1 km diameter asteroid. Assuming that the neutron yield is from 3% to 30% of the total yield, they conclude that energy from 100 kt to 1 Mt is required to divert a 1 km diameter asteroid.

For uniformity in the comparisons here, I assume a value of 10% neutron yield for all cases. The total required energy scales linearly with the inverse of that value, and linearly with the asteroid mass for other size asteroids. The energy is assumed to be absorbed uniformly in a surface layer of depth 20 cm, which for the 200 m standoff gives a maximum specific energy just below the device of 2 MJ kg^{-1}, which is just below the melt energy of silicate materials. A further crucial assumption made by Ahrens and Harris (1994) is that the heated material has no tensile strength, so the resulting expansion results in blow-off of the material.[3]

Shafer *et al.* (1994) consider a different case. They assume that the surface material is vaporized, and consider both X-rays and neutrons. To vaporize material requires a much closer detonation, and a resulting smaller area of deposition. They consider only the case of a one-dimensional deposition in the main body of the paper, and do a simplified analysis of the spherical case in their Appendix II using a different approach. There they estimate a blow-off velocity by assuming a perfect gas model, and assume a fixed average thickness of heated material for all standoff heights. They determine the impulse transmitted as a function of standoff distance, and find that the maximum momentum occurs when the standoff is about 25% of the asteroid radius, or 125 m for a 1 km asteroid. At that standoff, 1 Mt of neutron yield gives a velocity increment of 5.3 cm s^{-1}. They also give values for other size asteroids, down to a value of 1.5 cm s^{-1} for a 2 km diameter asteroid. A small extrapolation gives the asteroid size having a velocity increment of 1 cm s^{-1} as having a 2.4 km diameter. Thus, if a 2.4 km asteroid requires 1 Mt of neutron yield with 10% efficiency, it requires 10 Mt total yield. Scaling down to the asteroid diameter of 1 km gives the required total yield of 720 kt. The maximum specific energy just under the device is about 15 MJ kg^{-1}, sufficient to give an initial state in the melt region. The material will then vaporize as it expands (blows off) to low pressures.

For the X-ray case, Shafer *et al.* (1994) report that the peak velocity increment is about a factor of 3 less at a slightly closer standoff distance for all sizes, so it is less efficient by a factor of 3 than the neutron case for the same deposited energy. Simonenko *et al.* (1994) report a fraction of absorbed X-ray energy to be about 6%, which is 60% of the 10% assumed here for the neutron case. Using those two factors gives the estimate of 3.5 Mt total in the table.

2.3 Surface and sub-surface explosions

The third type of method uses the excavation of material in a cratering event from an explosion, either on or beneath the surface. Ahrens and Harris (1994) use the scaling results of Schmidt *et al.* (1986) for either of the gravity or strength regime, and obtain a requirement of 40–90 kt for a surface explosion.

---

[3] With strength, the heated layer will simply oscillate in thickness as the stress waves alternate in tension and compression.

For a deeply buried case, ranging from 11 m to 88 m burial depth for a 1 kt nuclear and conventional explosive, Solem and Snell (1994) report code cratering calculations. They determine impulse per unit energy ranging from $2 \ 10^9$ to $4 \ 10^{10}$ kg m/(s kt) for different geological materials and a nuclear bomb. Scaling to a 1 km asteroid the lower limit would require an 8 kt device at an actual burial depth of 22 m, and the upper value would translate to a 0.3 kt device at an actual depth of 58 m. They estimate the weight of a penetrator and conclude that the more shallow case and the larger 8 kt energy will be the most effective over the calculated range, but that an even better case is just to use a surface explosive. However, they do not analyze near-surface cases. Clearly, of all the cases in Table 6.1, these require the least energy, but require the weight penalties and uncertainties of reliability of a penetrator device. If their analysis is correct, the buried cases are not as efficient as a surface burst.

2.4 Disruption and dispersion

Finally, Ahrens and Harris (1994) determine that break-up and dispersal of a 1-km asteroid could be done with a nuclear weapon from 1–3 Mt yield, and Simonenko *et al.* (1994) obtain 1 to 100 Mt from a very simplified analysis. The 1–3 Mt energy provides an energy per unit asteroid mass in the range of $1–3 \ 10^3$ J kg^{-1}, while the 100 Mt estimate gives an energy per unit asteroid mass equal to 10^5 J kg^{-1}.

2.5 Summary of previous results

Taken at face value, these prior results show that the least energy required, a few kilotons, is for cratering by a buried explosive, which would require a penetrator design, but the weight of the penetrator device would more than offset the reduced explosive weight. The next smallest is the 40–100 kt of a surface explosion, followed by the 300–700 kt for a standoff nuclear weapon. The masses of larger nuclear weapons are less than 1 kg kt^{-1}, so the masses are on the order of only a few hundred kilograms. While the direct-impact cases require less energy, the corresponding masses are many orders of magnitude higher, so are not competitive. The disruption requires more energy than deflection, and raises questions about the fate and effects of multiple asteroid pieces, so is apparently not a good choice.

3 New analyses and porous asteroids

The prior analyses for the blow-off cases used some simple estimates for a net blow-off velocity of heated materials, and ignored any aspects of actual wave propagation. None of the previous analyses considers the effects of a porous asteroid. I now report

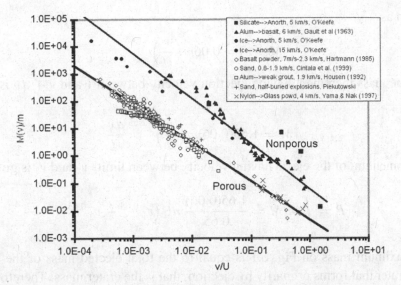

Figure 6.1 The mass of ejecta traveling faster than a velocity v as a function of v/U, where U is the impact velocity. In this scaled form, all porous materials fall on one curve and all non-porous and porous targets on another, but the curves are very different. (K. Housen, personal communication.)

on some new analyses: both analysis of solid asteroids using a wave code with a three-phase equation of state, and of porous asteroids using a porous model.

3.1 Impacts into non-porous asteroids

I first revisit the deflection by the impacts of large masses, which may be of interest for smaller asteroids. The momentum gained by the asteroid is the direct momentum transfer of the impacting mass, plus the momentum from all ejecta thrown back at speeds greater than the escape speed. The direct momentum is simply the impactor mass times its velocity. A straightforward way to calculate the momentum imparted from the ejecta is from a plot of the mass $M(v)$ of ejecta traveling faster than the velocity v as a function of that velocity. That can be determined by experiments or by code calculations. Figure 6.1 shows such results, where the ejecta mass is scaled by the impactor mass, and the ejecta velocity is scaled by the impactor velocity.

The notable feature of this plot is the great difference between the ejected mass for the non-porous materials compared to those for the porous materials. Note that the points for the non-porous materials are primarily from code calculations.

From these curves, the transmitted impulse can be calculated. First the non-porous cases are considered. The fit[4] shown to the data for the non-porous materials

---

[4] The point source scaling predicts this scaling exponent to be 3μ so this is consistent with $\mu = 0.55$; see Housen *et al.* (1983) and Holsapple (1993).

is given as

$$M(>v) = 0.06m\left(\frac{v}{U}\right)^{-1.65} \tag{6.4}$$

Then the mass increment having ejection velocity between v and $v + dv$ is given as

$$dM = 1.65(0.06)m\left(\frac{v}{U}\right)^{-2.65}\frac{dv}{U} \tag{6.5}$$

The momentum of the ejecta having velocity between limits v_1 and v_2 is given as

$$P = \int_{v_1}^{v_2} v\,dM = \frac{1.65(0.06)}{0.65}mU\left(v_1^{-0.65} - v_2^{-0.65}\right) \tag{6.6}$$

The maximum mass on Fig. 6.1 is equal to the total ejected mass of the crater. For a crater that forms primarily by ejection, that is the crater mass. Therefore, the lowest velocity is determined by where this fit intersects that upper cutoff, which is determined by either the gravity or the strength of the target. The uppermost data point on this plot should be that mass. In Fig. 6.1, the lowest ejecta velocity for the curve is at $v/U = $ 2e-4 at a mass ratio just above 10^4 so that is the lower velocity limit. It is in these cases determined by the strength of the target. The highest is at $v/U = 2$. Putting those limits into Eq. 6.6 shows that the upper limit is of no consequence, so that the momentum is dominated by the lowest-velocity ejecta.[5] Using the above equations, and assuming that even the slowest ejecta have a velocity greater than the escape velocity gives the momentum of the ejecta as

$$P = 13.6mU \tag{6.7}$$

or a factor of 13.6 times the momentum of the impactor. Assuming most of it is at a 45° angle, we must multiply by 0.71 to get the normal component. Adding then the direct momentum of the impactor, the total is 10.6 times that of the impactor.

Note that most of the data are not at the lower velocities that dominate the results. Since the slower velocity dominates the result, it might be better to just fit that portion of the data. That would reduce the constant of proportionality and the result by a factor of about 3. Then one obtains a total momentum of about 4 mU.

The nature of this result is well known for impact processes. There is a "momentum multiplication": the transmitted momentum can be many times greater than that of the impactor because of the ejecta thrown back along the path of the impactor.

With a momentum multiplication factor denoted as β, the required mass m at an impact velocity U to divert an asteroid with a mass M by a velocity increment Δv

---

[5] In contrast, the energy of the ejecta is dominated by the high-velocity ejecta.

is given as

$$m = \frac{M \Delta v}{\beta U} \tag{6.8}$$

For a 1 km diameter asteroid of density 3 g cm^{-3}, and $\Delta v = 1$ cm s^{-1}, and with $\beta = 10.6$, $U = 12$ km s^{-1}, the mass required is about 120 tons. At 40 km s^{-1}, the mass is about 37 tons. If the momentum multiplication is reduced to 4, the mass required at 12 km s^{-1} is 320 tons, while at 40 km s^{-1} it is 100 tons.

These results can be compared to those of Ahrens and Harris (1994). Their estimates of the mass required range from 250 to 450 tons at 12 km s^{-1} and from 50 to 100 tons at 40 km s^{-1}. Those have a momentum multiplication from 4 to 8. The values here are about the same, although better ejecta data at the lower velocities could easily change the results by a factor of several. In any case, the masses are probably not small enough to make this method attractive for kilometer-sized asteroids, but this method would for a smaller asteroid or by using multiple impacts.

Assuming that the same cutoff as apparent on these data holds for all asteroid sizes is essentially an assumption of strength (cube-root) scaling. Then the required mass increases linearly with the mass of the asteroid. Melosh *et al.* (1994) get much smaller required masses using gravity-scaled results. They assume the same gravity power-law form holds for all asteroids and all the way down to the escape velocity. It is uncertain whether gravity or strength scaling should hold, but I favor strength scaling for these small asteroids. Those who favor gravity scaling rely on results from code calculations that use a rate-dependent strength measure. Holsapple *et al.* (2002) discuss those results. For a 1-km asteroid, the escape velocity is less than 1 m s^{-1}, which, even for an impact velocity of 10 km s^{-1} is at an abscissa value of 10^{-4} on this plot. For the impact velocity of 40 km s^{-1} it is at the abscissa value of 2.5×10^{-5}. Therefore the lowest velocity on the plot is well above the escape velocity. Also, when the Melosh *et al.* (1994) equations (3) and (4) are combined, one gets the same form as Eq. 6.4, but with the proportionality constant of 0.147 compared to the value 0.06 here, an additional factor of 2.5. Further, if I take the factor of 3 reduction mentioned, there is a factor of 7.5 difference in the constant.

The analysis illustrates some major uncertainties of the calculation. The momentum is not commonly directly measured in experiments, so must be deduced from the ejecta measurements. Further, it is the very late-stage slow-moving ejecta that must be measured very carefully. The use of scaling relations for both crater size and ejecta separately may be suspect, especially when extrapolated to small bodies with very small gravity. Whether to use gravity or strength results is uncertain. In addition, experiments are not at the velocities of actual interest. Therefore, code calculations may be a better way to estimate these results in better detail, although only after the codes have been calibrated to the existing experiments.

Some code calculations using an SPH code have just been reported by Benz and Nyffeler (2004). Their model of porosity was based on simply including regions with no material particles. Even with 50% porosity, they still calculated a momentum multiplication factor of 4.3, much higher than the values noted above based on experiments. For a fully damaged but non-porous target, they got multiplication values as high as 35. Much more study of code approaches is necessary.

3.2 Impacts into porous asteroids

Few experiments have been made of impacts into porous materials. There are those reported by Housen et al. (1983) and by Shultz (2001) for moderately porous materials, and, more recently, those by Housen (2002) and Housen and Holsapple (2003) and those by Durda (2002) in highly porous materials. One should note the wide range of results these investigations produced, because there are many different types of "porous materials." The results by Shultz (2001) are not different in nature from those for conventional materials: craters form primarily by excavation and ejection. Shultz assumes that conventional gravity scaling defines extrapolations to low gravity, although that is a major extrapolation. Housen and Holsapple (2003) report that in highly porous materials craters are formed mostly by compaction, not by excavation, resulting in greatly reduced ejecta mass. Strength scaling is found for low gravity. Durda (2002) shoots projectiles entirely through open-pore foam materials with no cratering nor ejecta at all, albeit at an impact velocity lower than those of interest.

Figure 6.1 above also includes some ejecta data from Housen et al. (1983) in a moderately porous, weak grout material. The above calculations can be repeated using the curves[6] for the porous materials. The data for the grout data are fit[6] by

$$M(>v) = 0.005m \left(\frac{v}{U}\right)^{-1.4} \tag{6.9}$$

which holds to a lowest ejecta velocity of perhaps 3×10^{-3} times the impact velocity. At that value, the data clearly curve over to the upper mass cutoff. Assuming strength scaling in application to asteroids, this same cutoff is valid for all sizes. This is the total mass of the ejecta, which, for a porous asteroid is from a volume which may be only a part of the crater volume, the rest being compaction volume. Again integrating gives that the result that the ejecta momentum is only

$$P = 0.16mU \tag{6.10}$$

Multiply times 0.71 and add the momentum of the impactor to get the total of only 1.11 times the impactor momentum.

---

[6] The scaling exponent here is consistent with the point source scaling with the scaling exponent $\mu = 0.46$.

This result is very close to the perfectly plastic limit of impact physics, in which two colliding bodies simply stick together without the rebound of any material. That happens whenever no material is thrown backwards. In comparison to the solid asteroid case, for this material the momentum of the ejecta is almost negligible, and only that of the impactor matters. Thus, the required masses for a porous body for a given deflection increment are a factor of about 10 times greater than for a solid asteroid. Assuming strength scaling, that holds for all asteroid diameters. While again the numerical results are not terribly precise, it clearly shows that one cannot count on any momentum multiplication for impacts into very porous bodies, and diversion of porous asteroids or comets by impacts is even less attractive than for solid ones.

3.3 Explosions on or below the surface

Ahrens and Harris (1994) and Simonenko *et al.* (1994) analyzed a surface explosion scenario. Both nuclear weapons (NE) and a conventional high-explosives (HE) were considered. However, since the specific energy of an NE weapon is about 6 orders of magnitude higher than that of HE, even though its cratering efficiency is less by a factor of about 20 (Schmidt *et al.* 1986), it still has a great advantage over HE regarding the mass required. Ahrens and Harris (1994) use the cratering database given in Schmidt *et al.* (1986) and obtained an estimate of 40–90 kt required for the 1-km asteroid. Simonenko *et al.* (1994) use a much less sophisticated method, but obtain about the same result. Both results are probably roughly correct for a non-porous target.

For penetrator devices, the code approach by Solem and Snell (1994) should be calibrated against the terrestrial database as given, for example in Schmidt *et al.* (1986). There for kiloton field events the ejected mass is about ten times higher for a burst at the optimum depth of burst than for a surface burst, although the ejected velocities are lower. That optimum depth for a 1-kt weapon is at about 5 m, which is probably a feature of the effects of terrestrial gravity; on Earth deeper burial reduces the cratering. Therefore, they should include gravity to calibrate the code, then turn it off to obtain data for an asteroid. However, it is true that the factor of 10 applied to the surface estimates of Ahrens and Harris (1994) is roughly consistent with the Solem and Snell (1994) buried results, so there is really no reason to doubt the results.

What about porous bodies? Unfortunately, there are almost no data nor calculations for explosive cratering and ejecta in highly porous materials.[7] We do not yet know how cratering in porous bodies scales with size. Therefore, there is presently

---

[7] I am only aware of some HE field tests in snow and ice, but those data are strongly affected by the near-melt temperatures of the materials.

no basis on which to base estimates. The best we can do is note that the results
for impacts and for shallow buried explosions of equal energy are about the same
in conventional materials. Therefore, if the efficiency of diversion by impacts is
reduced by the porosity by a factor of, say, 10, then that for explosions might be
equally reduced.

Deeply buried explosives may just form a cavity by compaction, without any
ejected mass. That will depend on the crush strength of the material compared
perhaps to the lithostatic pressure. Clearly much more study and experiments are
required, although at present I would guess that surface and deeply buried cratering
are probably not viable method of deflection for porous asteroids.

3.4 Standoff nuclear blasts

It was noted above that both Ahrens and Harris (1994) and Shafer *et al.* (1994)
consider the deposition of energy into a surface layer of an asteroid by a nuclear
explosion detonated some distance above the surface, and that these analyses assume
two very different situations. Ahrens and Harris (1994) assume both that the layer
of heated material remains within the solid regime, and also that it lacks tensile
strength. Subsequently, solid material blows off the surface as it expands from an
initial heated state. The initial heated state must have a specific energy less than that
for melt (a value of perhaps 2 MJ kg^{-1}) and its resulting tensile strength must be
small compared to the initial stress (which is on the order of a few gigapascals). Their
analysis rests further on the estimate that there is from 3% to 30% of the weapon
yield in neutrons that will deposit energy in a layer of depth 20 cm along the line
of view for a spherical asteroid. That plus some spherical geometry considerations
determine the heated surface mass as a function of distance around the asteroid
surface from directly below the detonation point. They assume a blow-off velocity
$\Delta v = \frac{\Gamma Q}{c}$ where Q is the specific energy deposited, Γ is the Gruneisen coefficient,
and c is the sound speed. A numerical integration with respect to that distance from
just below the detonation point to the limit of exposed surface gives the imparted
impulse for a given yield and standoff distance. They report numerical results for
a standoff distance of 40% of the asteroid radius, which is about as close as their
assumption of solidity is valid.

Shafer *et al.* (1994) assume instead that the material is vaporized, so the specific
energy must be perhaps 5 MJ kg^{-1} or greater. It is consistent in this case to assume
no tensile strength. In the main body of their article they consider the physics of
energy deposition for one-dimensional cases, for the delivered impulse per unit area
arising from an energy deposition per unit area, as measured by their parameter
denoted by I^*. Then, in their Appendix II, they present a different analysis, for a
spherical asteroid. There they use a relation for an average thickness of the region

of deposition of energy, and a simple perfect gas blow-off velocity relation (rather than the results from their energy deposition analyses) to determine the impulse.

None of these analyses includes the wave physics of an actual energy deposition nor uses the complete equation of state behavior of the material. Here I present results from a calculation of the dynamics of the processes, using one-dimensional wave codes to study the initial state and subsequent motion of the heated layer and the underlying asteroid material. I then determine the three-dimensional effects for a spherical asteroid by a numerical method similar to that of Ahrens and Harris (1994). I used both the Lagrangian code WONDY (Herrmann *et al.* 1967) and the Eulerian code CTH (Thompson 1979). Both are available with complex three-phase equations of state and also with models for porous materials.

3.4.1 Standoff nuclear: solid asteroids

I first consider the case of a non-porous silicate asteroid, and only the case of the deposition of neutrons, which penetrate much deeper and therefore are more effective than the case of the deposition of X-rays. Following Ahrens and Harris (1994), I assume that the neutrons penetrate a depth of 20 cm instantaneously and uniformly.[8] That determines an initial state at the original mass density and at some high internal energy and pressure. Subsequent states are then determined by the equation of state for the asteroid material and by wave mechanics. For the calculations I took a three-phase equation of state for alpha-quartz (SiO_2) with an initial mass density of 2.65 g cm^{-3} using the ANEOS analytical EOS package (Thompson 1973), with the constants for this material attributed to J. Mclosh (Hertel and Kerley 1998). (It is also available as an equation of state in the SESAME tabular library used with CTH, but that tabular library is just a tabulation of the ANEOS model.)

The initial conditions for the code calculations have a layer of thickness 20 cm heated to some initial specific energy Q_1 and at the initial density of 2.65 g cm^{-3}, from which the EOS determines an initial high pressure P_1. The interior cold material beneath the heated layer is at its original zero pressure and low temperature state. Then a compression pressure wave develops that travels from the layer interface between the hot surface material and the cold interior into the cold interior, with magnitude approximately $P_1/2$, an unloading wave from that interface moving outward into the heated layer, and an unloading wave to zero pressure from the free surface moving into the hot layer. Thus, the interior is compressed to a higher density, and the hot layer expands (blows off) to zero pressure.

These states can be tracked on a plot of the equation of state. Figure 6.2 shows the EOS for quartz on the specific energy–density plot. The heated layer is initially at its original nominal density of 2.65 and at an elevated specific energy and temperature,

---

[8] Shafer *et al.* (1994) assume a depth of 10 cm.

K. A. Holsapple

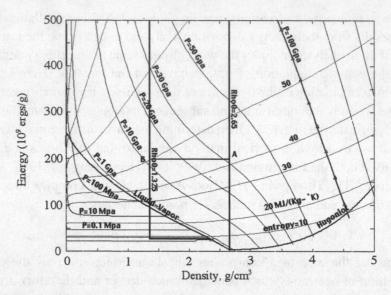

Figure 6.2 The equation of state plot for solid quartz with initial mass density 2.65 g cm^{-3}.

which may be in either of the solid or liquid phases. Those initial heated states are on the vertical line labeled "Rho0 = 2.65" at the appropriate initial energy deposition. An initial energy greater than about 3 MJ kg^{-1} (30×10^9 erg g^{-1}) is in the liquid state. Each material particle of the heated layer will then unload adiabatically along one of the constant-entropy curves to a reduced density as the surface layer expands to zero pressure. At some lower density, those adiabats enter the vapor dome shown, to a state that is an equilibrium mixture of vapor and melt.

The material points in the cold interior begin at their ambient zero pressure and very low temperature state, and are then loaded to a higher density as the compression wave from the heated layer reaches them. The shocked states are to the right along the Hugoniot curve shown, and those particles eventually unload from the compressed Hugoniot states along the adiabats back to zero pressure.

The code calculations determine these waves, and the transmitted impulse per unit area can be found either by adding all of the momentum per unit area of the heated layer as it expands, or by integrating the pressure vs. time curve at the interface. The results are shown in Fig. 6.3 as a plot of the momentum per unit area versus the initial specific energy. In these calculations there is no gravity so strength scaling holds. These results for a solid asteroid are the heavy curve with the diamond symbols. The result for porous asteroids on this figure will be discussed below.

These results for a non-porous asteroid can be compared to the approaches of the previous authors. Ahrens and Harris (1994) assumed an average blow-off velocity given as $\Delta v = \frac{\Gamma e}{c}$ where e is the specific energy e. The momentum per unit area

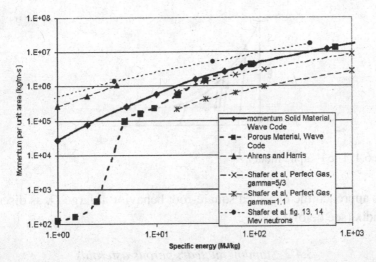

Figure 6.3 The momentum per unit area (impulse) imparted as a function of the initial deposited specific energy from neutrons in a 20-cm surface layer. The heavy curves are results from a wave code analysis for non-porous and porous asteroids, and can be compared to the diversity of previous results.

is that value times the mass per unit area ρt, where t is the deposition depth. Here $\rho = 2.65$ and $t = 20$ cm, so that the momentum per unit area is 53 times the average blow-off velocity. That curve is shown on the figure, but only for the lower specific energy values consistent with solid states. It is almost an order of magnitude above the present results.

Shafer *et al.* (1994: Appendix II) assume a blow-off velocity $\Delta v = \sqrt{(\gamma)(\gamma - 1)(e - e_v)}$, where e_v is the vaporization energy. They do not report the value they use, so I assumed a generic value of 10 MJ kg^{-1} for comparison purposes. Also, since they assume a neutron deposition depth of 10 cm, I used that value to get the momentum per unit area. They use a value for a perfect gas $\gamma = 5/3$, but they note it might be as low as 1.1. The quartz EOS does indeed have values as low or lower than 1.1 in the vapor region. Their results for both those values are shown on Fig. 6.3. Those results are both below the current results, for the $\gamma = 1.1$ case almost by an order of magnitude.

Shafer *et al.* (1994) in the main body of the chapter also present results of an energy deposition analysis for their I^* parameter, which is, to within a factor, the momentum per unit area divided by the energy per unit area. They do not consider the wave propagation aspects. The conversion to an energy per unit mass is again provided by multiplying the energy per unit area by the mass per unit area ρt. I show their results from their Fig. 13 for their "14Mev spike" neutron deposition. That result is distinctly above the present results, especially at low specific energy.

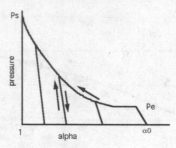

Figure 6.4 The P-alpha model.

All curves approach the expected square-root behavior at large Q, as discussed in the Appendix (see below).

3.4.2 Standoff nuclear: porous asteroids

Now the case of a porous asteroid structure is considered. Various models of the thermodynamical behavior of porous materials were developed several decades ago, partly in response to interest in porous materials as a method to protect weapons systems from the damaging influences of the X-ray deposition from nuclear bombs. One of the most used is the "P-alpha" model developed by Walter Herrmann at the Sandia Laboratory (Herrmann 1968), which is now a component of both the WONDY and CTH wave codes.

The mechanical behavior described by the model is like that of an elastic–plastic model, but for the dilatational (pressure–volume) component, not the deviatoric shear component of plasticity theories. The model assumes that the material is composed of small particles of normal "solid" density ρ_{solid} (say 2.65 g cm^{-3}) separated by intervening pore spaces. That gives a net mass density of a lower value ρ_{porous} (say 1.3). The ratio $\alpha = \rho_{solid}/\rho_{porous}$ is called the distension ratio. Then, as pressure P is applied, above a threshold pressure P_e the material can permanently crush to a smaller density as the pores collapse. If rapidly heated above that pressure, the solid particles expand into the voids as the overall mass density remains fixed. Above the pressure P_s the crush is complete, and the solid and porous densities are the same.

The introduction of the new distension parameter requires a new equation. That equation is provided by a curve of distension α versus pressure P. If the pressure is removed, there is some elastic recovery, but no change in the permanent crush. Figure 6.4 illustrates that crush behavior.

The thermodynamic assumptions are that the internal energy is contained in the solid particles, so that the energy per unit mass of the solid particles and the porous material is the same. The pressure of the porous material is a factor $1/\alpha$ of the pressure of the solid particles as required by thermodynamical consistency. Thus, in equations,

$$P_{porous} = \frac{1}{\alpha} P_{solid}(\rho_{solid}, e_{solid})$$

$$e_{porous} = e_{solid}$$

$$\rho_{solid} = \alpha \rho_{porous} \tag{6.11}$$

Inherent in this model is the assumption that this crush behavior is instantaneous. That is true only if the particles are small, so that the time scale of pressure equilibrium of the particles is negligible compared to time scales of a problem. For larger particles, this model would not be appropriate.

I performed a series of calculations with solid density $\rho = 2.65$ g cm^{-3} and an initial distension $\alpha = 2$, so that the initial porous density is 1.325 g cm^{-3}. I assumed that the neutrons are deposited into the same mass for both the porous and non-porous cases, so that the heated layer for the porous case was 40 cm. I assumed that crush began at the pressure $P_e = 1$ MPa and was complete at the pressure $P_s = 10$ MPa.

The nature of the porous cases can be best understood by a consideration of the initial states of the heated layer. There are three cases of interest. Suppose first that the heated solid material particles are below melt. These are the same as assumed by Ahrens and Harris (1994), but now for a porous material. Then, upon the instantaneous heating, the solid particles will expand into the void space to reach a reduced pressure consistent with the crush curve, with a pressure somewhere between $1/\alpha$ times 1 MPa and 10 MPa. Because of that expansion, the pressure is greatly reduced from that existent in the solid material. For example, the pressure in the solid material is given as $P = \rho \Gamma e$ which is several times the initial specific energy e. Figure 6.5 compares the pressure in the porous material to that in a solid as a function of the initial specific energy. In this regime, the pressure developed by heating is limited by the crush curve of the porous material. Those initial states in the EOS plot are just to the right of the vapor dome below the melt point labeled in the Fig. 6.2 EOS plot.

As a consequence of the greatly reduced pressure, the transmitted impulse is also reduced markedly. On Fig. 6.3 are the actual results; the cases where the initial specific energy is less than a few megajoules per kilogram are those conditions where melt is not induced. Those have an imparted momentum per unit area that can be more than two orders of magnitude below the non-porous cases.

Next are the cases where the initial specific energy is sufficient to cause melt. I assume that melt changes the nature of the porous crush: e.g., I assume that the crush pressure becomes zero and the melted material can expand to fill the void spaces entirely. As a consequence, the initial states in this case have a density of 1.325 g cm^{-3} and the initially solid but now melted particles expand to that density.

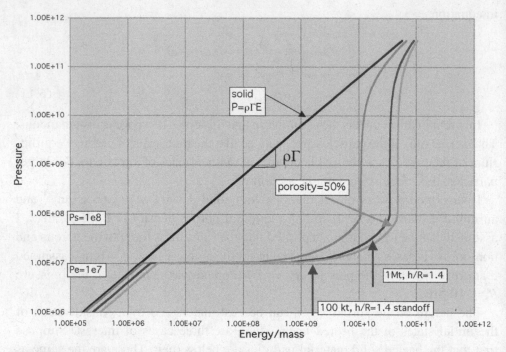

Figure 6.5 The pressure developed as a function of the initial specific energy for porous and non-porous materials. The expansion of solid particles into the void spaces reduces the pressure for the porous material by a substantial amount, until the pressures are well above the crush pressure.

Those particles have states on the vertical line labeled "Rho0 = 1.325" shown in Fig. 6.2. For specific energies above melt, greater than 2.6 MJ kg^{-1} but below about 10 MJ kg^{-1}, those initial states are well inside the vapor dome, so consist of an equilibrium mixture of melt and vapor. In this case, the expansion from the heating to the density of 1.325 g cm^{-3} allows vaporization,[9] and the initial pressures are extremely low compared to the case of a non-porous asteroid. However, the wave speeds in this regime are also greatly reduced, so the blow-off of the heated layer occurs at very low pressures but over much greater timescales, giving a transmitted momentum that is greater than the lower specific energy conditions having initial states below the melt transition. The code calculations in this regime are very difficult, since the sound speeds can approach zero. The curve is thus not well determined, but is roughly the curve shown in Fig. 6.3 for specific energies from about 2 to 10 MJ kg^{-1}.

Lastly, initial specific energy above about 10 MJ kg^{-1} gives initial states that are well above melt, but outside the vapor dome. The initial state is that of a liquid,

[9] One might question whether kinetic effects would postpone such phase changes, but the EOS here does not include any kinetic effects. Thus, this behavior might not be observed in experiments.

and are on the vertical line labeled "Rho0 = 1.325." As the heated layer expands, the unloading then leads to states inside the vapor dome. Figure 6.2 has two points, labeled as *A* and *B*. Point *A* is an initial state for a non-porous material with an initial specific energy of about 20 MJ kg^{-1}; it has an initial temperature of about 15 000 K and an initial pressure of about 30 GPa. In contrast, the porous initial state at that same initial energy is at point *B*. It has the same temperature, but a pressure of only about 11 GPa. The reduced pressure for the porous case results in reduced impulse compared to the non-porous case, but the reduced pressure effect is somewhat offset by a much lower sound speed which results in longer durations.

The entirety of results for the porous case are given in Fig. 6.3, as the dashed curve. At the lower specific energies that curve is much below the solid-body case and the transmitted momentum would not be effective at diverting an asteroid. At the specific energy of about 10 MJ kg^{-1} (temperature of about 8000 K) it is still below the non-porous curve by a factor of about 2, and finally approaches the non-porous curve as the specific energy reaches 100 MJ kg^{-1} (temperature of about 60 000 K) or greater. Therefore, it appears that, in order for a nuclear standoff method to divert a porous asteroid successfully, it is necessary to detonate the device very near the surface so that the resulting specific energy is extremely high. I will now look at the specifics of what the height of detonation should be and estimate some of the three-dimensional effects of a spherical asteroid shape.

3.4.3 Standoff nuclear: implications for asteroid deflection

The calculations of transmitted momentum per unit area provide the data to estimate the problem of asteroid deflection. The simplest case that illustrates the three-dimensional features is for a spherical asteroid, although clearly that may be a gross simplification. That case is presented here.

I borrow much of the analysis and notation of Ahrens and Harris (1994), specifically their geometry and angles defined in their Fig. 6.2. The problem is to determine the impulse transmitted to a spherical asteroid of radius *R* by the deposition of energy from a nuclear weapon detonated at a height *h* above its surface. The analysis is a quasi-one-dimensional one, in which circular rings of material around the surface in some angle range θ to $\theta + d\theta$ (Ahrens and Harris 1994: Fig. 6.2) have a deposition thickness *l*, a mass *dm*, a deposited energy *de*, and a specific energy $Q = de/dm$. All of these variables depend on the location on the asteroid, as measured by the angle θ. Then that heated mass blows off normal to the local surface at some velocity Δv. Ahrens and Harris (1994) assume a blow-off velocity Δv proportional to the specific energy Q; instead I will use the results of the code calculations given above to define an average blow-off velocity as a function of the specific energy.

The code calculations were done with a deposition depth of 20 cm, and at various specific energy values Q. One must scale those results to determine answers at other deposition depths. In the Appendix to this chapter, I consider that scaling problem, and show that for any other thickness, one simply takes the average blow-off velocity to get the momentum for any thickness. Implied in the analysis is that there is no effect of gravity, so strength scaling governs the process. If at a specific energy Q the one-dimensional momentum per unit area from a layer of material of mass density ρ and thickness l_0 is denoted by p, then the average blow-off velocity is given as the momentum per unit area divided by the mass per unit area:

$$\Delta v = \frac{p}{\rho l_0} \tag{6.12}$$

Now this velocity is a function of the specific energy Q that varies for the spherical asteroid with the angle θ. I used a distance variable $x = R \sin(\theta)$, which is the horizontal distance from the vertical axis (under the detonation point) to a point on the spherical surface. The thickness of deposition at any point on the surface is given as $l = l_0 \cos(\alpha)$ and the mass in an angle increment at that range is given as

$$dm = 2\pi \rho l R^2 \sin(\theta) d\theta \tag{6.13}$$

The deposited energy in a ring is given as

$$de = \frac{y}{2} \sin(\phi) d\phi \tag{6.14}$$

where y is the total (neutron) energy and α and ϕ are angles defined in Ahrens and Harris (1994: Fig. 6.2).

I generated an analytical fit to the curves of Fig. 6.3 above for both the porous and non-porous asteroids to define the blow-off velocity as a function of the specific energy Q. Using those, and various geometrical relations, it is easy to perform a numerical integration for the total momentum:

$$P = \int_0^{\theta_{max}} \Delta v \, dm \tag{6.15}$$

where θ_{max} is the value where the line of sight from the detonation point just grazes the asteroid surface, and defines the limits of the surface area that receives energy deposition.

I assume an asteroid of diameter 1 km, a 1 Mt nuclear bomb with 10% neutron yield, a solid mass density of 2.65 g cm^{-3}, a porous distension ratio of 2 for the porous case, a deposition depth of 20 cm for the solid asteroid, and a depth of 40 cm for the porous one. The numerical integration around the surface of the sphere of the

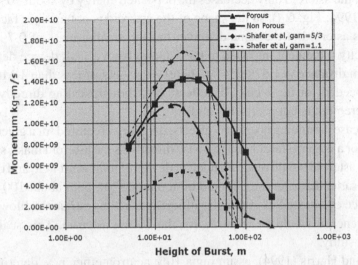

Figure 6.6 The total momentum imparted to a 1-km asteroid by a 1-Mt nuclear bomb as a function of the standoff distance. The solid asteroid is of density 2.65 g cm^{-3}, while the porous one is 1.325 g cm^{-3}. I also did the analysis using the perfect gas assumptions of Shafer *et al.* (1994) and find a requirement of much smaller standoff distance than they obtained.

incremental mass times the blow-off velocity for various standoff distances gives the imparted momentum.

The results are shown in Fig. 6.6. The maximum momentum is obtained for a standoff much less than the 40% radius value suggested by Ahrens and Harris (1994), and even much less than the 25% noted by Shafer *et al.* (1994). For the solid case it is at the distance of only 23 m, and for the porous case at about 15 m, only 5% and 3% of the asteroid radius, respectively. For large standoff distance, the porous case momentum is negligible, but approaches the non-porous asteroid case as the standoff distance becomes quite small. That is, of course, a consequence of the fact that the specific energies of deposition become much larger at the closer distances.

The mass of the solid 1 km asteroid is 1.4×10^{12} kg, so the velocity increment imparted at the optimum distance is just about 1 cm s^{-1}, so the 1 Mt bomb is just what is required. The maximum specific energy just under the detonation is about 1.2×10^3 MJ kg^{-1}, with an initial temperature of 3.5×10^5 K (28 eV), and an initial pressure of 1.5×10^3 GPa. Approximately 35% of the bomb neutron energy is absorbed over only 2.2% of the asteroid surface, over a 300 m diameter circle. The area-averaged deposition depth is 3.25 cm, ranging from 20 cm just under the device to zero at the extremities of the deposition circle where the line of sight from the detonation point just grazes the surface. On this size deposition area, the curvature of the asteroid only decreases the deposited energy by about 30% (Ahrens and Harris 1994: Fig. 6.4), so the shape of the asteroid is not a major factor.

A porous asteroid of the same diameter has the smaller mass of 0.7×10^{12} kg so the velocity increment imparted to the asteroid from a 1 Mt device detonated at the optimum distance is 1.67 cm s^{-1}. Therefore a yield of less than 1 Mt, perhaps 600 kt, is needed for the 1 cm s^{-1} deflection (although scaling directly with the velocity increment is not precisely valid).

In either case, assuming no gravity effects, the energy required for a given velocity increment for a larger asteroid will increase directly as the asteroid mass, so a 10 km non-porous asteroid would require a nuclear device of total yield 1 Gt, and the porous one of that same diameter a total yield of about 600 Mt, both about 100 times the largest device ever tested. If gravity is sufficient to act to restrain the blow-off, then even more energy would be required for the 10 km asteroid. This scaling issue needs more study.

Ahrens and Harris (1994), assuming a 10% neutron efficiency, determined that a 300 kt total yield would suffice for the solid asteroid at the distance of 200 m. The current results give that, at that distance, almost a factor of 10 more yield would be required for the non-porous asteroid, and a porous asteroid would not be deflected by any noticeable velocity increment with that standoff. Shafer et al.'s (1994) results gave that, again assuming 10% neutron efficiency, a 700 kt device would work at the standoff distance of 125 m, and that required energy would be greatly increased if the effective gas constant were only 1.1. Here the effectiveness at that standoff distance is negligible, but moving to a much closer distance gives the desired effect.

I compared the details of my numerical integration method with the analysis of Shafer et al. (1994) to discover the reason for this large discrepancy. I found that their approximation for the thickness of energy deposition is only valid for very large standoff distances, and is greatly in error for the standoff distances giving the most effective deflections. In particular, they estimate the average thickness to be 41% of the maximum. At the 23 m standoff, it is only 16%.

Note that if the standoff distance is reduced to these very small distances, the specific energy and the initial pressure are much larger than for a greater standoff distance, which might create questions about disrupting the asteroid. For the 23 m standoff, the total energy deposited is 35% of the neutron energy, or a total of 35 kt $= 1.5 \times 10^8$ MJ. The energy per unit mass of asteroid material (which is the standard measure of disruption criteria) is about 100 J kg^{-1}. As presented below, this is just about equal to estimates of the energy required to disrupt an asteroid, but a factor of 10 or more below that required to also disperse it. Therefore, there is probably no danger of dispersing it into many pieces, but it may be severely fractured by the deflection action.

3.5 *Laser, microwaves, and solar energy deposition*

These types of energy deposition have also been suggested. They differ in the deposition of neutron energy because they have a much more shallow penetration depth, and the heating of material to a blow-off state requires thermal diffusion to some depth. The energy deposition generally occurs in a heated plasma layer near the surface, whose calculation requires radiation-coupled wave codes. For that reason, the calculation by wave codes is much more difficult, and has not been done here. However, some questions about of the effects of a porous body should be raised. The primary result here for nuclear weapon deposition is that, although at specific energies near melt the porous material greatly diminishes the effectiveness of imparting momentum, much of that effect disappears at very high specific energies. Since these other deposition methods generally have a very high specific energy, it may be that the porosity has little effect. However, the mechanics of energy absorption are very different for these other methods. Clearly calculations and experiments to clarify these questions are needed.

3.6 *Disruption and dispersion*

Finally, I revisit the question of the energy needed if one wishes to blow up the asteroid. The disruption or dispersion of an asteroid by an impact is generally measured by the total energy averaged over the entire mass of the asteroid. Holsapple *et al.* (2002: Fig. 6.6) review the literature. There is a large range of estimates, although the more recent ones are tending to converge. For a 1 km diameter asteroid, the energy per unit asteroid mass for disruption is perhaps $100 \, \text{J} \, \text{kg}^{-1}$, while that to both disrupt and disperse is perhaps $5 \, \text{kJ} \, \text{kg}^{-1}$. Thus to disperse an asteroid requires an impactor at $20 \, \text{km} \, \text{s}^{-1}$ to have a diameter of about 30 m, a mass of about 3.7×10^4 tons, and an energy of about 2 Mt. An impactor deposits its kinetic energy in a target volume approximately equal to its own volume, so the initial specific energy is on the order of $1/2$ of the square of its velocity: $200 \, \text{MJ} \, \text{kg}^{-1}$ for a $20 \, \text{km} \, \text{s}^{-1}$ impact.

High-energy (HE) explosions at a depth of burial of about equal to their diameter act like impacts of the same energy (Holsapple 1980). Buried nuclear energy (NE) explosions are about half as effective as equal-energy HE explosives (Schmidt *et al.* 1986). Furthermore, a hot X-ray nuclear device initially vaporizes a sphere of the order of $2.2 \, \text{m} \, \text{kt}^{-1/3}$ (Schmidt *et al.* 1986), giving an initial deposited specific energy on the order of $100 \, \text{MJ} \, \text{kg}^{-1}$, on the same order as an impact. Therefore, a nuclear device with about two times the energy of an impactor and buried at a depth equal to the impactor diameter should be roughly comparable for disrupting or dispersing an asteroid. That equates to an energy of about 4 Mt buried to a depth of about 30 m, just above the 1–3 Mt range estimated by Ahrens and Harris (1994).

The energy required for both disruption and dispersion is definitely thought to be gravity scaled for asteroids above about 200 m. Figure 6.6 in Holsapple *et al.* (2002) has the required energy per unit asteroid mass increasing with the 1.65 power of the asteroid radius. That implies that the disruption and dispersion of a larger asteroid would scale as the 4.65 power of the radius, or the 1.55 power of the mass. Thus, a 10 km asteroid would require almost 5×10^4 times the energy as a 1 km asteroid. It would not be possible to disrupt a 10 km asteroid with a nuclear weapon. For smaller asteroids, the energy to disrupt a 100 m diameter asteroid is about the same as to disperse it, about $150 \, \mathrm{J \, kg^{-1}}$, implying about 50 tons of NE yield. That is also just about the energy to divert it by surface cratering, so a surface burst intended to divert it may also disperse it. For a 100 m asteroid, the standoff case looks more favorable, to avoid possible unwanted dispersions.

For porous asteroids, there are only a few experiments or calculations that can be used to estimate its disruption or dispersion energy (see the summary in Holsapple *et al.* 2002). Generally they are found to require much more energy per unit asteroid mass because of their ability to absorb energy. Much more information is needed to establish a coherent scaling theory that can be applied to large bodies.

4 Summary and final remarks

It is probably out of the question to use mass impactors against 1 km or larger porous asteroids. Surface explosions are probably not viable for a 1 km or larger porous asteroid. A penetrator device would require only a few kilotons to divert a solid asteroid if it could penetrate on the order of a few tens of meters, but the weight penalty is probably too high to compete with a surface explosion. It is not likely to be effective for a large porous asteroid, but that is presently unknown. A solid 1 km asteroid could be dispersed with maybe a 4 Mt weapon if it could be planted at about 30 m depth. A 10 km asteroid could not be dispersed with any conceivable energy. For either non-porous or porous asteroids, nuclear standoff methods would be ineffective at the heights found by prior researchers, but regain effectiveness at much closer ranges. A 1 Mt nuclear weapon detonated 23 m above the surface of a 1 km solid asteroid will impart a velocity increment of about 1 cm $\mathrm{s^{-1}}$. That is also about the energy required to disrupt it, but not to disperse it into separate pieces. About 700 kt at a height of 15 m will give that velocity increment to a porous asteroid, and probably would not disrupt it. In either case, these requirements will increase at least linearly with increasing mass of an asteroid, but the governing scaling is unknown. The effects of a porous structure on other surface energy deposition methods are also presently unknown.

Perhaps some of these methods could be used multiple times and in creative ways. For example, a first surface explosion might compact a region so that following explosions became more effective. Maybe the wasted energy of a standoff nuclear weapon could be focused towards the asteroid. A second explosion over the same point on a porous asteroid as an earlier one may encounter compacted material. Material on a porous asteroid at a pre-existing impact crater may be much more compact.

There is much more to be learned about the mitigation of threatening asteroids. We now know that the effects of a porous structure can be very significant, to the point of ruling out some of the suggested methods, and requiring a different use of others. What about the scaling issues for large asteroids? What other effects have we overlooked? While we know something about the mineralogy of asteroids and comets, we know almost nothing of their structure, or of their mechanical and thermodynamical behavior. We have very limited experiments in the porous types of materials that may be of interest, and there are almost no calculations for them.

A focused synergistic program of laboratory experiments and code calculations could clarify many of the questions raised here. Missions to asteroids and comets can determine more about their actual structure, if their surface is whacked or zapped. Such programs must be a component of programs to study the issues of the mitigation of the effects of large-body impacts into the Earth.

As Don Yeomans observed at the recent workshop, which was the precursor to this book, "We know much less about asteroids now than we did ten years ago." Hopefully we will know much more before the one with our name on it is sighted.

Appendix I Scaling of energy deposition in a surface layer

For the analysis of energy deposition in a spherical body from any standoff source such as a nuclear weapon or a laser, it is necessary to know the momentum contribution as a function of various initial specific energy Q and thickness l in some given material. A local one-dimensional approximation was used in the text.

For a one-dimensional energy deposition in a finite thickness surface layer, the transmitted momentum per unit area p is a result of complex wave dynamics and material properties. However, the final result can be written as some function:

$$p = F(Q, l, \rho, c, \mathcal{M}) \qquad (6.A1)$$

of the specific energy Q, thickness l, the two material properties: ρ the mass density and c the sound speed, and a set $\mathcal{M}$ of all other material properties, assumed to have only units that are combinations of mass density and velocity units.[10] It has also been assumed that there is no other length scale in the problem, which is valid

---

[10] There can be no material properties with time or length scales. That is true of the material descriptions commonly used.

for an infinite half space or if the deposition thickness t is small compared to a finite problem total thickness.

The number of variables can be reduced by three, using dimensional analysis. For example, the function can always be written as

$$\frac{p}{\rho c l} = G\left(\frac{Q}{c^2}, \overline{\mathcal{M}}\right)$$

(6.A2)

where now the $\overline{\mathcal{M}}$ are dimensionless ratios of material properties normalized to combinations of ρ and c. Define

$$\overline{v} = \frac{p}{\rho t}$$

(6.A3)

the momentum per unit area divided by the mass per unit area, it defines the average blow-off velocity. Thus for a fixed material the results have the simple form

$$\frac{\overline{v}}{c} = G\left(\frac{Q}{c^2}\right)$$

(6.A4)

which is independent of the thickness t.

To determine this average blow-off velocity, calculations can be made in a given material with any thickness t desired, for various specific energy values Q, and then plotted in this form. The results will then apply to any other thickness in that same material.

Others use special forms of this relation. Ahrens and Harris (1994) assume that

$$\overline{v} = \frac{\Gamma Q}{c}$$

(6.A5)

which is the linear form of this relation. They assume that the constant of proportionality is the Gruneisen constant Γ.

Other useful forms can be determined. While a solid has a material wave speed c, there is none for a perfect gas. Thus, for a perfect gas Eq. 6.A4 must have a form where the sound speed c cancels out. In addition, for large Q (and pressure) the sound speed should be of no consequence, which gives that the function G must be a square root and therefore that

$$\overline{v} = K\sqrt{Q}$$

(6.A6)

Shafer et al. (1994: Appendix II) use this form, where the constant is given in terms of the gas constant γ as $\sqrt{\gamma(\gamma - 1)}$.

If one divides by Q/c^2 an alternate dimensionless form in a fixed material is given as

$$\frac{pc}{Q\rho l} = H\left(\frac{Q}{c^2}\right)$$

(6.A7)

where the function H is the function G divided by its argument. The combination $f = Q\rho l$ is the energy per unit area, or the "fluence." If we use the fluence rather than the specific energy, the relation for the transmitted momentum per unit area becomes

$$\frac{pc}{f} = H\left(\frac{f}{\rho l c^2}\right) \tag{6.A8}$$

Shafer *et al.* (1994) denote the momentum per unit area divided by the fluence as I and define I^* using an arbitrary reference velocity v_{ref}:[11]

$$I = \frac{p}{f}, \quad I^* = \frac{p v_{ref}}{f} \tag{6.A9}$$

Therefore results can also be given in the form:

$$I^* = \frac{v_{ref}}{c} H\left(\frac{f}{\rho l c^2}\right) \tag{6.A10}$$

In a given material, and for a given radiation source type, the penetration depth l is also a material property, so that results can be written as

$$I^* = \overline{H}(f) \tag{6.A11}$$

which is the form Shafer *et al.* (1994) use. Since the G function approaches a square root for large values of its argument, the H functions approach a negative square root for large values of their argument. The Shafer *et al.* (1994) Figs. 11 and 13 show the G functions for different sources, and their Fig. 15 shows the H function. They do approach the expected square root limits at large argument values.

Acknowledgments

This research was supported by Grant NAG5-11446 from the National Aeronautics and Space Administration under the Planetary Geology and Geophysics Program. I thank Kevin Housen for Figure 6.1 and many useful discussions.

References

Ahrens, T. J. and Harris, A. W. 1994. Deflection and fragmentation of NEAs. In *Hazards due to Comets and Asteroids*, ed. T. Gehrels, pp. 897–928. Tucson, AZ: University of Arizona Press.

Benz, W. and Nyffeler, B. 2004. Impact sur asteroide. *Pour la Science*. In press.

---

[11] Since v_{ref} is not a problem parameter, it would seem better to me to just let $v_{ref} = c$ to get non-dimensional forms.

Durda, D. 2002. Impacts into porous foam targets: possible implications for the disruption of comet nuclei and low-density asteroids. Poster presentation, workshop on *Scientific Requirements for Mitigation of Hazardous Comets and Asteroids*, Arlington, VA, September 3–6, 2002.

Gehrels, T. (ed.) 1994. *Hazards due to Comets and Asteroids*. Tucson, AZ: University of Arizona Press.

Herrmann, W. 1968. Equation of state of crushable distended materials. Sandia Laboratories Report SC-RR-66-2678.

Herrmann, W., Holzhauser, P., and Thompson, R. 1967. WONDY, a computer program for calculation problems of motion in one dimension. Sandia Laboratories Report SC-RR-66-601.

Hertel, E. S and Kerley, G. I. 1998. CTH Reference manual: the equation of state package. Sandia Laboratories Report SAND-98-0947.

Holsapple, K. A. 1993. The scaling of impact processes in planetary sciences. *Ann. Rev. Earth and Planet. Sci.* **21**, 333–373.

    1980. The equivalent depth of burst for impact cratering. *Lunar Planet. Sci. Conf.* **11**, 1980.

Holsapple, K. A. and Schmidt, R. M. 1982. On the scaling of crater dimensions. II. Impact processes. *J. Geophys. Res.* **87**, 1849–1870.

Holsapple, K., Giblin, I., Housen, K., *et al.* 2002. Asteroid impacts: laboratory experiments and scaling laws. In *Asteroids III*, eds. W. F. Bottke, Jr., A. Cellino, P. Paolicchi, and R. P. Binzel, pp. Tucson, AZ: University of Arizona Press.

Housen, K. R. 2002. Does gravity scaling apply to impacts on porous asteroids? *Lunar Planet. Sci. Conf.* **33**,

Housen, K. R. and Holsapple, K. A. 2003. Impact cratering on porous asteroids. *Icarus* **163**, 102–119.

Housen, K. R., Schmidt, R. M., and Holsapple, K. A. 1983. Crater ejecta scaling laws: fundamental forms based on dimensional analysis. *J. Geophys. Res.* **88**, 2485–2499.

Melosh, H. J., Nemchinov, I. V., and Zetzer, Y. I. 1994. Non-nuclear strategies for deflecting comets and asteroids. In *Hazards due to Comets and Asteroids*, ed. T. Gehrels, pp. 1111–1134. Tucson, AZ: University of Arizona Press.

Schmidt, R. M., Housen, K. R., and Holsapple, K. A. 1996. Gravity effects in cratering. Report no. DNA-TR-86-182. Washington, DC: Defense Nuclear Agency.

Shafer, B. P., Garcia, M. D., Scammon, R. J., *et al.* 1994. The coupling of energy to asteroids and comets. In *Hazards due to Comets and Asteroids*, ed. T. Gehrels, pp. 955–1012. Tucson, AZ: University of Arizona Press.

Shultz, P. H. 2001. Cratering on a comet: expectations for deep impact. American Astronomical Society, DPS meeting no. **33**, Abstract no. 31.10.

Simonenko, V. A., Norgin, V. N., Petrov, D. V., *et al.* 1994. Defending the Earth against impacts from large comets and asteroids. In *Hazards due to Comets and Asteroids*, ed. T. Gehrels, pp. 929–953. Tucson, AZ: University of Arizona Press.

Solem, J. C. and Snell, C. M. 1994. Terminal intercept for less than one orbital period warning. In *Hazards due to Comets and Asteroids*, ed. T. Gehrels, pp. 1013–1033. Tucson, AZ: University of Arizona Press.

Thompson, S. L. 1973. Improvements in the CHARTD energy flow – Hydrodynamics code V: 1972/1973 modifications. Sandia Laboratories Report SLA-73-0653.

    1979. CSQII: an Eulerian finite difference program for two dimensional material response. Sandia Laboratories Report SAND-77-1339.

7

Scientific requirements for understanding the near-Earth asteroid population

Alan W. Harris

DLR Institute of Planetary Research, Berlin

1 Introduction

The known NEA population contains a confusing variety of objects: there are many different "animals in the zoo" of near-Earth asteroids. Some NEAs are thought to be largely metallic, indicative of material of high density and strength, while some others are carbonaceous and probably of lower density and less robust. A number of NEAs may be evolved cometary nuclei that are presumably porous and of low density but otherwise with essentially unknown physical characteristics. In terms of large-scale structure NEAs range from monolithic slabs to "rubble piles" and binary systems (asteroids with natural satellites or moons). An asteroid that has been shattered by collisions with other objects may survive under the collective weak gravitational attraction of the resulting fragments as a cohesionless, consolidated, so-called rubble pile. A rubble pile may become a binary system if it makes a close approach to a planet and becomes partially disrupted by the gravitational perturbation. More than 20 NEAs in the currently known population are thought to be binary systems and many more are probably awaiting discovery.

The rate of discovery of NEAs has increased dramatically in recent years and is now seriously outstripping the rate at which the population can be physically characterized. The NEA population is still largely unexplored.

Which physical parameters are most relevant for mitigation considerations? Preventing a collision with an NEA on course for the Earth would require total destruction of the object, to the extent that the resulting debris poses no hazard to the Earth or, perhaps more realistically, deflecting it slightly from its catastrophic course. In either case accurate knowledge of the object's mass would be of prime importance. In order to mount an effective mission to destroy the object knowledge of its density, internal structure, and strength would also be required. Deflection of

---

Mitigation of Hazardous Comets and Asteroids, ed. M. J. S. Belton, T. H. Morgan, N. H. Samarasinha, and D. K. Yeomans. Published by Cambridge University Press. © Cambridge University Press 2004.

the object from its course would require the application of an impulse or continuous or periodic thrust, the magnitude and positioning of which would depend on the mass and its distribution throughout the (irregularly shaped) body and on the spin vector. In either case mitigation planning takes on a higher level of complexity if the Earth-threatening object is a rubble pile or binary system.

A very important question is how remotely sensed parameters relate to the physical properties relevant to mitigation scenarios. For instance, what can we learn about the shape, mass, or structure of an asteroid from optical photometry, thermal–infrared photometry, reflectance spectroscopy, radar observations, etc.? The main techniques used in the remote sensing of asteroids that are most relevant to mitigation considerations are discussed.

2 Optical photometry

Optical photometry is a very important source of information on the rotation rates and shapes of asteroids. Repeated sampling of the rotational lightcurve of an NEA during one or more apparitions can lead to a very accurate rotation period and an estimate of the spin-axis orientation. Given sufficient resolution of lightcurve structure the basic shape of the body can be derived via lightcurve-inversion techniques (Kaasalainen *et al.* 2002).

A rapidly developing field in which optical lightcurve studies have played a major role is that of binary NEAs. Multiple periodicity in lightcurves has been observed in a number of cases, which cannot be attributed to the effects of shape or a complex rotation state. The most plausible cause is the presence of a companion satellite or moon with mutual eclipses and occultations giving rise to extra dips in the lightcurve (Fig. 7.1). Analysis of such lightcurves can reveal the period of revolution of the system and, via Kepler's Third Law, the mean bulk density of the bodies. Kepler's Third Law relates the orbital period, p, of the system to its semi-major axis, a, via:

$$p^2 = 4\pi^2 a^3 / G(M_P + M_S) \tag{7.1}$$

where M_P and M_S are the masses of the primary and secondary bodies, respectively.

Assuming the primary is spherical (see below) and its mass dominates (typically $D_S/D_P = {\sim}0.3$), we can substitute (density × volume) for mass and rearrange to obtain:

$$\rho_P = 24\pi a^3 / (G D_P^3 p^2) \tag{7.2}$$

where ρ_P and D_P are the density and diameter of the primary body, respectively.

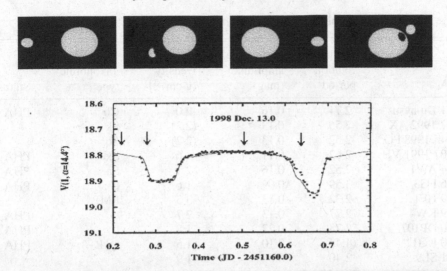

Figure 7.1 1998 FG$_3$ computer-generated model (upper four frames) and residual lightcurve after subtraction of the contribution from the rotation of the primary asteroid. The residual lightcurve structure is due to the eclipses and occultations illustrated in the upper four frames. The arrows in the lightcurve plot indicate the approximate points corresponding to the orbital phases depicted in the four frames taken in order from left to right. Graphics and lightcurve from S. Mottola (DLR). For more details see Mottola and Lahulla (2000).

Table 7.1 lists the currently known or suspected binary NEAs and the rotation periods of their primaries, lightcurve amplitudes, and densities calculated using Eq. 7.2. It is interesting to note that the bulk densities are generally low compared to the reference (solid) densities of the mineral mixtures of which asteroids are composed. For instance, silicate-based minerals have densities of 3–4 g cm^{-3} and nickel–iron has 7–8 g cm^{-3}. The low densities of binary primaries suggest that they have high porosities. Furthermore, the rotation periods are all just a few hours and the lightcurve amplitudes are nearly all less than 0.2 mag. These data suggest that binary NEA primaries have significant internal cavities, rotate at a rate just below the threshold for break-up via centrifugal force of a strengthless body, and are nearly spherical in shape, characteristics indicative of consolidated rubble piles.

A further observation from Table 7.1 is that the binary population includes a wide range of taxonomic classes. Although the database is still small it seems that taxonomic class cannot be used as an indicator of the probability of an asteroid being a rubble pile.

Recent radar observations have confirmed the presence of binary systems in the NEA population (Margot *et al.* 2002, and references therein). The rubble-pile hypothesis offers a convincing explanation for the formation of binary NEAs: a close approach of a rubble pile to a planet may result in partial disruption of the

A. W. Harris

Table 7.1 *Binary near-Earth asteroids[a]*

NEA	Primary rotation period	Lightcurve amplitude (mag)	Density[b] (g cm^{-3})	Taxonomic type[c]	Status[d]
3671 Dionysus	2.71	0.16	2.0	Cb	PHA
5407 1992 AX	2.55	0.13	(2.3)	Sk	
31345 1998 PG	2.52	0.13	(2.2)	Sq	
35107 1991 VH	2.62	0.11	1.6	Sk	PHA
1994 AW1	2.52	0.16	2.1	Sa	PHA
1996 FG3	3.59	0.09	1.6	C	PHA
1999 HF1	2.32	0.13	3.5	EMP	
1999 KW4	2.77	0.13	2.7	Q	PHA
2000 DP107	2.78	0.22	1.8	C	PHA
2000 UG11	(4.44)	0.10	1.5	Q, R	PHA
2001 SL9	2.40	0.09	1.9	–	

[a] Based on the web list of P. Pravec: http://www.asu.cas.cz/~asteroid/binneas.htm. Values in brackets are based on less secure data.
[b] Typical uncertainty in density ~ ±50%.
[c] Taxonomic types are from Bus and Binzel (2002), Binzel *et al.* (2003), or the above website.
[d] PHA, potentially hazardous asteroid (Minor Planet Center definition).

asteroid via a sudden increase in its spin rate. A large fragment that drifts away from the rubble pile in this manner may then remain gravitationally bound to the resettled pile as a moon.

If this model of binary asteroids is correct it is vital to establish the probability that the next large Earth-threatening asteroid will be a binary rubble-pile system, since it is hard to imagine a more difficult scenario for mitigation planners to deal with. Investigating the physical characteristics of rubble piles and NEA binary systems, about which very little is known at present, and devising appropriate mitigation strategies, should be assigned a high priority on the mitigation agenda.

Since shape, rotation rate, spin vector orientation, possible binary and/or rubble-pile nature, and density are all crucial parameters for space-mission and mitigation planning, lightcurve observations are a relatively cost-effective way of obtaining mitigation-relevant information about a large number of objects. Much can be done with 1-m-class telescopes and modest CCD-cameras, although a large amount of observing time over several years is required to obtain reliable data, e.g., spin vector orientation, on some objects, depending on their lightcurve amplitudes and orbital geometries. Furthermore, accurate work on the 1-m-class telescopes typically used for such observations is limited to objects brighter than $V \sim 18.5$. Larger telescopes are rarely used for such observations so information on the large number of small, but still potentially dangerous, NEAs with diameters less than 300 m is lacking.

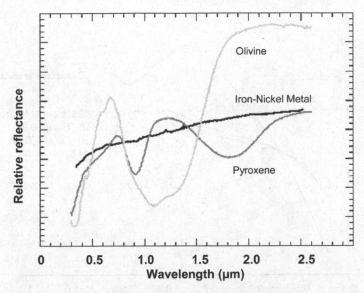

Figure 7.2 Reflectance spectra of common minerals found in meteorites and observed in asteroids. The spectral reflectance "fingerprints" of different minerals in the visible to near-infrared region form the basis of our knowledge of the mineralogical composition of asteroids. Adapted from NEAR–Shoemaker NIS Fact Sheet, Johns Hopkins University, Applied Physics Laboratory.

3 Reflectance spectroscopy

Spectroscopy is the main source of information on the mineralogy of asteroid surfaces. Analysis of asteroid spectra of modest resolution in terms of absorption band depth, spectral slopes, positions of maxima and minima, etc. reveals details of mineralogical composition and allows asteroids to be classed into taxonomic types according to their spectral features. While currently used taxonomic classification systems are based on optical spectra, investigations of the mineralogical composition of asteroids are greatly aided by extending the spectral range through the near infrared to cover the broad absorption bands of the metal-bearing silicate minerals olivine and pyroxene around 1.0 and 1.9 μm (Fig. 7.2).

Spectra of asteroids are compared with those of mineral mixtures and meteorite material to identify their most probable composition. A serious problem in the interpretation of reflectance spectra in terms of composition is the mineralogical ambiguity of featureless spectra. Minerals such as enstatite, and other iron-free silicates, metals, and dark, organic-rich carbonaceous material all display similar, relatively featureless spectra but have very different compositions and albedos. It may not be possible on the basis of reflectance spectroscopy alone to establish whether an Earth-threatening asteroid is a massive metallic object or a fragile, porous cometary nucleus of relatively low density. Such objects would presumably

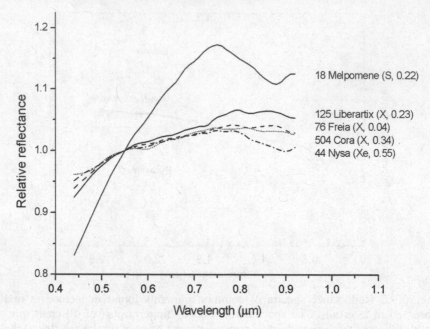

Figure 7.3 Reflectance spectra of main-belt asteroids with similar featureless spectra, compared to the spectrum of the S-type asteroid 18 Melpomene (from the
SMASS database at smass.mit.edu). The spectra have been normalized at 0.55 μm.
The spectral types (Bus and Binzel 2002) and geometric albedos (Tedesco 1992)
are given in brackets after the asteroid names. Despite the four X-type asteroids
having very different albedos, and presumably very different compositions, their
spectra are very similar. This illustrates the importance of albedo for distinguish-
ing between different mineralogical compositions giving rise to similar reflectance
spectra.

require very different mitigation approaches. In order to distinguish between the
different mineralogical compositions giving rise to featureless spectra the geometric
optical albedo of the object is required (Fig. 7.3). Approximate albedo ranges of
asteroids of different taxonomic types and mineralogical compositions are given
in Table 7.2. Albedos are most often determined via a combination of optical and
thermal–infrared observations.

4 Thermal–infrared spectrophotometry

The optical brightness of an asteroid depends on the product of its geometric albedo
and projected area; these parameters cannot be individually determined from optical
photometry alone. If, however, observations of the optical brightness are combined
with measurements of the object's thermal emission, both its albedo and size can
be individually derived. The largest database of asteroid albedos and diameters

Table 7.2 *Typical geometric albedos and probable mineral compositions associated with important asteroid taxonomic types*

Taxonomic type	Range of typical albedo	Probable mineral composition
D, P	0.03–0.06	Carbon, organic-rich silicates
C, B, F, G	0.03–0.1	Carbon, organics, hydrated silicates
M	0.1–0.2	Metals, enstatite
S	0.1–0.3	Olivine, pyroxene, metals
Q	0.2–0.5	Olivine, pyroxene, metals
V	0.2–0.5	Pyroxene, feldspar
E	0.3–0.6	Enstatite, other iron-poor silicates
X	0.03–0.6	Unknown (X signifies otherwise unclassifiable featureless spectrum)

compiled to date is based on thermal–infrared photometry with the IRAS satellite of some 1800 (mostly main-belt) asteroids (Tedesco 1992).

While thermal–infrared measurements have provided the vast majority of asteroid albedo determinations to date, the extension of the thermal–infrared, or radiometric, technique to NEAs is not straightforward. Complications arise in the thermal modeling of NEAs due to their often irregular shapes, compared to observed main-belt asteroids, apparent wide range of surface thermal inertia, presumably reflecting the presence or absence of a dusty, insulating regolith (small objects may have insufficient gravity to retain collisional debris), and the fact that they are often observed at large solar phase angles.

The problem of irregular shape can be largely overcome by combining thermal–infrared observations with lightcurve-tracing optical photometry obtained at about the same time. The optical photometry allows the infrared fluxes to be corrected for rotational variability.

One approach to the thermal modeling of NEAs that appears to be successful in most cases is that of Harris (1998), who modified the application of the so-called standard thermal model (STM) of Lebofsky *et al.* (1986) to allow a correction for the effects of thermal inertia, surface roughness, and rotation vector. The STM was designed for use with large main-belt asteroids and incorporates parameters that apply to asteroids having low thermal inertia and/or slow rotation observed at solar phase angles of less than 35°. In the near-Earth asteroid thermal model (NEATM) of Harris (1998) the model temperature distribution is modified to force consistency with the observed apparent color temperature of the asteroid, which depends on thermal inertia, surface roughness, and spin vector, and the solar phase angle is taken into account by calculating numerically the thermal flux an observer would detect from the illuminated portion of a sphere visible at a given solar phase angle, assuming no emission originates on the night side:

$$F_\lambda = \frac{\varepsilon D^2}{\Delta^2} \frac{hc^2}{\lambda^5} \int_0^{\pi/2} \int_{-\pi/2}^{\pi/2} \frac{1}{e^{\frac{hc}{\lambda k T(\theta,\phi)}} - 1} \cos^2\phi \cos(\theta - \alpha) d\theta d\phi \qquad (7.3)$$

where D is the diameter of the sphere, Δ is the geocentric distance, α is the solar phase angle, and θ is measured from the sub-solar point in the same plane as α and ϕ is analogous to latitude. Other symbols have their usual meanings in relation to blackbody radiation. The temperature distribution is given by

$$T(\theta, \phi) = T(0) \cos^{1/4}\phi \cos^{1/4}\theta \qquad (7.4)$$

For surface elements not visible to the observer at phase angle α, or on the night side, $T(\theta, \phi)$ is set to zero. $T(0)$, the maximum (sub-solar) temperature, is given by

$$T(0) = [(1 - A)S/(\eta\varepsilon\sigma)]^{1/4} \qquad (7.5)$$

where A is the bolometric Bond albedo, S the solar flux at the asteroid, η the beaming parameter, and σ Stefan's constant. The beaming parameter is treated as a variable for the purposes of fitting the model continuum to the observed thermal–infrared fluxes obtained at several wavelengths around the thermal peak in the range 5–20 μm. For further details see Harris (1998), Harris and Lagerros (2003), and Delbo and Harris (2002).

The NEATM has been applied to thermal–infrared spectrophotometric data from the Keck, UKIRT, ESO 3.6 m, and IRTF telescopes of some 40 NEAs to date and appears to give reasonable diameters and albedo values judging from comparison data derived from radar or space mission results and reflectance spectra, where available (two examples are given in Fig. 7.4). In general, the accuracy of NEATM appears to be better than ±15% in diameter and ±30% in albedo. Ongoing efforts to probe the accuracy of the NEATM, especially at solar phase angles beyond 50°, should lead to improvements in its reliability.

Given sufficient observing time on suitable telescopes, thermal–infrared observations offer a means of obtaining the sizes and albedos of a significant sample of the NEA population. Since mass is proportional to diameter cubed, the size–frequency distribution of the NEA population is of critical importance to the impact-hazard issue. Since asteroids have albedos, p_v, ranging from a few percent to about 60%, measurement of absolute brightness (H-value) alone allows the diameter to be determined to an accuracy no better than a factor of 2, and the mass to an accuracy no better than a factor of 8.

There is some evidence of a size dependency of asteroid albedos that might reflect the effects of space weathering (Binzel et al. 2003; Harris and Lagerros 2003). Collisional fragments from the break-up of larger bodies would be younger

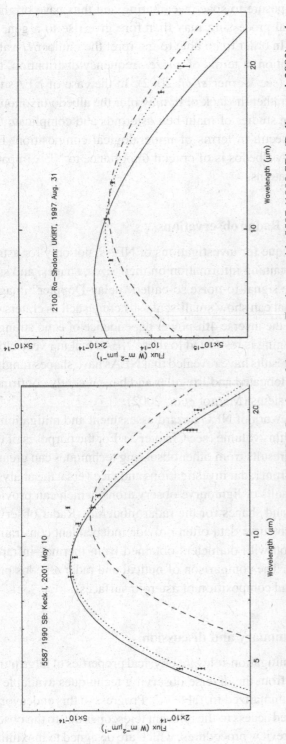

Figure 7.4 Thermal-model fits to Keck-1 and UKIRT flux data of the near-Earth asteroids 5587 (1990 SB) (Delbo *et al.* 2003) and 2100 Ra-Shalom (Harris *et al.* 1998). The error bars are 1σ uncertainties in the photometry and do not include absolute calibration uncertainties or errors due to variable atmospheric absorption. In the case of 5587 the STM fit (continuous curve) is nearly as good as that of the NEATM (dotted curve). Best-fit model thermal continua from the fast rotating model (FRM) (see Lebofsky and Spencer 1989), which assumes high thermal inertia and/or rapid rotation, are also plotted for comparison (dashed curves). In the case of 2100 only the NEATM fit is acceptable. The FRM best fits are very poor in both cases. The resulting NEATM effective diameter (corrected to lightcurve mean) and albedo for 5587 (Sq type) are $D = 3.6$ km, $p_V = 0.32$, respectively, and for 2100 (Xc type), $D = 2.5$ km (lightcurve maximum), $p_V = 0.13$, respectively. The STM results for 5587 are very similar to those of the NEATM; in the case of 2100, however, the STM diameter is too low compared to radar results. In many cases the FRM gives diameters that are too high compared to radar results and/or albedos that are below the ranges normally associated with the taxonomic classifications. The taxonomic classifications quoted here are from Bus and Binzel (2002).

and therefore have had less exposure to space weathering and thus have brighter surfaces. Continuous collisional processing may therefore give rise to a general dependence of albedo on size. In order to be able to interpret the number/*H*-value distribution of the NEA population in terms of a size–frequency distribution, the albedo distribution is required (see Werner *et al.* 2002). In the case of NEAs the effects of space weathering, or rather the lack of it, may blur the albedo/taxonomic class associations familiar from studies of main-belt asteroids and complicate the interpretation of reflectance spectra in terms of mineralogical composition. For these reasons, the study of NEA albedos is of crucial importance to NEA impact-hazard and mitigation considerations.

5 Radar observations

Radar is a very powerful technique for investigations of NEAs, not only for astro-metric purposes but also for obtaining information on their sizes, shapes, and sur-face properties. Given adequate signal-to-noise so-called "delay-Doppler" images of objects can be constructed that can show small-scale structure such as craters on their surfaces. However, due to the inverse 4th-power dependence of echo strength on distance, delay-Doppler imaging is restricted to those NEAs making very close approaches to the Earth. Radar results have revealed that NEAs have shapes ranging from almost spherical to very elongated and irregular and have recently confirmed the existence of NEA binary systems (Margot *et al.* 2002).

The use of radar in the framework of NEO hazard assessment and mitigation is described in detail elsewhere in this volume (see Chapter 3). For the purposes of the present discussion we note that results from other observing techniques can greatly assist the interpretation of data from radar investigations and vice versa: the analysis of radar data is aided by the results of lightcurve observations, which can provide valuable input on spin vectors and shapes for the radar modeling. Radar observa-tions combined with optical lightcurve data often provide independent constraints on asteroid sizes for comparison with diameters obtained from thermal–infrared spectrophotometry (see above). The comparison of optical and radar albedos pro-vides insights into the nature and composition of asteroid surfaces.

6 Summary and discussion

A wealth of information on the mitigation-relevant physical properties of large num-bers of NEAs can be obtained from the diverse observing techniques available at Earth-based observatories, as summarized in Table 7.3. Progress in this endeavor is limited, however, by the restricted access to the relevant telescopes due to the obser-vatories' observation-proposal review procedures, which are designed to maximize

Table 7.3 *Summary of mitigation-relevant information obtainable from Earth-based observations of NEAs*

Optical photometry (lightcurves)	Reflectance spectroscopy	Thermal–infrared spectrophotometry	Radar
• Rotation rates • Shape estimates • Spin-axis orientation • Identification of binaries • Densities (from modeling of binary-system lightcurves)	• Mineralogical composition	• Sizes • Optical geometric albedos (p_v) (composition) • (Future: thermal inertia; mineralogical composition from wavelength-dependent emissivity?)	• Accurate astrometry • Sizes • Radar albedos (composition) • Shapes • Identification of binaries (densities)

overall *scientific* return in astronomy in general and do not, at present, make allowance for the special circumstances of the NEA hazard assessment/mitigation task.

An important aspect of the application of astronomical remote-sensing techniques to the investigation of NEAs is the *interdependency* of the various techniques described above for the interpretation of observational data. A comprehensive understanding of the nature of NEAs requires a combination of virtually all observing techniques available. Rendezvous missions can provide "ground truth" for a few objects and thus aid in the accurate interpretation of Earth-based remote-sensing data, but for a thorough understanding of the diverse physical characteristics of the overall population of NEAs, which is vital for the development of an effective mitigation strategy, collaborative international Earth-based observation programs are essential.

References

Binzel, R. P., Lupishko, D., Di Martino, M., *et al.* 2003. Physical properties of near-Earth objects. In *Asteroids III*, eds. W. F. Bottke, Jr., A. Cellino, P. Paolicchi, and R. P. Binzel, pp. 255–271. Tucson, AZ: University of Arizona Press.

Bus, S. J. and Binzel, R. P. 2002. Phase II of the small main-belt asteroid spectroscopic survey: a feature-based taxonomy. *Icarus* **158**, 146–177.

Delbo, M. and Harris, A. W. 2002. Physical properties of near-Earth asteroids from thermal-infrared observations and thermal modelling. *Meteor. and Planet. Sci.* **37**, 1929–1936.

Delbo, M., Harris, A. W., Binzel, R. P., *et al.* 2003. Keck observations of near-Earth asteroids in the thermal infrared. *Icarus* **166**, 116–130.

Harris, A. W. 1998. A thermal model for near-Earth asteroids. *Icarus* **131**, 291–301.

Harris, A. W. and Lagerros, J. S. V. 2003. Asteroids in the thermal infrared. In *Asteroids III*, eds. W. F. Bottke, Jr., A. Cellino, P. Paolicchi, and R. P. Binzel, pp. 205–218. Tucson, AZ: University of Arizona Press.

Harris, A. W., Davies, J. K., and Green, S. F. 1998. Thermal infrared spectrophotometry of the near-Earth asteroids 2100 Ra-Shalom and 1991 EE. *Icarus* **135**, 441–450.

Kaasalainen, M., Mottola, S., and Fulchignoni, M. 2002. Asteroid models from disk-integrated data. In *Asteroids III*, eds. W. F. Bottke, Jr., A. Cellino, P. Paolicchi, and R. P. Binzel, pp. 139–150. Tucson, AZ: University of Arizona Press.

Lebofsky, L. A. and Spencer, J. R. 1989. Radiometry and thermal modeling of asteroids. In *Asteroids II*, eds. R. P. Binzel, T. Gehrels, and M. S. Matthews, pp. 128–147. Tucson, AZ: University of Arizona Press.

Lebofsky, L. A., Sykes, M. V., Tedesco, E. F., *et al.* 1986. A refined "standard" thermal model for asteroids based on observations of 1 Ceres and 2 Pallas. *Icarus* **68**, 239–251.

Margot, J. L., Nolan, M. C., Benner, L. A. M., *et al.* 2002. Binary asteroids in the near-Earth object population. *Science* **296**, 1445–1448.

Mottola, S. and Lahulla, F. 2000. Mutual eclipse events in asteroidal binary system 1996 FG_3: observations and a numerical model. *Icarus* **146**, 556–567.

Tedesco, E. F. (ed.) 1992. The IRAS minor planet survey. Technical Report no. PL-TR-92–2049. Hanscom Air Force Base, MA: Phillips Laboratories.

Werner, S. C., Harris, A. W., Neukum, G., *et al.* 2002. The near-Earth asteroid size–frequency distribution: a snapshot of the lunar impactor size–frequency distribution. *Icarus* **156**, 287–290.

8

Physical properties of comets and asteroids inferred from fireball observations

Mario Di Martino

INAF–Osservatorio Astronomico di Torino

Alberto Cellino

INAF–Osservatorio Astronomico di Torino

1 Introduction

Understanding the inventory and size distribution of those bodies which during their orbital evolution can intersect the orbit of the Earth with a non-zero probability of collision is a high priority task for modern planetary science. Apart from obvious considerations about mitigation of the impact hazard for the terrestrial biosphere, this is also a challenging theoretical problem, with important implications for our understanding of the orbital and physical evolution of the minor bodies of our solar system.

Several mechanisms have been discovered and analyzed in recent years to explain the steady influx of bodies from different regions of the solar system to the zone of the terrestrial planets. Several unstable regions in the orbital element space have been identified in the asteroid main belt, which can lead bodies to be decoupled from the belt and evolve into near-Earth object (NEO) orbits (Morbidelli *et al.* 2002). Both conventional collisional mechanisms and dynamical non-gravitational mechanisms (Yarkovsky effect) can be responsible for a steady injection of main-belt asteroids into these unstable orbits. Even in the absence of unstable regions, the Yarkovsky effect can cause a continuous drift in semimajor axis and eccentricity, eventually leading to a close encounter with a terrestrial planet, and removal from the main asteroid belt (Spitale and Greenberg 2001, 2002). The NEO production rate and the resulting NEO inventory and size distributions can be theoretically estimated and compared with observations (Bottke *et al.* 2002). It should be noted that the effectiveness of the different supply mechanisms is eminently size dependent. This is trivially true in the case of the Yarkovsky effect, since the acceleration induced by the thermal radiation force becomes progressively less relevant as the mass of the object increases, but also the average collisional lifetime of the possible

Mitigation of Hazardous Comets and Asteroids, ed. M. J. S. Belton, T. H. Morgan, N. H. Samarasinha, and D. K. Yeomans. Published by Cambridge University Press. © Cambridge University Press 2004.

parent bodies in the main belt in the case of direct collisional injection of fragments is size dependent. An improved understanding of the NEO inventory and size distribution, therefore, would have important implications for our understanding of the inventory and size distribution of the possible parent populations (main-belt asteroids and comets) down to sizes which are mostly not observable, as well as on the effectiveness of the supply and transport mechanisms in different size ranges. A major problem, however, is still our lack of knowledge of the size distribution of NEOs at small diameters. The interval of sizes for which the known NEO population can be considered essentially complete is above 1 km (the goal of current telescopic surveys being that of discovering 90% of existing NEOs larger than 1 km before 2008), and includes only a minor fraction of the whole NEO inventory. Therefore, it would be very useful to obtain more information on the existing bodies smaller than the current completeness limit. In particular, the size range between about one and some tens of meters corresponds to the least-known objects of the solar system. There is practically no hope of detecting objects in this size range except in the case of their close approach to the Earth. In particular, with current technology these bodies can be efficiently detected only when they actually collide with the Earth. In this size range, the usual outcome of a collision is a very bright fireball or bolide, delivering an energy typically in the range between 10^{-2} and 10^4 kt of TNT. These events typically occur at a sufficiently high rate to justify a systematic observational effort. The observations of fireballs can provide invaluable information on the inventory and size distribution of the NEO population in a very important interval of sizes, and can at the same time be used to derive data on the physical properties of these bodies, and on their likely origin. In particular, it is important to determine the relative ratio between the asteroidal and the cometary components of the NEO population at these sizes. In the following sections, we will give a short summary of the observational properties of fireballs, and how they can be used to infer information on the physical properties of the bodies. The problems to be faced in fireball detection will also be discussed, and we will explain why we are convinced that a dedicated space-based observing facility would be the most efficient tool to obtain a decisive improvement of the current situation.

2 The influx of cosmic material on the Earth

The Earth is subject to a steady bombardment by interplanetary bodies generally called meteoroids. The vast majority of these bodies are tiny and do not hit the ground, being mostly visible as meteors in the night sky. When the impacting bodies are sufficiently big, very bright fireballs are produced, and in some cases it can happen that the body is not destroyed during its path across the atmosphere, and some solid survivors, called meteorites, can reach the ground. Exceptionally, the impacting body may be so big as to produce a local or regional devastation. In the worst possible cases, global devastations are produced, which can trigger events of mass extinction in the biosphere.

About 75% of the meteor events are sporadic, whereas one-fourth of the observed meteors are genetically associated in a number of meteoroid streams producing meteor showers in well-defined epochs of the year. This is due to the fact that each meteoroid stream is formed by bodies having very similar orbits, likely produced by low-velocity ejection from a common parent body. Many known meteor showers (like the Leonids, Lyrids, and the Perseids to mention only a few of them) are known to be associated with a parent comet. In the case of the Geminids, the parent body is an object previously classified as an asteroid, 3200 Phaeton. In what follows, we give a brief summary of the observable features of meteoric events. For an excellent and exhaustive review of the subject, including a vast body of bibliographic references, the reader is referred to Ceplecha *et al.* (1998).

2.1 Detectable effects

Meteoroids hit the Earth's atmosphere at hypersonic velocities ranging mostly between 11 and 72 km s^{-1}. Depending on the entry velocity and the mass of the meteoroid, different phenomena are then produced. All of them, in practice, result from the conversion of the kinetic energy of the impacting meteoroid into other forms of energy. As a general rule, there is always the release of ions and free electrons along the meteoroid path in the atmosphere, which are produced by the collisions of the material on the body's surface with atoms and molecules of the atmosphere. As a consequence, the events can be detected by means of radar echoing. In many cases, visible light is produced during a process of ablation experienced by the body during the passage through the atmosphere (see below). In less frequent cases, corresponding to the most energetic events, detection of acoustic and infrasonic blast wave effects is also possible.

The most common outcome of the entry of a meteoroid into the Earth's atmosphere is the meteor phenomenon, in which the body is heated up to temperatures higher than about 2200 K. After a preliminary heating at heights between 200 and 130 km (preheating phase), a phase called ablation follows, in which the surface material starts to sublimate and a layer of hot vapor is produced around the body. At temperatures around 2500 K evaporation from the melted material starts. The impact of atmospheric molecules on the body and the meteoric vapor creates ions and leads to collisional excitation. Excited ions in this surrounding layer emit light at characteristic lines while they lose energy and are de-excited. This process continues until the body is completely ablated. Ions and free electrons are produced in an ionized column along the path of the body in the atmosphere, and this makes it possible to detect these events also by means of radar techniques.

When the impacting body is larger than some limit depending on the entry velocity vector (about 20 cm for a velocity entry of 15 km s^{-1} from the zenith direction), a very bright event can occur. When the apparent brightness of the meteor reaches a magnitude around -4 or brighter at visible wavelengths, it is generally called *meteoric fireball*. The term *bolide*, or *fireball*, is also generally used for events reaching magnitude -14 or brighter. When the magnitude reaches -17, the term *superbolide* is often used. Superbolides constitute the main subject of the present article, since they are produced by bodies with sizes mostly within the range of about 1 and some tens of meters, which is particularly interesting for the reasons mentioned at the beginning.

The ablation of the meteoroid can be incomplete when it is decelerated below hypersonic velocity, and in this case a phase called dark flight follows, in which some part of the meteoroid remains and is further decelerated while it cools down without emitting further visible radiation. The velocity can reach the free-fall limit, and a remnant mass may eventually hit the ground as a meteorite. When the impacting body is even larger and/or sufficiently strong and some of the cosmic velocity remains, an explosive impact occurs. The body hits the ground before being decelerated below hypersonic velocity, and emits light all the time until contact with the ground, where an explosive crater is produced.

Meteoric matter can also survive if brought in as very small particles. When the size of the meteoroid is small, of the order of some micrometers or tens of micrometers, and it radiates heat at sufficiently high rate, it can be substantially slowed down without reaching melting temperature. When this happens, the ablation process does not occur, and the meteoroid is led to softly reach the ground. Ions and free electrons are still produced, however, and these events are detectable by means of radar sensing. Samples of such interplanetary dust particles have been collected in the stratosphere by means of dust collectors mounted on aircrafts. According to some analyses (e.g., Kortenkamp and Dermott 1998), the chemical composition of these samples suggests that they are mostly of asteroidal origin, being probably related to the dust bands discovered by IRAS in 1984. The major sources of these bands are thought to be the dynamical families from major and/or recent break-ups in the asteroid main belt, like Eos, Veritas, and Koronis. According to the same authors, probably less than 25% of the collected dust particles should be of likely cometary origin. However, one should also take into account that the collected dust particles are preferentially those entering the Earth's atmosphere at relatively slow velocities, then a bias against the collection of cometary material is likely present. We should mention that other models predict that interplanetary dust particles at an heliocentric distance of 1 AU should be predominantly of cometary origin (Ishimoto 2000).

In addition to the above-mentioned observable effects in the interaction of meteoroids with the Earth's atmosphere, some additional phenomena are also observed in some cases. Meteor flares are a sudden increase in the apparent brightness of the object during the ablation process. In the case of high-entry velocities, flares are fairly common. When they occur at the end of the terminal point of the visible track, they can be due to the explosive disintegration of the body. Observations in NASA's Leonid Multi-Instrument Aircraft Campaign have shown that when this happens solid debris can be deposited even by meteors that enter at top speeds of 72 km s^{-1} (Borovička and Jenniskens 2000; Russell *et al.* 2000). This debris may be linked to micrometeorites found in ocean sediments. It should be taken into account, in fact, that the ablation is a complex phenomenon, which can be severely affected by many physical circumstances, such as (trivially) the direction of the entry velocity vector, which determines the length of the path of the meteoroid across the different atmosphere layers. Other common phenomena are *meteor wakes (meteor afterglow)* and *meteor trains*, a radiation emitted behind the meteoroid along its path. Wakes can last over several tens of seconds. Spectral observations indicate that meteor wakes are due to forbidden green O I line emission. The meteor afterglow shows spectral lines of the same elements producing the main "head" radiation, but at lower excitation energies (Borovička and Jenniskens 2000). Recombination line emission is also observed in bright fireballs on timescales of tens of seconds. Meteor trains have durations of up to minutes and hours, but a substantial part of the radiation consists of iron oxyde airglow due to the catalytic recombination of oxygen atoms in the train with ozone molecules in the environment (Jenniskens *et al.* 2000; Kruschwitz *et al.* 2001). Some meteor showers have been found to exhibit an increased presence of the meteor train phenomenon, compared to sporadic meteors.

2.2 Physical and orbital characterization of meteoroids

Several techniques can be used to detect observable features of the events, in order to derive from them information about the physical properties of the meteoroids impacting the Earth. A list of the most interesting properties includes the original orbit of the meteoroid, its size and composition, its mass and density, and the overall strength of the body. The above properties can be used to understand the origin and likely thermal, collisional, and dynamical history of the meteoroid. Also an assessment of the occurrence frequency of events produced by meteoroids of different sizes is very useful for obtaining information of the inventory and size distribution of meteoroids of these sizes, to be compared with predictions coming from models and extrapolations of the observations of kilometer-sized NEOs detected by telescopes.

In principle, the pre-impact orbital motion of the meteoroid can be derived from the entry velocity vector. This can be inferred from the visible trajectory of the meteor, in the case of optical observations. Since a single observing station can only give a single projection of the actual three-dimensional path, simultaneous observations from different stations are needed. For this purpose, dedicated networks of stations equipped with photographic cameras were established in the past in Europe, in the USA, and in Canada. Due to the limited sensitivities of the photographic detectors, these networks could be effective mostly for the detection of fireball events (although fainter meteors have also been and are observed by Superschmidt and small-camera networks). Only the European network is currently still in operation (Oberst et al. 1998).

The measurement of radar echoes is another powerful tool to determine meteoroid orbit data. A very important facility is the Advanced Meteor Orbit Radar (AMOR) system, operative since the early 1990s near Christchurch, New Zealand (Baggaley 2001). In very general terms, the orbits of many meteors turn out to be of likely cometary origin. This is certainly true for meteors belonging to the major showers known to be associated with comets, but a large contribution of likely cometary orbits is also found in analyses specifically devoted to study the case of sporadic meteors. In that case, however, a non-negligible asteroidal contribution is also found (Jones and Brown 1993).

When meteoroids of masses larger than 100 kg are considered, the asteroidal contribution seems to be increasingly important. According to Voloshchuk et al. (1997), in that mass range meteoroids of asteroidal origin would be dominant, accounting for 75% of "stream" meteoroids, and 37% of sporadic meteoroids. These conclusions are in qualitative agreement with results from the observations of the major events recorded by the European Fireball Network (Oberst et al. 1998). The role of asteroids is dominant when events producing recovered meteorites are concerned. For instance, all meteorite falls with recovered pre-impact orbits, namely Lost City, Innisfree, Pribram, Peekskill, and Tagish Lake, had orbits of a likely asteroidal origin, with the aphelia in the asteroid main belt. It is believed that meteoroids of cometary origin have generally no chance of hitting the ground due to their weak strength. Only extraordinarily large bodies of this kind can survive the ablation process. It is known that no meteorite collected so far has been found to be of probable cometary origin, whereas the interplanetary dust particles collected at high heights by aircraft-based experiments suggest a cometary contribution of around 25% to the interplanetary dust in the vicinity of the Earth (Kortenkamp and Dermott 1998), although, as previously mentioned, a bias against the recovery of cometary dust particles is likely at work here.

Further information can be obtained through an analysis of other observational features. Spectroscopic observations of the light radiated at visible wavelengths during meteoroid ablation has revealed the presence of a number of well-identified emission lines, the most intense ones being due to O I, N I, Na I, Mg I and II, Ca I and II, Fe I, Si I, Mn I, and Cr I. Relative abundances of different elements as derived from observations of the spectral lines in actual events, and also from laboratory experiments, have been obtained by several authors (again, see Ceplecha *et al.* (1998) for an extensive review), and they have been compared with analyses of collected meteorites of different types. The general results of these investigations are that it is generally possible to identify an iron or achondrite meteoroid from an observed good-quality spectrum. However, it turns out also that it is generally impossible to distinguish chondritic meteoroids (of likely asteroidal origin) from cometary meteoroids. The reason is that cometary and chondritic meteoroids seem to exhibit the same elemental ratios for elements that can be identified in meteor spectra. The problem here is that the most volatile elements, for which we could expect large differences for asteroidal and cometary meteoroids, are hardly measurable in meteor spectra. The bottom line is that cometary meteoroids cannot be easily distinguished in practical terms from chondrites (especially carbonaceous chondrites) purely on the basis of the spectra observed during meteor ablation in the atmosphere, a possible exception being the phenomenon of so-called differential ablation, in which sodium is seen to be released earlier than magnesium in the case of some meteoroids of likely cometary origin (Borovička *et al.* 1999). In any case, meteor spectroscopy seems to be a field of investigation deserving further efforts, in order to improve the overall observational data set.

Additional information can be obtained, however, from other observable parameters. They include ablation coefficients, and bulk densities derived from the photometric properties of the events after comparison with the predictions of physical modeling of the ablation process. Moreover, differences in beginning and terminal heights have long been used to distinguish among different classes of meteor behavior. In the case of bright meteoroids (fireballs), four classes have been originally defined by Ceplecha and McCrosky (1976), and they have been substantially confirmed by more updated investigations. The four classes have been named I, II, IIIa, and IIIb, respectively. They differ in terms of beginning height of the ablation process, bulk density, and ablation coefficient (which describes the rate of ablation experienced by the meteoroid). They are also preferentially associated with different orbit types, although it turns out that it is not possible to separate the four classes in terms of orbital elements alone. Class I fireballs have mostly asteroidal properties, with the highest bulk densities (beyond 3.5 g cm^{-3}) and relatively small ablation coefficients. They tend to have orbital semimajor axes close to the $3/1$ mean/motion resonance with Jupiter at 2.5 AU, large eccentricities (beyond 0.6), and small inclinations. Class II fireballs have a smaller density, around 2.0 g cm^{-3}, and a slightly larger ablation coefficient. The corresponding orbits tend again to be asteroidal. The common interpretation is that these objects should mostly correspond to carbonaceous chondrite parent bodies, which much less frequently are able to survive beyond the ablation process, and to hit the ground. They could in principle be of either asteroidal or cometary origin. Classes IIIa and IIIb fireballs are of evident cometary origin. They have bulk densities below 1 g cm^{-3}, and high ablation coefficients. Cometary shower meteors belong to class IIIa. It should be noted also that a corresponding classification has been developed (using other terms to indicate the classes) for fainter meteors on the basis of the observations made by networks of photographic cameras and television detectors.

2.3 Current estimates of the flux

Photographic and television systems, as well as simple visual observation campaigns, have been developed and organized for the general purpose of assessing the frequency of meteor events (and in particular to detect the changes in the intensities of several meteor showers during different apparitions), in addition to more detailed physical investigations of single events.

In turn, the observed influx of meteoroids in different size ranges corresponding to the maximum efficiencies of different meteor detecting devices (radar meteors for masses between 10^{-9} and 10^{-6} kg and visual/television/photographic observations between 10^{-6} and 10^4 kg), to be linked with observations of bodies in different size ranges, like NEOs observed with optical telescopes, has been used to derive an estimate of the total influx of interplanetary material on the Earth per year. The results obtained by different authors (Ceplecha 1992; Ceplecha *et al.* 1998) indicate that the Earth collects on the average something of the order of a few times 10^8 kg per year of interplanetary material. This average flux, however, is dominated by very large bodies which hit the Earth over very long timescales (tens and hundreds of millions of years). Over time-spans corresponding to a human life (100 years), one should disregard the contribution from bodies beyond 10^8 kg in mass, and the expected influx becomes a few times 10^7 kg. According to Ceplecha *et al.* (1998), the most important contribution would come from bodies between 10 and 100 m in size, corresponding to very bright fireballs and superbolides. The situation is not completely clear, however. Bodies in the 10–100 m size range have long been expected to hit the Earth on timescales of hundreds or thousands of years, the most impressive recent example being the Tunguska event in June 1908. According to a more recent analysis of a set of satellite detections covering a time-span of 8.5 years, however, the frequency of impacts with bodies between 50 and 100 m in size should be around one event every 1000 years (Brown *et al.* 2002). What is certain is that any extrapolation of the available data on the mass influx is still uncertain due to poor statistics. It is clear, then, that the observation of meteors and fireball events deserves further efforts to reach a more solid body of statistical evidence.

3 The observation of fireballs

In the past there was a much larger gap between the typical masses and sizes of the objects detectable by the fireball networks, and the typical sizes of NEOs accessible to telescopic observation. Now, this gap is narrower due to progress in sensor technology. Thus it seems now that the most interesting size range to be explored by fireball observations should be moved to slightly smaller values in the 10–100 m range.

In the next sections we will focus on the use of space-based observing devices. Here, we would like to mention first a few other possible options for ground-based fireball observations.

The first obvious possibility is to use ground-based optical facilities for spectroscopic investigations of the events at visible wavelengths. The importance of spectroscopic observations to derive information on the meteoroid compositions has already been summarized in Section 2.2. Possible instrumental improvements could conceivably be made in order to improve the performances of these observations, although this is in any case a challenging field, due to the peculiar nature of the events to be recorded (very short, and unpredictable).

Apart from the very rare and hugely destructive explosive impacts producing craters on the Earth's surface and regional or global devastations, fireballs and superbolides represent the most spectacular events of impact with extraterrestrial material. They are associated with the fall of meteorites. For several decades a big effort has been made to detect and record the maximum possible number of fireball events, with the general purpose of being able to determine the three-dimensional entry velocity vector of the bodies, in order to derive their pristine heliocentric orbits, and to determine their path in the atmosphere in order to determine also the likely regions of fall of possible associated meteorites. This task requires that a single event must be recorded by more than one observing station. For this reason, different networks of observing stations have been developed, as mentioned above, and the European Fireball Network is the last one still operating. The observing stations consist typically of very wide field cameras (including all-sky and fish-eye cameras). Typical performances of the European Fireball Network in the late 1990s, when it included 34 major observing stations mainly located in central Europe, was a total of about 10 000 image exposures per year, corresponding to an average of 1200 hours of clear-sky observations (Oberst *et al.* 1998). The detection rate was an average of around 50 fireballs per year, having peak magnitudes brighter than −6. Among them, about 50% could be actually observed by at least two of the observing stations of the network. In the period between 1959 and 1995, 13 objects were found to have (computed) pre-atmospheric masses over 1000 kg, and/or peak brightness over magnitude −15. Five events had magnitudes brighter than −17 (superbolides). These 13 events were classified according to the Ceplecha criteria based on the beginning heights (see Section 2.2), and only six of them were found to belong to class I, of clear asteroidal origin. The remaining events were of either class II or class IIIb (three cases). We recall that class II consists of bodies of likely carbonaceous chondritic composition, compatible with either an asteroidal or cometary origin, whereas class IIIb includes bodies of clear cometary origin.

Large networks of observing stations are needed for visible detection of the events, but they suffer from a number of inherent problems. First, the covered sky area is in any case forcedly limited in practice (for instance, the potentially conceivable development of an all-sky system would face the problem that most of the Earth's surface is covered by oceans). Second, these observations can be made only during the night. Third, the efficiency is affected by the varying weather conditions. Moreover, in spite of the fact that large observing networks had the capability of predicting the likely location fall of meteorites, the results have been largely insufficient. In the case of the European Fireball Network, after the first successful recovery of the Pribram meteorite in 1959, one more meteorite has been recovered since then, in Bavaria following a bright bolide on April 6, 2002, which was registered by eight camera stations of the network (Spurny *et al.* 2002).

Another possibility is given by infrasonic and acoustic techniques. Bright fireballs and superbolides deposit part of their kinetic energy into the atmosphere in the form of blast waves, which propagate through the atmosphere, and decay to amplitudes that can be detected by ground-based infrasonic sensor arrays (ReVelle 1976). The acoustic signals are subject to severe perturbations by the atmosphere, due to the presence of vertical temperature gradients and horizontal winds. Moreover, this technique should be intrinsically biased against weak bodies which are not capable of penetrating the atmosphere down to great depths (cometary meteoroids). According to some computations (ReVelle 1997) several thousands of events should occur every year with a kinetic energy sufficient to produce detectable signals. Several detections of very large bolides have been made in the last 15 years, and results have been summarized by ReVelle (1997). The detection of infrasonic signals from a fireball associated with the Leonid meteor shower on September 17, 1998 is also reported by ReVelle and Whitaker (1999).

4 Bolide detection from space-based sensors

The observable phenomena exhibited by fireball events are best detectable from space-based facilities, like satellites. This includes observations both of the light spike at visible wavelengths, and of the infrared radiation produced by the heating of the meteoroid material all along its path in the atmosphere. In fact, after the flash the heated air and debris take some time to cool down, during which they are detectable in the mid-infrared spectral range. Moreover, the trail and cloud of debris left in the atmosphere by the incoming body, when illuminated by sunlight, creates a localized source detectable in the near infrared. Infrared observations are also possible from the ground, but it is known that at these wavelengths space-based detectors work better and more efficiently. This is trivially true when sources above the atmosphere are concerned, but also in cases like this in which the infrared emission takes place in the high layers of the atmosphere, space-base detectors have in any case a much reduced atmospheric extinction at these wavelengths with respect to ground-based facilities. Moreover, fireball events also include phases during which the meteoroid material is heated up to very high temperatures, and emits detectable radiation in the ultraviolet spectral region. The emission of the ionized material along the fireball track (fireball train) is expected to include spectral lines in the ultraviolet region, and space-based sensors can be useful to record also this kind of radiation.

Space-based sensors also have a number of other obvious advantages with respect to ground-based observing stations. They can cover wide areas of sky, and a relatively small number of satellites orbiting at sufficiently high orbits (20 000 km or more) could conceivably provide a full-sky coverage. Moreover, being located above the atmosphere, a satellite is not limited by weather conditions, and, even more important, can also operate in daytime.

Given the very large number of satellites currently in orbit, it may seem strange that fireball detections are not very frequently reported. The simple reason is that current satellites are dedicated to other purposes, and it is not usual that fireball events are detected and recorded. In many cases satellites are equipped with optical sensors that work in scanning mode with a small field of view for high-resolution observations, and that limits the chances of detection of fireball events. Most satellites currently equipped with sensors useful for fireball detections have military purposes. Unless the operators have some reason to record an observed event, this is discarded and no data remain of it. Even recorded events, moreover, can be not made public, being included in the records of classified activities. Moreover, even in cases of properly equipped satellites devoted to civilian activities, like systematic monitoring of large areas of the Earth for various purposes, what happens is that generally the data pipeline has been designed to record data having very different characteristics with respect to fireball events. For instance, satellites aimed at monitoring the long-term evolution of some ecological environments are not supposed to record events with typical durations of a couple of seconds. As a consequence, data showing sudden changes like fireball events are generally simply automatically discarded as a source of high-frequency noise, for these systems.

In this respect, technologies developed for military purposes are much more helpful. For instance, the techniques of so-called transient radiometry developed for detecting very short duration events like missile engine ignitions and nuclear bomb detonations are very useful for fireball detection purposes (Tagliaferri *et al.* 1994).

The most recent estimate of the flux of meteoroids in the size range between 50 and 100 m has been based on data on about 300 events recorded by satellites of the US Department of Defense between 1994 and 2002 (Brown *et al.* 2002). Some discrepancy between the flux derived from these observations, which are in a qualitative good agreement with recent telescopic data on NEOs, and the estimates based on infrasound/acoustic sensors is apparent. Extrapolations of the numbers of events recorded by infrasound sensors would lead to predict a higher influx of meteoroids delivering energies around 1000 kt of TNT, corresponding to sizes between 20 and 50 m, approximately. This discrepancy might be due to poor statistics, or to the models used to compute the energy released in the events. The data seem in any case to suggest that most of the influx of meteoroids is made up of objects of likely asteroidal origin. But since it is true that in any case a large fraction of the real events are essentially lost, it is clear that some effort should be made in order to have at our disposal a *dedicated* system of fireball detection from space.

5 Future perspectives

Fireball observations are important if we are to derive important physical information on the population of meteoroids orbiting in the vicinity of the Earth. After decades of excellent ground-based activities, which have been able to obtain the best conceivable results based on ground-based detectors, the time now seems ripe to plan the development of a new generation of dedicated, space-based observing facilities. Visible and infrared cameras should be the primary sensors for these facilities, which do not require the launch of independent satellites, since they could be carried on board orbiters designed to carry out other activities, to reduce the overall costs of the deployment of a fireball-observing network. As a purely speculative example, for instance, one could think about the possibility of adding fireball-dedicated cameras to about ten satellites of the fleet that will be launched in the near future for the European Global Navigation Satellite System (Galileo). This is expected to consist of a fleet of about 30 satellites, orbiting at altitudes of about 24 000 km, and should be operative starting in 2008.

High satellite orbits are better for sky-covering issues, but there is also a corresponding intrinsic reduction in sensitivity performances, and in precision in the determination of the entry trajectory of the bodies. For this reason, an ideal space-based network would include also a number of satellites on lower-altitude orbits. In addition, the International Space Station could include some of the sensors of a more general network.

In practical terms, proposals are also being addressed to the European Space Agency by the authors of the present work and other colleagues (i.e., SFERA: Space-based sensors for Fireball Explosion Recognition in the Atmosphere). Very recently, ESA issued an announcement of opportunity (AO/1-4229/02/NL/CP) for the development of systems devoted to the detection of sudden events in the atmosphere like lightning and meteors, or the observation of similar events on the dark side of other planetary objects. The detectors to be designed to fulfill these tasks are precisely of the kind needed for any project of systematic space-based detection of fireballs; thus there are reasons to hope that the studies carried out in the framework of this ESA project will soon lead to the development of a new generation of detectors optimized for these specific purposes.

Apart from the purely scientific issues, a space-based system for fireball surveillance could have also some more immediately practical purposes. It is known, for example, that the clouds of dust released by these events can be an hazard for aircraft, and moreover the impact of a sufficiently big meteorite in an ocean or sea could produce a dangerous tsunami wave. For these and other reasons, prompt detection of these events could be of the highest importance for mitigation of possible danger for human beings in particular circumstances.

Acknowledgments

The authors wish to thank G. Valsecchi and G. Cevolani for useful suggestions and comments. The detailed comments by an anonymous referee are also gratefully acknowledged.

References

Baggaley, W. J. 2001. The AMOR radar: an efficient tool for meteoroid search. *Adv. Space Res.* **28**, 1277–1282.

Borovička, J. and Jenniskens, P. 2000. Time resolved spectroscopy of a Leonid fireball afterglow. *Earth, Moon and Planets* **82/83**, 399–428.

Borovička, J., Stork, R., and Bocek, J. (1999). First results from video spectroscopy of 1998 Leonid meteors. *Meteor. and Planet. Sci.* **34**, 987–994.

Bottke, W. F., Jr., Morbidelli, A., Jedicke, R., *et al.* 2002. Debiased orbital and absolute magnitude distribution of the near-Earth objects. *Icarus* **156**, 399–433.

Brown, P., Spalding, R. E., ReVelle, D. O., *et al.* 2002. The flux of small near-Earth objects colliding with the Earth. *Nature* **420**, 294–296.

Ceplecha, Z. 1992. Influx of interplanetary bodies onto Earth. *Astron. Astrophys.* **263**, 361–366.

Ceplecha, Z. and McCrosky, R. E. 1976. Fireball end heights. *J. Geophys. Res.* **81**, 6257–6275.

Ceplecha, Z., Borovička, J., Elford, W. G., 1998. Meteor phenomena and bodies. *Space Sci. Rev.* **84**, 327–341.

Ishimoto, H. 2000. Modeling the number density distribution of interplanetary dust on the ecliptic plane within 5 AU of the Sun. *Astron. Astrophys.* **362**, 1158–1173.

Jenniskens, P., Lacey, M., Allan, B. J., *et al.* 2000. FeO "orange arc" emission detected in optical spectrum of Leonid persistent train. *Earth, Moon and Planets* **82/83**, 429–438.

Jones, J. and Brown, P. 1993. Sporadic meteor radiant distributions: orbital survey results. *Mon. Not. R. Astron. Soc.* **265**, 524–532.

Kortenkamp, S. J. and Dermott, S. F. 1998. Accretion of interplanetary dust particles by the Earth. *Icarus* **135**, 469–495.

Kruschwitz, C. A., Kelley, M. C., Gardner, C. S., *et al.* 2001. Observations of persistent Leonid meteor trails. II. Photometry and numerical modeling. *J. Geophys. Res.* **106**, 21525–21542.

Morbidelli, A., Bottke, W. F., Jr., Froeschlé, C., *et al.* 2002. Origin and evolution of Near-Earth Objects. In *Asteroids III*, eds. W. F. Bottke, Jr., A. Cellino, P. Paolicchi, and R. P. Binzel, pp. 409–422. Tucson, AZ: University of Arizona Press.

Oberst, J., Molau, S., Heinlein, D., *et al.* 1998. The "European Fireball Network": Current state and future prospects. *Meteor. and Planet. Sci.* **33**, 49–56.

ReVelle, D. O. 1976. On meteor-generated infrasound. *J. Geophys. Res.* **81**, 1217–1230.

1997. Historical detection of atmospheric impacts by large bolides using acoustic-gravity waves. *Ann. New York Acad. Sci.* **822**, 284–302.

ReVelle, D. O. and Whitaker, R. W. 1999. Infrasonic detection of a Leonid bolide: November 17, 1998. *Meteor. and Planet. Sci.* **34**, 995–1005.

Russell, R. W., Rossano, G. S., Chatelain, M. A., *et al.* 2000. Mid-infrared spectroscopy of persistent Leonid trains. *Earth, Moon and Planets* **82/83**, 439–456.

Spitale, J. and Greenberg, R. 2001. Numerical evaluation of the general Yarkovsky effect: effects on semimajor axis. *Icarus* **149**, 222–234.

2002. Numerical evaluation of the general Yarkovsky effect: effects on eccentricity and longitude of periapse. *Icarus* **156**, 211–222.

Spurny, P., Heinlein, D., and Oberst, J. 2002. The atmospheric trajectory and heliocentric orbit of Neuschwanstein meteorite fall on April 6, 2002. In *Asteroids, Comets, Meteors 2002*, ESA SP-500, pp. 137–140.

Tagliaferri, E., Spalding, R., Jacobs, C., *et al.* 1994. Detection of meteoroid impacts by optical sensors in Earth orbit. In *Hazards due to Comets and Asteroids*, eds. T. Gehrels, M. S. Matthews, and A. Schumann, pp. 199–220. Tucson, AZ: University of Arizona Press.

Voloshchuk, Y., Vorgul', A. V., and Kashcheev, B. L. 1997. The meteor complex near the Earth's orbit: sporadic background, streams, and associations. III. Sources of streams and sporadic meteoric bodies. *Astronomichevskii Vestnik* **31**, 306.

9

Mitigation technologies and their requirements

Christian Gritzner

Dresden University of Technology, Institute for Aerospace Engineering

Ralph Kahle

DLR Institute of Planetary Research, Berlin

Protection of the Earth from undesirable impacting bodies is not just a
science fiction project for some improbable future. The cost might be
comparable to, even smaller than, the world's current military expenditures. We could choose to do it now. We could choose to protect ourselves
from asteroids and comets rather than from each other.

Fred L. Whipple, *The Mystery of Comets*, 1985

1 Introduction

Impacts of near-Earth objects (NEOs) onto our planet are natural events where
the effects of each single impact mainly depend on NEO size, structure, relative
velocity, and impact location. To determine if a newly discovered object might
impact on Earth one day, the object's orbit has to be numerically computed into
the future. The accuracy of this orbit prediction basically depends on the optical
and eventually radar measurements available for that object, and of course on the
completeness and precision of the numerical perturbation models. Care has to be
taken when handling few observations, long prediction periods, and unendorsed
NEO properties, which might lead to large uncertainties in collision probability
prediction (e.g., Giorgini *et al.* 2002).

NEOs larger than 150 m in diameter and approaching Earth's orbit closer than
7.5 million km (0.05 AU) are called Potentially Hazardous Objects (PHOs). Due
to their susceptibility to small orbit disturbances on short timescales they are can-
didates for future collisions with Earth. Typical near-Earth asteroid (NEA) impact
velocities onto the Earth range from 11.18 to about 25 km s^{-1}, whereas comets
typically impact at higher velocities up to 73.65 km s^{-1} if on a retrograde orbit
(e.g., Gritzner 1996).

Mitigation of Hazardous Comets and Asteroids, ed. M. J. S. Belton, T. H. Morgan, N. H. Samarasinha, and
D. K. Yeomans. Published by Cambridge University Press. © Cambridge University Press 2004.

In Section 2 we assess the risk of NEO impacts and compare expenses for NEO hazard mitigation with expected costs due to the damages. Section 3 gives an overview on current, near-future, and futuristic technologies that could be used for NEO hazard mitigation. In Section 4 warning and preparation time demands as well as scientific requirements for NEO mitigation missions are discussed. The solar concentrator concept seems to be a very promising mitigation technology and is therefore described in more detail in Section 5. Here we also give numerical orbit computation results for the diversion of two PHOs. Finally, in Section 6 several steps to prevent future NEO impacts are defined.

2 Risk assessment

To assess the risk of NEO impacts the risk management approach is applied, which is an integral part of project management. Risk management is a formal process that allows evaluating and comparing different risk events in order to identify the ones that should better be mitigated in advance, i.e., before they occur. Solving risks in advance ("pro-active approach") requires fewer resources than solving problems, i.e., risks that already have occurred. For the case of NEO impacts solving risks means protecting humanity from a (global) catastrophe. The qualitative definition of the risk level depends on its probability of occurrence and its consequence if occurring. In general the NEO impact risk is a "high-consequence – low-probability" level risk. Therefore, it is categorized as a "moderate" level risk which requires certain precautionary actions (Gritzner and Fasoulas 2002).

The current worldwide expenditure for NEO hazard analysis and mitigation of about $5 million per year is negligible when compared to the efforts justifiable by the risk management approach, which would be on the order of several hundred million US dollars per year. Current activities (e.g., NEO detection, follow-up observations, etc.), and the proposed worldwide search campaigns and mitigation system studies are fully justified, even if financial aspects only are considered.

The benefits of deflection missions can be defined simply as the costs of the damages that would occur in the event of a NEO impact. Canavan (1994) derived the losses from impacts with global effects from an interruption in Earth's gross product (2×10^{13} per year) for a 20-year period (4×10^{14}) due to damages and evacuation of large regions. Further, Canavan calculates the average losses of impacts of all sizes to be about $400 million per year, dominated by the largest objects. From this approach it follows that the requested $50 million of investment costs and $15 million per year for operation costs for a worldwide search program (Morrison 1992) are fully justified. This should be extended by $5–10 million per year for the development of mitigation systems. In total, the average annual benefits of preventing impacts of NEOs larger than 2 km are about 20 to

80 times higher than the annual costs for planned and current detection activities, respectively.

If an actual threat and a successfully conducted mitigation mission are considered the relation of benefits to cost is even more promising. Assuming benefits (saved damages) from a large impact of 4×10^{14} and costs for a large-scale deflection mission of 10^{10} the benefit/cost relation would equal a factor of 40 000.

A recent study conducted by a business economist at Dresden University of Technology pointed out that the approach of Canavan (1994) uses some modeling assumptions that have to be redefined in order to get results closer to reality (Duerfeld 2002). Therefore, the following statements were made: (a) a homogeneous spread of production over the whole Earth is not given; (b) Earth's gross product is gained in relatively small areas (industrialized nations); and (c) sufficient warning time for the determination of the impact site and the evacuation is possibly not available in many or most cases.

Duerfeld (2002) showed that several common cost-analysis methods, such as cost–benefit analysis, human capital approach, and cost-efficiency analysis, are not applicable for this purpose. Thus, the cost-prevention approach was applied to calculate the mean annual costs for NEO detection and mitigation depending on the number of affected people in several scenarios. Four categories were defined in order to represent similar regions on Earth ranked according to the number of inhabitants. In comparison, for a homogeneously populated Earth the overall average would be about 12 people km^{-2} (including the oceans). The following categories were defined:

(1) 2500 people km^{-2} on 0.1% of Earth's surface (e.g., industrial centers).
(2) 50 people km^{-2} on 15.7% of Earth's surface (e.g., typical countryside).
(3) 10 people km^{-2} on 13.5% of Earth's surface (e.g., deserts, mountains, etc.).
(4) No inhabitants on 70.7% of Earth's surface (e.g., oceans).

Impact scenarios for S-type asteroids with diameters of 60 m, 150 m, 400 m, 750 m, 1.5 km, 3 km, and 10 km were analyzed with regard to the expected casualties and damages for the different categories. For NEOs ranging from below 750 m to 3 km the results strongly depend on the location of the impact area. Here, the values for casualties and damage costs differ by up to two orders of magnitude for the different categories. Larger NEOs that would cause global catastrophes were assumed to affect all regions of our planet, so that no differentiation of the impact location was necessary.

For example, in the 150 m scenario about 73 215 people are expected to die due to the impact and secondary effects. This is an area-specific average value derived from the casualties for the different inhabitant categories. For categories 1, 2, 3, and 4 the expected casualties are 15 664 000, 313 000, 63 000 and 0, respectively.

Further, the expected area-specific average materially losses are calculated to be about $58.68 million. This is a sum of losses due to the impact itself and losses due to 20 years of production failure in the affected area. The underlying losses per category are $52 193 billion, $41 billion, $54 million, and $0, respectively. Note that casualties and losses for category 4 are set to 0. For larger NEOs these numbers go up and become larger than the ones for category 2. This is because of the large destruction capabilities of tsunamis, which would result from an impact into an ocean. A 1.5 km asteroid would probably devastate an area 100 times larger than for the 150 m asteroid. The same factor is expected for the number of casualties and the amount of materially losses.

The resulting cost estimates can be compared with other common expenditures, e.g., the military budget of a nation, which shall be used to protect the lives and security of the people. Clearly, the absolute annual mitigation costs (including NEO search) are well below other comparable expenditures such as the military budget, which was about $750 billion worldwide in 1997 (CIA 2001). This applies for the annual mitigation costs per world population or per potential casualty, too. For the smallest NEOs the annual mitigation costs are the highest, because these impact events are more frequent. Duerfeld (2002) estimated the annual costs to prevent an Earth impact of a 150 m asteroid to be $10.2 million, which is equivalent to $0.002 per capita of world population, or $139.32 per potential casualty on average. For larger NEOs with global consequences both per capita values drop below $0.002, and annual mitigation costs would be about $3 million. Thus, it can be concluded that NEO detection and mitigation activities are not too expensive at all and that they should be greatly intensified.

Despite all cost analyses we should not forget that an impact of a 1 km NEO would probably cause a global catastrophe killing millions or even billions of people, and innumerable animals and plants. If human lives are considered in these statistical calculations it is an ethical question of whether to state that each human life is invaluable or to assume a theoretical amount of money for each lost life such as a compensation payment. However, if we are not able to avoid a large impact then modern civilization, most cultural values, and our standard of living would be irretrievably lost.

3 Review of NEO mitigation systems

Currently, various ideas for the deflection and disruption of hazardous asteroids and comets are under consideration. Among them are systems that might be technologically feasible at present, such as chemical rocket engines, kinetic energy impacts, and nuclear explosives. Others are currently under development and might be available with some development effort in the near future, such as solar concentrators.

But there are also systems that seem to be too far off to be realized in the required dimensions for the task of NEO deflection within the next decades, such as solar sails, laser systems, and the utilization of the Yarkovsky effect. Besides, there are also futuristic technologies such as eaters and the use of antimatter.

The most important characteristics are briefly described in the following. For more detailed information we refer to the sources cited at each system description. Further references can be found in a database established for the European Space Agency (ESA), which includes papers on NEO mitigation from 1967 to 2000 (Gritzner 2001).

3.1 History of NEO mitigation

The first technical approach to define a NEO mitigation system was the students' system project "Project Icarus" at the Massachusetts Institute of Technology (MIT) in 1967 (Kleiman 1967). This interdepartmental students' project in systems engineering took place during spring term 1967. The team developed a concept to avoid a collision of asteroid 1566 Icarus (provisional designation: 1949 MA) with the Earth. They recommended the use of six nuclear explosive devices each delivered to Icarus by a Saturn V launcher. The spacecraft bus was assumed to be a modified Apollo command module with an attached service module.

The first conferences on the problem of asteroid and comet impacts on Earth took place in the early 1980s ("Snowmass conferences"), but the publication of conference proceedings was refused in order to avoid a possible controversy about nuclear explosives in space. As recommended in the 1990 position paper of AIAA, the US Congress asked NASA to conduct workshops on NEO detection and interception. Both conferences (NEO Detection Workshop and NEO Interception Workshop) took place in 1992 and international collaborations began.

Seven conferences with major contributions on NEO mitigation have taken place up to now:

- 1992, January 14–16, Near-Earth Object Interception Workshop, Los Alamos, USA
- 1994, September 26–30, SPE-94, 1st International Conference on Space Protection of the Earth, Chelyabinsk, Russia
- 1995, April 24–26, UN International Conference on Near-Earth Objects, New York, USA
- 1995, May 22–26, Planetary Defense Workshop, Livermore, USA
- 1996, September 23–27, SPE-96, 2nd International Conference on Space Protection of the Earth, Snezhinsk/Chelyabinsk, Russia
- 2000, September 11–15, SPE-2000, 3rd International Conference on Space Protection of the Earth, Evpatoria, Ukraine
- 2002, September 3–6, NASA Workshop on the Scientific Requirements for Mitigation of Hazardous Comets and Asteroids, Arlington, USA.

Besides these conferences many authors have published results of their research activities related to NEO mitigation. A total of 288 scientific papers and conference abstracts on NEO mitigation were published between 1967 and 2000 (Gritzner 2001). The main work (about 60%) was done in the USA and in Russia, where most papers were published after 1992. Apart from the conference papers there is an average of about ten publications per year on NEO mitigation strategies and systems.

3.2 Mitigation strategies

Generally, mitigating the impact hazard could be either done by deflecting the threatening object from its collision course or by destroying the object itself. In case of NEO deflection one can further distinguish between the long-term application of a continuous small thrust, and the sudden application of a large impulsive thrust. In the first case, the NEO would be propelled for a period of several months. Ideally, the force would be applied through the object's center of mass parallel or antiparallel to its velocity vector, which would increase or decrease the semimajor axis of the NEO's orbit, respectively. Thus, its arrival at the intersection point with Earth's orbit would be delayed or advanced, respectively. But, this kind of orbit alteration can be applied only if sufficient warning time is given, e.g., a decade. Here, one could possibly deal with solar concentrators, attached thrusters, mass drivers, or laser irradiation.

If less warning time is available, the orbit has to be changed rapidly. Dealing with the worst case we would have to apply our mitigation means in the very last orbit before the collision. Here, the correction force should be applied in an almost perpendicular direction with respect to the NEO velocity vector. The only possible means for conducting such a high-energy interaction would be either nuclear explosives or kinetic energy impacts. For both techniques the danger of an uncontrolled fragmentation of the NEO has to be considered. The size of the fragments that remain on a collision course with the Earth is a critical factor. Only fragments smaller than 30 to 50 m burn up in the atmosphere. Larger objects could penetrate the atmosphere and if many, could cause a series of impacts over a large territory (firestorm).

Destroying the NEO might not guarantee that the resulting fragments will be very small and therefore harmless. Due to existing fractures inside a NEO it is possible that some of the resulting fragments could be large enough to cause even more damage on Earth than a single object because of the larger area affected. Therefore, this strategy should only be applied to smaller NEOs, where an atmospheric protection against the largest fragments exists. Exceptions are the iron–nickel asteroids, which experience only slight losses during their atmospheric entry because of a higher density and strength. Fortunately they are rare (about 3% of all meteorite falls). The diversion of a NEO orbit may be the only practicable way to prevent the impact of an object larger than a few hundred meters.

When dealing with threatening objects that cross the sphere of influence of other celestial bodies, such as Venus, Mars, or main-belt asteroids, a very slight diversion of the threatening body would result in a different geometry for the natural "gravity assist" at the celestial body. Then, the overall effect on the object orbit would be much larger than the one due to the pure mitigation interaction. When considering that a threatening object will probably perform a close encounter with Earth several orbits in advance of its collision, very good chances for orbit diversion would exist during that flyby. An example of such a mitigation maneuver is discussed in Section 5.

3.3 Attached thrusters

When applying thrusters to a NEO the heliocentric velocity vector could be adjusted to avoid a collision with Earth. A problem accompanied with propulsion systems is the attachment to the object that rotates at spin rates of a few to tens of revolutions per day. Several anchoring techniques are under consideration for small space probes (e.g., drills, harpoons), but the selection of the optimal method strongly depends on the predominating local NEO surface properties. Depending on spin rate and possibly precession, longer waiting periods might exist between periods of thruster operation to ensure an optimal alignment of the thrust vector. Several types of propulsion systems are under consideration for the purpose of NEO hazard mitigation.

3.3.1 Chemical propulsion

Typical values for the specific impulse I_{sp} of conventional chemical propulsion systems range from 200 s for monopropellants to 450 s for bipropellants, e.g., LO_2/LH_2. Here, the specific impulse is defined as the effective gas exhaust velocity c_{eff} of a rocket engine divided by the Earth's mean gravitational acceleration g (=9.83 m s^{-2}): $I_{sp} = c_{eff}/g$. The propellant mass m required to change the velocity of an object with diameter D and density ρ by Δv can easily be derived from conservation of momentum:

$$m = \frac{\rho D^3 \Delta v}{19 I_{sp}} \tag{9.1}$$

Assuming a 150 m object with a density of 2500 kg m^{-3} the required mass of propellant for a 0.05 m s^{-1} velocity change at 450 s of specific impulse would be 49 tons. For a 300 m object an eight-fold amount of propellant would be required.

Because of these very high propellant masses, the chemical propulsion system could only be delivered to the target as a modular device when considering the currently available launchers. Thus, the application of such a system would demand very expensive logistic flights. The utilization of *in situ* resources could be an alternative, but mission complexity would increase drastically due to the required devices for resource extraction and fuel production.

The rotation of the NEO is not expected to become a serious problem for chemical propulsion since the total engine burning time would be only about 1 h for the 150 m object.

3.3.2 Super high energetic propulsion (SHEP)

A variation of the conventional chemical propulsion system is the application of special high energetic fuels in order to increase the specific impulse (Gritzner 1996). When using meta-stable triplet helium in combination with helium or hydrogen, specific impulses of about 540 s or 800 s could possibly be achieved, respectively. Compared to standard chemical propulsion systems the achieved advantage is less than a factor of 2. The technology to produce, handle, and store meta-stable helium is very demanding and costly. Operating such a system in space would raise enormous costs that probably would not justify the small advantage compared to chemical propulsion systems.

3.3.3 Nuclear propulsion

In nuclear thermal propulsion systems, hydrogen propellants are heated to a high temperature (e.g., 3000 K) by a nuclear reactor. These were tested in the 1960s and 1970s but never reached operational status. A significant drawback is the radioactive contamination especially during systems development, which cannot be excluded as known from former tests. Conducting tests with nuclear propulsion systems on ground (and even in space) will possibly face strong resistance for political and ecological reasons. When operating the nuclear propulsion at the hazardous object, a radioactive contamination would probably pose no problem. As for the chemical propulsion option, the *in situ* utilization of resources to gather hydrogen or other propellants could be an alternative to decrease the overall system mass, but again at the expense of complexity. An advantage of the nuclear propulsion systems would be, as for chemical propulsion systems, the comparably short operation time in the order of minutes to hours.

Theoretically, the specific impulse of nuclear propulsion systems could be as high as 134 100 s for solid core fission reactors and 374 200 s for gas core fusion reactors. But these values are clearly indicators of an out-of-reach limit. Radiation losses, cooling requirements, temperature limits of materials, etc. decrease realistic specific impulses to comparably low values. In ground-based tests, the solid core fission reactor NERVA XE only achieved 715 s of specific impulse (Durham 1969). A new approach towards lightweight nuclear propulsion systems for planetary missions has been outlined by Powell *et al.* (1999). Their miniature reactor engine (MITEE) is to be developed, integrated, and tested within a 7-year period. The baseline MITEE concept is planned to achieve a specific impulse of 1000 s, while the maximum performance of their concept "Hybrid Electric Thermal MITEE" could reach 1600 s, which would be four times more than conventional chemical · propulsion systems. When applying Eq. 9.1 this would be equivalent to a propellant mass reduction of 75% in comparison to chemical propulsion. Anyway, the launch mass still would be too high. Picking up the example from Section 3.3.1, 14 tons of fuel (LH$_2$) would have to be carried to divert the 150 m object. If the problems accompanied with the radioactive contamination in ground-based tests could be overcome, nuclear propulsion could be an attractive means for the diversion of those objects where *in situ* resource utilization for propellant production is possible.

3.3.4 Nuclear pulse

Nuclear pulse propulsion systems were analyzed as early as the 1950s. Here, frequent nuclear explosions act onto a buffer plate delivering an impulse for a quasi-continuous medium-level thrust. To increase the efficiency a certain amount of "fuel" (e.g., H$_2$) can be added. Nuclear pulse mitigation systems combine the advantage of high-energetic nuclear explosions with a continuous thrust, but have a high system mass of up to 100 tons and require technologies that have never been tested before (Project Orion 1964). The development time for such a system would surely be more than 10 years. For all these reasons the nuclear pulse option will not be available for NEO mitigation at least for the next decades.

3.3.5 Electric and plasma propulsion

Electric and plasma propulsion systems are operational today but only in low-thrust configurations for attitude and orbit control tasks or as main propulsion system for small space probes such as Deep Space 1. They provide a very high specific impulse (e.g., 3200 s for Deep Space 1), but they require enormous power to provide a considerable thrust level (about 25 kW N^{-1}). Typical values for the specific mass of power supply units are 4.5 kg kW^{-1} for nuclear reactors (e.g., SNAP 50) and 30 kg kW^{-1} for solar-electric systems (Gritzner 1996). When applying the solar electric propulsion to divert the 150 m object that was discussed in Sections 3.3.1 and 3.3.3, an operation period of 1 year might be realistic. Assuming that 3000 s of specific impulse applies, the requirement of 7.4 tons of fuel is found from Eq. 9.1. A thrust level of 7 N was derived from the product of mass flow rate (7400 kg per year) and exhaust velocity (30 000 m s^{-1}). Thus, about 175 kW of continuous power would be required, where the solar electric power-supply mass would be about

5.2 tons. When using a nuclear reactor instead, power-supply mass could be reduced to 800 kg. Thus, a "high-level" thrust electric or plasma propulsion system for the purpose of NEO mitigation seems unrealistic today, because enormous power has to be supplied, which could not be delivered by solar-electric means within the limits of space transportation systems. The use of a nuclear reactor could reduce the total mass but at the expense of increased system complexity. Because of the low thrust a long operation time of between some months to several years would be required, where the autonomous operation over such a long period would imply a high risk to mission success.

Further, the system operation time would be much longer than the average NEO rotation period. Therefore, a pulsed operation would be necessary, i.e., the engine has to be turned on when the thrust vector points towards an optimal direction. Assuming an acceptable thrust angle of $\pm 6°$, which corresponds to 1/30 of the NEO rotation period, the interaction period would increase by a factor of 30. Precession would further increase the period of interaction. Possibly the use of a Cardan joint device could solve this problem but at the expense of increased system complexity (Gritzner and Fasoulas 2004). Another alternative could be the use of several smaller electric thrusters, which would be spread over the NEO for a successive operation. Of course, this would increase system mass and complexity, too. Thus, electric propulsion systems are not considered as a promising alternative for NEO mitigation in near future.

From the examples given within this section it is obvious that attached thrusters might only be applicable to the diversion of very small objects (e.g., up to 200 m in extent). Clearly, these thrusters are not the best choice and further techniques for the diversion of small objects have to be considered (see Section 3.5).

3.4 High energetic interaction

When dealing with large objects (e.g., larger than 1 km in extent), mass and power budgets for conventional techniques will soon approach unrealistic numerical values. Therefore, mitigation options with a much higher energy density have to be applied. Here, energy coupling to the target object is much more complex than in the previous examples. The interior structure of the hazardous object could drastically decrease the efficiency of a kinetic energy impact or nuclear explosion. In the following, only a rough overview on these techniques is given; for a detailed discussion see Chapter 6.

3.4.1 Kinetic energy impact

The technologically simplest method to correct an orbiting object's velocity vector might be the high-speed collision of a projectile (spacecraft). The impacting process goes along with the ejection of crater material. The total momentum change of the target object is the momentum of the escaping ejecta plus the translational momentum carried with the projectile. For non-porous targets the ratio between ejecta momentum and projectile momentum can be as large as 13, whereas for porous targets this could be decreased to 0.2, yielding a total momentum of 14 and 1.2, respectively (Holsapple 2002). Several articles that deal with the kinetic energy interaction have been published based on theoretical considerations, e.g., Melosh *et al.* (1994), hydrocode simulations, e.g., Asphaug *et al.* (1998), orbital computations, e.g., Conway (Chapter 13), and high-speed projectile experiments. Nevertheless, the opinions of experts diverge about the outcome of the impact on comet P/Tempel-1 by the Deep Impact mission. A first attempt towards such fundamental tests was the cancelled Clementine II mission where small probes were to impact asteroids.

The outcome of the impact of a spacecraft with a relative velocity of 10 km s^{-1} can be estimated from the momentum ratios given above. Applying the impacting spacecraft to the reference asteroid from Section 3.3, a moderate 1600 kg spacecraft would work for the non-porous object, whereas a 22 ton spacecraft would be required to accelerate the porous target by 0.05 m s^{-1}. Clearly, the target material properties (and of course modeling uncertainties) demand a careful analysis of the object's nature before an interaction. Otherwise, several medium-sized spacecrafts would have to be launched to impact in time-shifts on the target. Then, each individual effect could be studied to control the sequencing impacts.

Another mitigation idea based on the principle of kinetic energy interaction arose during the planetary defense workshop in 1995: the cookie cutter. Here, small tungsten bullets weighing a few tens of grams would penetrate the NEO surface. Further bullets following directly behind the first ones would excavate a deeper hole. With a whole string of bullets, a hole might be drilled through the object. Hundreds of thousands of tungsten bullets strung together by lightweight fiber to form a three-dimensional lattice could be used to cut the hazardous object into small boulders.

3.4.2 Nuclear explosives

Nuclear explosives offer the highest energy per unit mass, which is about 4 × 10^6 MJ kg^{-1}. The impulse delivered to the NEO is mainly due to neutrons and X-rays. Explosion debris can contribute to the momentum, too. Extensive studies of nuclear explosive interactions can be found for example in Shafer *et al.* (1994) and Hammerling and Remo (1995). Although nuclear explosives seem to offer the best answer to a threatening object, testing of nuclear deflection systems poses technical, legal, and environmental problems that have to be considered carefully (Harris *et al.* 1994).

The scaling force of nuclear weapons becomes obvious in the following example. In Remo (2000) the momentum coupling coefficient C for X-ray and neutron interaction for standoff explosions is between 5×10^{-5} s m^{-1} and 10^{-3} s m^{-1}. Further, the required energy E to change the velocity of an object with diameter D and density ρ by Δv is

$$E = \frac{\pi}{6} D^3 \rho \cdot \Delta v / C \qquad (9.2)$$

From this equation and with the help of the energy density, the payload mass to divert our 150 m example object by nuclear explosives is found to be between 1 kg and 50 g for the lower and upper coupling coefficient, respectively.

Clearly, nuclear explosives seem to offer the best cost effectiveness. But, "the mere existence of . . . nuclear weapons itself constitutes a terrifying threat to civilization that may overshadow the danger posed by comets and asteroids" (Melosh et al. 1994). "Hazards during launch as well as construction are unacceptable if alternatives exist" (Phipps 1992).

3.4.3 Chemical explosives

Due to the detonation of a single kilogram of chemical explosives about 6 MJ of energy are released. The impulse coupled into the NEO arises from the direct impingement of detonation products (Shafer et al. 1994). Thus, a buried detonation is much more effective than a surface or standoff explosion. Nevertheless, the use of chemical explosives as a direct mitigation technique seems unrealistic since the energy density of nuclear explosives is a million-fold larger. But chemical explosives could be relevant for mining, e.g., in supplying a mass driver with fresh material.

3.5 Mitigation technologies in reach

3.5.1 Solar sails

Solar sails use the solar radiation pressure on large reflecting sails for low-thrust propulsion. Therefore, no propellant is required for orbit control. They are most efficient in the vicinity of the Sun because radiation pressure depends on the inverse 2nd power of distance. The coupling of a solar sail is aggravated by the spin state and rate of the object. Either a large and complex Cardan joint or the deceleration of the object's spin would be required to allow for several months of operation.

Picking up the example from Section 3.3, 7 N of continuous thrust would be required to change the velocity of the 150 m object by 0.05 m s^{-1} within a 1-year operation phase. For an optimistic scenario at 1 AU solar distance, the radiation pressure acting on the sail would be roughly 8×10^{-6} N m^{-2}. Thus, a reflecting area of 875 000 m^2 would be required. Such a structure would be 100 times larger than the International Space Station. Clearly, the expenses for the diversion of a small object by solar sails would be very high.

Although solar sails might never become a realistic means for mitigation, the technology could become of great importance for the exploration of small solar system bodies (e.g., Bernasconi and Zurbuchen 1995).

3.5.2 Mass driver

The idea of mass drivers goes back to 1953, when Arthur C. Clarke described an electromagnetic catapult for a lunar material launch application. When surface material is accelerated via electromagnetic coils, the impulse acting on the asteroid could be used to change the orbital energy of the object and thus to deflect it. Since 1979 several engineering models of mass drivers have been built and tested at the Space Studies Institute (Dyson *et al.* 2002). Their latest system mainly consists of a drive coil, wherein a bucket coil is accelerated by a traveling magnetic wave. Melosh *et al.* (1994) determined ejection velocities of several hundred meters per second for an efficient conversion of supplied electrical energy into orbital energy. Both authors prefer the use of a solar collector instead of a nuclear reactor as energy supplier.

Although the development of the mass driver itself seems to have reached an advanced state, several other difficulties have to be overcome. For example, the continuous operation of a mass driver demands a continuous supply of material, too. Assuming optimal conditions a small vehicle collecting and catapulting loose surface material could be sufficient. In a more realistic scenario mining equipment would be required to process pieces of rock into adequate boulders that fit into the mass driver bucket coil. In any case, the predominant microgravity would require an expensive anchoring device, which would allow for a movement of the complex. The vacuum environment and the thermal gradient during day and eclipse additionally aggravate the use of mining tools.

Other difficulties might arise from the use of a large solar collector in combination with the mass driver. Slower material could be captured by the object's gravity and might hit the collector. Anchoring of the collector would also mean phases of restricted operation during eclipse. Otherwise heavy energy-storage devices would be additionally required. Thus, the use of mass drivers for the deflection of hazardous objects might become very complex and expensive.

3.5.3 Pulsed lasers and microwaves

Beamed energy has the advantage of a high-intensity and small-divergence beam, which enables a wide and thus safe distance for operation, e.g., as compared to solar concentrators. The energy absorbed by the target heats the surface and an ablation jet forms. The momentum transferred by this jet is much larger than the momentum due to light pressure. The coupling coefficient C, which is the ratio of total momentum and beamed energy, strongly depends on target material composition, energy intensity at target surface, laser wavelength, and pulse duration. From experiments C was estimated to be between 10^{-5} s m^{-1} to 10^{-4} s m^{-1} (Shafer et al. 1994; Phipps, 1995).

First, a ground-based laser is considered, where the laser light is beamed on an Earth-based mirror to focus the spot on the NEO. The mirror must be several meters in diameter to ensure the spreading of the beam over many square meters to avoid air breakdown. Assuming an optimistic maximum beam range of 300 000 km and a relative velocity of the hazardous object of only 15 km s^{-1}, the interaction period would amount to 5.5 h. Further, we assume that the huge laser, which is currently under construction at the National Ignition Facility (NIF) at the Lawrence Livermore National Laboratory, would be available for the deflection of NEOs. The NIF will be able to produce laser pulses with an infrared energy of 4 MJ. To avoid self-destruction of the facility, the maximum shot rate is limited by the cooling time between shots and would be less than 100 per day (D. Dearborn, personal communication). Thus, only 23 shots with a total energy of 92 MJ could be fired on the NEO, provided that the laser facility is located at the corresponding hemisphere. If a moderate coupling coefficient of 2×10^{-5} s m^{-1} applies, a NEO 1 m in extent could be deflected by 1 m s^{-1}, or a NEO 12 m in extend by 1 mm s^{-1}, respectively. Since objects of this size usually burn up during their entry into the atmosphere, the laser would be useless.

Clearly, much higher beam intensities and larger beam areas (e.g., the gas lens concept discussed in Phipps (1992)) would be required for the deflection of larger objects. But, even if this technology was available, ground-based laser would still suffer from a very restricted control of thrust vector orientation. When irradiating the object during Earth passage, the thrust will mainly be applied transverse to its trajectory. Applying the same impulse collinear with the NEO trajectory would yield a larger change of orbital parameters. To achieve an optimized thrust vector alignment a space-based laser would be required. In addition, the additional expenditure for assembling and maintenance in space, size, and mass of the power supply would be critical parameters.

Melosh et al. (1994) analyzed the use of beamed microwaves for the diversion of NEOs. Although higher coupling efficiencies than for lasers are expected, this technology suffers from the huge phased antenna array required for focusing microwaves into beams of high intensity and small divergence. Melosh et al. (1994) estimated that the array has to be 160 km in diameter for a suitable energy flux at a 450 000 km distant object. Clearly, such a system cannot be built within the near future.

3.5.4 Utilization of the Yarkovsky effect

Bodies in space are exposed to solar radiation, where a certain portion of this energy is absorbed while the rest is reflected. The absorbed energy is a function of distance from the Sun, albedo, and area exposed to the Sun. Additionally, the object emits energy as a function of the 4th power of surface temperature according to the Stefan–Boltzmann law. Thereby, thermal photons leave the surface with a very small momentum. Nevertheless, the reaction on the asteroid is large enough to affect its long-term motion. This Yarkovsky effect was observed first for spacecraft motion and is a function of mass, size, shape, spin state, and global distribution of surface optical and thermal properties of an object (Giorgini *et al.* 2002).

Recently, Spitale (2002) proposed several ideas for the utilization of the Yarkovsky effect for NEO hazard mitigation. All of them deal with alteration of the upper surface layer (to a depth of a few centimeters) to change albedo and/or surface thermal conductivity. From Spitale's point of view this could be achieved by covering the object with a 1 cm dirt layer or by painting it white. Another more realistic approach is to shatter the surface to a certain depth by detonations, or to remove a pre-existing surface layer by the same means.

Besides the problems accompanied with such immense efforts, the interaction would only slightly affect the course of the asteroid. Spitale estimated an orbital displacement of about 15 000 km in 100 years for a 1 km spherical stony asteroid at 1 AU solar distance. Conducting such a mission would require full knowledge of all object parameters related to the Yarkovsky effect before the interaction and of course thereafter. Slight uncertainties in orbit prediction and in those parameters could possibly turn a former harmless object into a collision candidate. Further, it has to be mentioned that a change in albedo would entail a change in solar radiation pressure too, prompting an even more careful analysis.

The importance of the right or wrong selection of force model parameters was demonstrated by Giorgini *et al.* (2002). They analyzed the numerical computation of an individual asteroid orbit on the sensitivity to several gravitational and non-gravitational forces. As an outcome the largest uncertainty in predicting the orbit arises from the Yarkovsky effect, especially due to the choice of thermal conductivity and spin rate.

Thus, we conclude that utilization of the Yarkovsky effect could only become an option for mitigation in a very restricted case where the orbit is known at high precision for a time-span of a few hundred years in advance. Further, if that were the case, the change of the spin state of the object could possibly be achieved more easily and more effectively than the alteration of the complete object's surface.

3.5.5 Solar concentrators

The application of solar concentrators for NEO mitigation was first discussed by Melosh and Nemchinov (1993). The basic idea of this technology is to concentrate solar radiation onto the NEO surface with a lightweight (parabolic) reflector. Depending on duration and intensity of illumination, the material within the spot will be heated up and will vaporize, delivering a low but continuous thrust. We believe that this technology would be a very good choice for a long-term mitigation mission. Therefore, this system is treated in more detail below (see Section 5).

3.5.6 Mini-magnetospheric plasma propulsion

The idea of the mini-magnetospheric plasma propulsion is related to the solar sail concept, in that both tap the ambient solar energy to provide thrust to a spacecraft. Winglee *et al.* (2000) invented this revolutionary propulsion concept for interplanetary space missions. The system is based on an electromagnetic coil, which generates a magnetic field. Due to the injection of plasma this field is enlarged and will intercept the solar wind. At 1 AU solar distance the solar wind has a particle density of about 6 cm^{-3} and moves at speeds of 300 to 800 km s^{-1}. The resulting constant dynamic pressure on the magnetic bubble is about 2 nPa.

If the magnetic field cross-section is large enough, a continuous force of a few to tens of newtons could be provided. We propose to use such a system for NEO diversion. Although the generated thrust is low, this system could be operated for the duration of several months to divert a small threatening object. When operating with small thrust over long periods the force should be aligned with the NEO's velocity vector. But for mini-magnetospheric plasma propulsion, the largest portion of the generated force will point in radial direction (Sun–NEO direction). To increase the performance, the magnetic axis could be tilted by 45° into the solar wind, which would result in a larger magnetosphere (because the magnetic field is stronger at the poles) and a tangential force component (Winglee *et al.* 2000).

A hypothetical system has been scaled based on the system characteristics given by Winglee *et al.* (2000) Here, we assume a 60 km artificial magnetosphere. To maintain the electromagnetic field and to generate the plasma would require about 7 kW of power, which could be provided by a radioactive isotope power system weighing about 1.2 tons. Together with the equipment (electromagnetic coil and radio-frequency antenna) and the "fuel" for 1 year of operation one would probably end up with a 3 ton spacecraft. Assuming an efficient coupling to the object a tangential force of 11 N could be delivered at 1 AU solar distance. Again, the example of the 150 m object from Section 3.3 is used to quantify the problem.

When applying the mitigation system to that object for the duration of 8 months, the postulated 0.05 m s^{-1} velocity increment could be achieved.

One significant problem is the attachment of such a system to the NEO. When landing the system onto the NEO, a continuous operation might be impossible because of spin state and rate. Anchoring a complex Cardan joint would allow for an optimum force vector alignment. Apart from these difficulties the magnetospheric plasma propulsion system has not been fully developed yet and further problems will arise when dealing with a functioning device. But, even if coupling to the NEO for deflection purposes should not be feasible, this propulsion system would be of great importance for reaching the hazardous object to apply a suitable mitigation technique. For example, kinetic energy impacts would profit highly from the enormous relative speed available by magnetospheric plasma propulsion. Small spacecrafts of 100 kg or 200 kg mass could be accelerated to speeds of 75 km s^{-1} or 50 km s^{-1} respectively (Winglee *et al.* 2000). This would enable us to conduct exploration missions to asteroids and comets much faster, too. Objects could be explored that are not accessible with current propulsion systems and the additional help of gravity assists.

3.6 Futuristic mitigation technologies

3.6.1 Antimatter

Accumulation and storage of antimatter are unsolved problems today. Nevertheless, if feasible one day, antimatter might be the perfect means for NEO mitigation. Antimatter could be used for NEO mitigation in two ways, either in an explosive device or in a propulsion system. For the first case, the payload mass of antimatter would annihilate an equivalent mass of matter and generate 1.8×10^{11} MJ kg^{-1} of energy (Remo 2000). In the case of a propulsion system, the working principle would be similar to thermal nuclear propulsion, but the system complexity would be increased additionally.

3.6.2 Eaters

Mechanical, biological, or chemical NEO "eaters" could fragment the NEO into smaller boulders, dust, or even gas, which would be harmless to Earth in case of a collision (Urias *et al.* 1996). Depending on its definition, the above-mentioned kinetic impact cookie cutter could belong to the group of mechanical eaters, too. If complete disintegration of an object is impossible, a partial fragmentation could possibly be sufficient to cause a destruction of the NEO when entering Earth's atmosphere. One might also think of an NEO surface modification to influence the atmospheric entry flight path of the object (D. A. Mitchell, personal communication).

Yet nothing is known so far about any biological or chemical eating processes. The NEO material would probably play an important role for the efficiency of the interaction. And there may be safety issues associated with the accidental release of potentially toxic or otherwise dangerous biological or chemical eaters (Urias *et al.* 1996).

3.6.3 Neutral particle beams

Besides concentrated sunlight, microwaves, and pulsed lasers, particle bombardment could be used for surface ablation. The NEO surface material could be vaporized by a particle bombardment to form a gas jet for impulse delivery. When applying such a technology over large distances, the particles have to be neutral since charged particles (ions and electrons) would be deflected by terrestrial and solar magnetic fields (Letteer *et al.* 1991).

A particle gun to be stationed in Earth's orbit was studied for purposes of the "White Horse" project in the 1980s by the Defense Advanced Research Projects Agency (DARPA) in Los Alamos. Here, the gun was thought to deflect approaching nuclear weapons. A system mass of 60 tons and beam energy of 200 MeV in conjunction with currents of up to 1 million A was considered. Ions are strongly accelerated in this apparatus and subsequently neutralized. The radio-frequency quadrupole (RFQ) method, which was developed in the USSR, could be used to improve acceleration and focusing. The RFQ method employs electric fields instead of magnetic ones (Horeis and Liebig 1985).

Many problems are expected from the use of neutral particle beams for NEO deflection. Focusing the particle beam over long distances (hundreds of thousands of kilometers) is problematical, given that only a few 1000 km were targeted as the range in the Strategic Defense Initiative (SDI) program. To avoid the Earth's atmosphere, neutral particle beam systems have to be installed and operated in space, which imposes high costs. As with the laser mitigation system, the neutral particle beam system requires a very high pointing accuracy, which may not be achievable with current technology. A solution could be to operate this system only during a close encounter of the target object with Earth. But in such a case the available operation time is clearly limited to few days.

4 Requirements for NEO mitigation

4.1 Warning time demand

The available warning time is of great importance for a successful mitigation mission. If more time is available the chances of a successful outcome of NEO mitigation measures will increase. With current search activities there may be sufficient warning time in a few cases, but mostly the warning time will be too short or not available at all. For example, in June 2002 the 100 m wide asteroid 2002 MN passed within 120 000 km of our planet. It was discovered 3 days after its near miss.

While a certain warning time becomes available only in an actual threat situation, the time for mission preparation for a certain mitigation system cannot be decreased. If the preparation time is higher than the warning time, there is no chance for a successful mitigation. But the preparation time could be reduced in advance since some parts of the preparation tasks could already be done today. Typical values for the duration of preparation tasks are as follows:

(a) NEO detection, hazard recognition, and decision-making   >1 month
(b) Analysis and selection of mitigation strategy and system   >2 months
(c) Concept development of selected mitigation system   >6 months
(d) System development, production, and testing   >3 years
(e) Launch and flight to target NEO   >3 months
(f) Operation of mitigation system   seconds to years
(g) Coasting phase   days to years

Assuming optimal conditions for the development and operation of the selected mitigation system, as mentioned in points (a) to (g) above, a minimum warning time of about 5 years is required. Small satellite missions based on known technology could be accomplished within this short period, but large space systems such as the International Space Station, or huge space probes such as Rosetta require at least 10 years between project start and the beginning of operations. Thus, some NEO hazard mitigation activities should be done in advance of a threatening situation. We propose several tasks to minimize the required warning time:

(1) The analysis of potential mitigation systems should be done immediately. This partially includes point (a) in preparing the decision-making process, and completely includes point (b). This could save at least 3 months at low cost.
(2) Additionally to (1) several suitable mitigation concepts could go through a development process now, as represented by point (c). Plans for a mitigation system would be completed, so the final development, production, etc. could start immediately on demand. The preparation time saved would be 6 months or more; together with point (1) this would save more than 9 months.
(3) A further step could be to develop and test one or more prototype mitigation systems. This does not include the production of the operational mitigation system, which should be done on demand. The time for production cannot be saved here, so the saved preparation time would be about 50% or 18 months for point (d), and 27 months altogether.
(4) Completing point (d) would save the whole 36 months, or 45 months altogether. But a mitigation system being ready for launch without facing a real threat would need to be maintained all the time at high costs. Additionally, the deflection system would have to be guarded to prevent misuse. A compromise could be to develop and test one or more

complete deflection systems. Later these systems would be disassembled and stored at several locations. In the case of an emergency the parts could be assembled and supplied with propellants, batteries, etc., and then be launched.

Currently, point (3) seems to be the best compromise between minimizing the preparation time and respecting political and financial requirements. But this strategy would only make sense if worldwide NEO detection and tracking capabilities were significantly increased at the same time.

4.2 Scientific requirements for NEO mitigation

Although the working principles of the mitigation techniques mentioned above might differ greatly, they all have one thing in common: a detailed knowledge of the target object's properties is required. For example, for the efficient application of any mitigation technique the dynamic properties of the object must be known, e.g., orbit, albedo, size, shape, mass, and state of spin. These data are generally gathered by means of remote sensing, including measurements by ground-based and Earth-orbiting telescopes, or spacecraft flyby and rendezvous missions (Huebner and Greenberg 2001).

Mitigation systems that are based on the generation of thrust, e.g., propulsion systems, mass drivers, or magnetospheric plasma propulsion, have to be mounted onto the hazardous object in some way. When the device is anchoring to the asteroid, the mechanical structure of the surface layer plays an important role. While the presence of loose regolith might demand deep penetration, drilling or hammering devices would be required for solid rock. Further, it has to be ensured that the level of thrust does not exceed the cohesion of a rubble-pile object.

For mitigation techniques that depend upon NEO surface ablation by radiation, e.g., pulsed lasers, microwaves, solar concentrators, and standoff nuclear explosives, the surface material composition influences the portion of radiation energy coupled into the object. Further, thermodynamic properties of the surface layer such as thermal conductivity, melting, and vaporization energy are of great importance with respect to the required irradiation time and vapor production rate.

When dealing with high-energy interactions, such as surface and sub-surface nuclear explosions or kinetic energy impacts, the interior structure of the object will dramatically affect the efficiency of the interaction. Porosity (rubble-pile structure or porous regolith) could decrease the efficiency of a kinetic energy impact and a surface nuclear explosion by a factor of 5 and 100, respectively (see Holsapple, Chapter 6 in this book). To gather the relevant information about the interior of the object, exploratory missions are essential. Seismic investigations would reveal material strength parameters for dense objects such as metallic or stony asteroids;

on the other hand, radio tomography could detect the state of fracture (for detailed descriptions of these techniques see Chapters 10 and 11).

If a short warning time permits a precursor mission, the mitigation mission itself should include capabilities for the assessment of relevant parameters. Clearly, for kinetic energy impacts this could hardly be realized, but instead of a single large impacting spacecraft several smaller ones could be sent there. Due to temporal separation the effects of each single impact could be analyzed to correct the remaining impacts. Such a procedure might be applicable to nuclear explosives as well. In either case system redundancy would be improved at the expense of mission complexity.

If the required interaction data cannot be gained and a staggered arrangement was not possible, then short-term deflection methods would imply a very high risk and a low probability for a successful diversion maneuver. To prevent such a dilemma, the necessity of conducting short-term deflection missions should be reduced, i.e., the number of unknown NEOs and especially PHOs has to be reduced. The best effort would be an intensified search for NEOs down to 100 or 200 m. Since these small objects are both more numerous and fainter than the 1 km NEOs of NASA's current goal, such a survey may have to last several decades. In the meantime, we cannot rule out the possibility of an Earth impact. This situation demands simultaneous physical characterization of NEOs and an analysis of suitable mitigation technologies along with the surveys. In the following section, a solar concentrator mitigation option will be discussed, which could become important for both mitigation and exploration. In an early development phase, this technology could be used to analyze surface properties of NEOs and to demonstrate deflection capabilities at the same time.

5 Solar concentrators for NEO diversion

5.1 History

The idea of using solar concentrators to avert hazards at large distances is attributed to the Greek philosopher Archimedes, who possibly constructed such devices to burn Roman ships attacking the city of Syracuse, Sicily, in the year 212 BC. Unfortunately, no contemporary documents are available to confirm if Archimedes really applied such mirrors. Later, the Greek engineer Ioannis G. Sakas conducted several experiments to prove that story. On November 6, 1973, he gathered 70 Greek Navy sailors on a pier at Piraeus, the harbor of Athens, each holding a plate of polished copper, to retest whether these ancient mirrors could set a 55 m distant ship on fire. The 2.3 m long wooden boat covered with tar went up in flames in less than 3 minutes (Temple 2000).

The first space application was described by Oberth (1923) in his book *Die Rakete zu den Planetenraeumen*, where he proposed the use of solar mirrors or concentrators to illuminate parts of the Earth such as industrial areas at night, to keep shipping channels free of ice, or even for military purposes. His later book *Der Weltraumspiegel* described these ideas in greater detail (Oberth 1978). First experiments with solar reflectors in space started in 1993 when the Russian Znamya-2 experiment was carried out. This mission showed that a simple reflector may be used to illuminate regions on Earth, although the illuminated spot was not stationary. The reflector consisted of rotator-like foils of 20 m in diameter and a supporting structure. In 1999 this experiment was followed by the Znamya-2.5 mission, featuring a 25 m reflector. In 1996 the US Spartan 207/IAE experiment showed that inflatable structures can be used in space to create lightweight beam-focusing devices.

Solar concentrators for the purpose of NEO deflection were first proposed by Melosh and Nemchinov (1993). This idea was also presented at the Livermore Planetary Defense Workshop by Melosh (1995). In Gritzner (1996) the potential high efficiency of this system compared to other NEO hazard mitigation options was confirmed on the basis of a mass model. Another mass model was presented by Mikisha *et al.* (1996) at the 2nd Space Protection of the Earth (SPE) Conference. Finally, various solar concentrator concepts were studied in detail at Dresden University of Technology (Voelker 2002).

5.2 Working principle

Depending on duration and intensity of the illumination, solar radiation concentrated with a lightweight (parabolic) reflector will heat up and vaporize NEO surface material within the focused spot. The evaporated material accelerates to speeds of about 1 km s^{-1} and thus transfers an impulse to the NEO. Although the generated thrust is small (tens to hundreds of newtons) it will suffice to deflect the NEO from its collision course with Earth if sufficient lead time is available (years). Such a system could be operated for the duration of several months, which would lead to a slight increase in the semimajor axis of the NEO when the thrust vector is aligned with the orbital velocity vector of the NEO.

The following equations and numbers are applicable for a parabolic solar collector. Simplified equations for the estimation of vapor production rates and thrust are adopted from Melosh *et al.* (1994). Here, the area specific mass flow rate dm/dt of the ejected material is given by light intensity P and heat of vaporization of surface material H,

$$\frac{\mathrm{d}m}{\mathrm{d}t} = \frac{P}{H} \qquad\qquad (9.3)$$

Table 9.1 *Thermal properties of stony and FeNi meteorites*

	Stony meteorites	FeNi meteorites
Melting point	1350 ... 1800 K	1770 K
Boiling point	2960 K	3510 K
Average specific heat	900 J kg^{-1} K^{-1} (solid)	<700 J kg^{-1} K^{-1} (solid, liquid)
	1100 J kg^{-1} K^{-1} (liquid)	<400 J kg^{-1} K^{-1} (gaseous)
Heat of vaporization	6.05 MJ kg^{-1}	6.4 MJ kg^{-1}
Heat of fusion	0.27 MJ kg^{-1}	0.27 MJ kg^{-1}
Estimated energy for vaporization H	9.3 MJ kg^{-1}	9.1 MJ kg^{-1}

Source: Remo (1994).

Clearly, this simplified equation assumes a 100% conversion of beamed energy into vaporization energy. Losses due to thermal conduction and heat radiation are not explicitly considered, but have to be taken into account. Conductive losses are expected to appear in the range of kilowatts per square meter. But total losses are dominated by radiation, which soon increases to megawatts per square meter. Both effects will be treated in more detail at the end of this section.

The light intensity P in Eq. 9.3 is a function of the solar constant S at a distance R from the sun, the diameter of the solar mirror D_{mirror}, the spot diameter D_{spot}, and the reflectance ψ (about 85% for aluminum-coated foils),

$$P = \psi \cdot S(R) \frac{D_{mirror}^2}{D_{spot}^2} = \frac{\psi}{R^2} \frac{D_{mirror}^2}{D_{spot}^2} 3 \cdot 10^{25} W \qquad (9.4)$$

The heat of vaporization H depends on the thermodynamic properties of a specific material. Values of H have been derived from meteorite data given in Remo (1994) and are summarized in Table 9.1. Note that data were gathered under terrestrial conditions (atmospheric pressure). Melting and boiling points might differ for vacuum conditions.

The ejection velocity v is supposed to be the molecular speed of vapor molecules at evaporation temperature and is on the order of 1 km s^{-1}. Finally, the generated specific thrust f (N m^{-2}) normal to the NEO surface is given as the product of specific vapor mass flow rate, ejection velocity, and a numerical factor $\beta = 0.5$ that accounts for a hemispherical spread of the vapor,

$$f = \beta \frac{dm}{dt} v \qquad (9.5)$$

Melosh *et al.* (1994) examined the influence of the light intensity on the start-up time for evaporation. They found that this time t increases for decreasing intensities P as $t \sim P^{-2}$ (e.g., 10^{-4} s for 10^9 W m^{-2} and 1 s for 10^7 W m^{-2}). This evaporation time has to be shorter than the residence time of the NEO surface within the spot. Depending on NEO spin rate, spot latitude, and NEO shape the velocity at which the surface moves beneath the spot can be determined. In Section 5.4 we derive two reference scenarios where a 100 m concentrator is applied to a 210 m and a 580 m diameter object. Typical spin rates of NEAs larger than 200 m in diameter are between 1 and 10 revolutions per day (e.g., Pravec and Harris 2000). Thus the maximum surface velocities at the object's equator would be 8 to 76 mm s^{-1} for the 210 m object and 21 to 210 mm s^{-1} for the 580 m object, respectively.

The evaporation process is strongly influenced by the thermal conductivity λ of the surface material. Here, the difference between stony (1.5 to 2.4 W m^{-1} K^{-1}) and iron–nickel asteroids (about 40 W m^{-1} K^{-1}) becomes obvious. One-dimensional heat transfer computations affirmed the dependency that was found by Melosh *et al.* (1994). As a result, a heat flux of 4 MW m^{-2} (this corresponds to a 100 m diameter collector at 1 AU Sun distance and 160 m distance towards the asteroid) would vaporize surface material of a stony asteroid within 20 to 40 s (lower limit: density $\rho = 2000$ kg m^{-3}, $\lambda = 2$ W m^{-1} K^{-1} and upper limit: $\rho = 3000$ kg m^{-3}, $\lambda = 2.4$ W m^{-1} K^{-1}). Applying the same concentrator to a metallic object would most probably end in an insufficient result. After 250 to 300 s of illumination the object's surface would approach an equilibrium temperature of 2850 K and 2780 K, respectively ($\rho = 4700$ kg m^{-3}, $\lambda = 40$ W m^{-1} K^{-1} and $\rho = 7700$ kg m^{-3}, $\lambda = 40$ W m^{-1} K^{-1}). At these high temperatures almost all of the incoming heat flux is lost in radiation heat, which depends on the 4th power of surface temperature. For the given examples an emission coefficient of 0.8 was assumed. Although vaporization is likely to occur at a temperature below the 3510 K given in Table 9.1, the temperatures achieved here might not suffice. Thus, for a metallic object one would probably choose a larger collector since a further reduction of the distance to the object might induce operational problems in the close proximity to the object. Extending the concentrator to 160 m diameter the boiling temperature of about 3510 K could be achieved within 60 to 100 s for the low-density and high-density metallic objects, respectively. Here, the incoming heat flux would amount to 10 MW m^{-2}. These examples show the importance of the thermal properties of the object's surface. While thermal conductivity affects how fast the boiling and vaporization processes elapse, thermal radiation decides whether the vaporization level will be achieved at a given beam intensity. In the case of objects with regolith layers, thermal conductivity would be further reduced. Thus, a regolith layer would probably increase the concentrator performance, especially for fast-rotating objects.

5.3 Solar concentrator properties

Focusing solar radiation onto a spot on the surface of a NEO can be done in two ways: by reflection or by refraction. For refraction a Fresnel lens could be used, which might consist of a lightweight polymer foil with engraved concentric refracting rings. A reflecting solar collector could be either a large parabolic mirror or an array of small parabolic or plain facets. A typical value for the reflectance is 0.85 for aluminum-coated foils. Higher values of up to 0.95 are technically feasible.

When operating a solar collector system close to an NEO the pointing accuracy is an important issue, because of the orbital movement of the NEO and the collector system towards each other, as well as the NEO's rotation. Due to the generally non-spherical shape of the NEO and the properties of the collector surface, focusing mismatches will occur that have to be considered in advance in the system layout. Another disadvantage with the large concentrator is the disturbing force due to the solar pressure that acts on the mirror surface. For example, a 100 m diameter concentrator implies 0.07 N of thrust at 1 AU solar distance when directly pointing towards the Sun. To partly counteract that force it could possibly be aligned with the gravitational force of the NEO, but a much larger portion of pressure might result from the upstreaming vapor. This vapor stream poses a key problem during operation. The pollution of the collector surface due to the evaporated NEO material (vapor and debris) would result in a degeneration of the optical components. This implies a threat to the mirror when partly exposed to the vapor jet. Particles moving at high speed could penetrate the mirror foil while slower particles could deposit on the mirror surface. This would be in addition to the vapor deposition on the reflecting surface. These processes would impair the efficiency of the mitigation system, in the same way as the degradation of solar cells on a spacecraft.

Most probably the vapor plume would expand towards a hemispherical vapor cloud, whose dust concentration would depend on the distance and direction from the spot. To avoid too many losses the mirror should be positioned outside of the main evaporation plume whose main elongation is assumed to be normal to the surface tangent. Therefore, Melosh *et al.* (1994) proposed a free-flying secondary mirror that redirects the focused beam and allows the location of the primary mirror far away from the evaporating site. The secondary mirror would be much smaller than the primary mirror and could be exchanged if polluted. But the use of secondary and tertiary mirrors to protect the primary one does not necessarily improve the situation, since the degradation problem is then transferred to those mirrors. The extent of the degradation of reflectivity is not well known as of now. Further analysis and simulations are needed to gain data close to reality. Nevertheless, it has to be guaranteed that the mirror foil will not rupture if hit by debris.

This could be achieved through a honeycomb cell structure of the foil where the propagation of fissures stops at the boundary of the cell. If these cells are small enough, damage by penetration could be minimized. Besides, some simple and therefore low-cost options to avoid or slow down degradation could include the following:

(1) The use of double-sided primary and secondary mirrors for simply switching to a new second mirror surface, which doubles the operating time.
(2) The use of several transparent foils to cover the mirrors, which can be stripped off when polluted. This approach extends the operating time depending on the number of removable foils, but each foil reduces the efficiency of reflectivity by a small factor.
(3) The use of a movable transparent protecting foil, which covers the mirror surface like a blind. If the active area is polluted the foil will be rolled to one side while a fresh foil is pulled out of a protecting drum on the other side. The efficiency of reflectivity is only reduced by one foil. The operating time increases according to the amount of protecting foil available.

These mechanical protection concepts will increase the overall system mass slightly, but they might be essential for a successful operation. Further, the use of secondary mirrors allows for optimal thrust vector orientation. Several secondary mirrors on the same orbit around the NEO would enable a quasi-constant operation. Another important aspect could be the use of non-imaging mirrors or a Cassegrain-type arrangement as proposed by Melosh *et al.* (1994).

5.4 Systems performance simulation

At the Dresden University of Technology various solar concentrator concepts have been analyzed (Voelker 2002). To give an example of the concentrator capabilities, a small system has been chosen that would meet the limited payload capacity of currently available launchers. Here, we consider a 100 m diameter concentrator system, which was the basis for our vaporization process computation in Section 5.2. Figure 9.1 shows system components and specific masses of the device. The total weight of concentrator and spacecraft is expected to be less than 2000 kg.

At 1 AU solar distance such a concentrator would collect about 8 MW of solar power. When beamed onto a 1.6 m diameter spot an intensity of 4 MW m^{-2} could be achieved on the NEO surface. When dealing with silicate material and applying Eq. 9.3–9.5 the specific thrust equals 215 N m^{-2} for the spot area. Assuming that losses due to thermal conduction, radiation, and spot displacement decrease the overall efficiency by a factor of 2, a continuous thrust of 215 N would be produced by the vapor jet at the spot surface.

From the NEA database, which is maintained at the European Asteroid Research Node by Gerhard Hahn (Hahn 2002), two potentially hazardous asteroids have been

	Specific mass	Masses for 100 m mirror
Mirror foils (aluminum coated, 90% reflectivity)	10.5 g/m² PEN 12.4 g/m² Kapton 18.9 g/m² Mylar	80 kg (min) 160 kg (max)
Telescopic rods	~ 0.6 kg/m	240 kg (8 rods, 50m each)
Centre boom		50 kg
Ropes	0.04 kg/m	16 kg
Winches	~ 4 kg each	32 kg
Total (collector)		500 kg (max)
Estimated total S/C		< 2000 kg

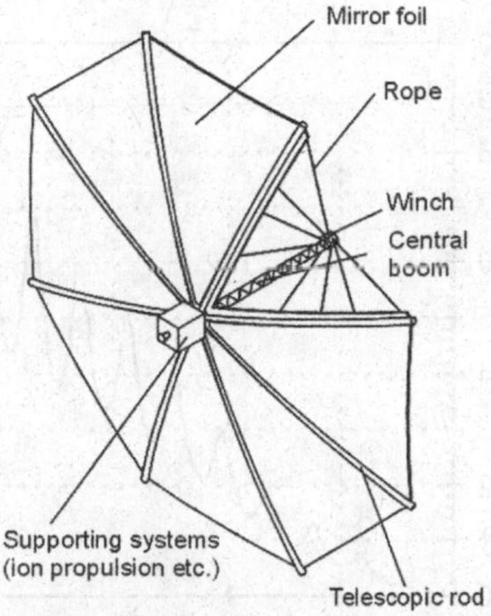

Figure 9.1 System mass breakdown and model for solar concentrator "umbrella."

selected for demonstration: 2000 WC1 and 1999 AQ10. Assuming both objects belong to the C-type group (albedo of 0.04), their diameters are approximately 210 m (absolute magnitude, $H = 22.5$) and 580 m ($H = 20.3$), respectively. Further, their densities are assumed to be 3000 kg m^{-3}. Note that this density is relatively high and has been chosen to give a worst-case scenario.

Figure 9.2 is a plot of the orbital evolutions of the objects. Here, the unperturbed orbits have been computed until the year 2010. Then, the solar concentrator interaction (215 N parallel to the orbit velocity vector) has been applied for the duration of 400 days followed by a coasting phase of several years. The orbital evolution has been compared to the unperturbed orbit to determine the position difference, which is the distance between diverted and unperturbed NEO positions at given epochs. As a result of the interaction, the semimajor axis is slightly enlarged causing a delay of the perturbed object along its orbit path with respect to its unperturbed position. As a consequence the displacement of the NEO from its unperturbed orbit increases constantly with time. This confirms the importance of an early discovery and interaction. The variations in the plots are due to perihelia (ups) and aphelia passes (downs). After 10 years of coasting phase the miss distance for 2000 WC1 would be about 100 Earth radii or 6.4×10^5 km (Fig. 9.2a).

In the case of 1999 AQ10, the object is 2.76 times larger in diameter and thus 21 times heavier than 2000 WC1 when assuming spherical bodies. Therefore, the effect of the same 100 m concentrator is about 21 times less. This can be seen in Fig. 9.2c when following the orbital evolution to the year 2020. Thereafter, a sudden increase in the slope of position difference occurs. The reason for this is a close approach of 1999 AQ10 with Venus in 2019 (Fig. 9.2b). Due to the previous orbit alteration, the flyby additionally diverts the object when compared to its unperturbed orbit.

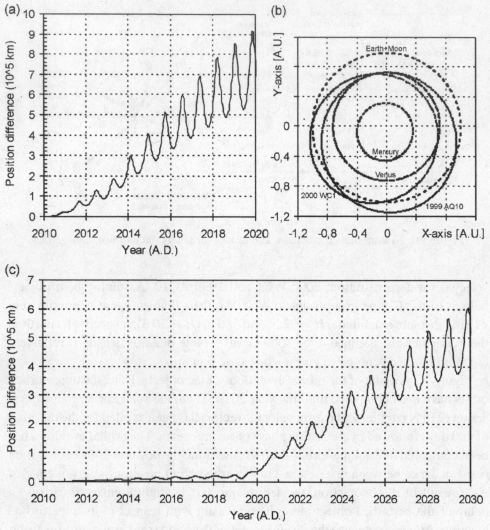

Figure 9.2 Simulation of orbital evolution of (a) 2000 WC1 and (c) 1999 AQ10 after 400 days of solar concentrator interaction beginning in 2010; (b) is a projection of planetary and asteroids orbits into the ecliptic plane of the heliocentric system.

The position difference achieved after 20 years of coasting phase is about 80 Earth radii.

Clearly, this example shows the relevance of precise orbit determination and the importance of radar measurements for orbit refinement, too. Only when based upon reliable orbit data can a successful mitigation be ensured. Unreliable data might even lead to an unintended diversion causing an Earth collision.

5.5 *Solar concentrators for* in situ *experiments*

For technology demonstration a small space probe could be built at low cost within a short time. A deployable 4 m diameter concentrator would probably weigh less than 4 kg and could be easily attached to a small NEO space probe. When equipped with instruments (e.g., a mass spectrometer) material properties of the target NEO could also be studied. At 1 AU solar distance, such a small concentrator would excavate a 0.2 m deep hole when applied to a stony asteroid within 20 min of operation. Thus, the deflection capabilities of solar concentrators could be proven and demonstrated along with the conduction of scientific experiments. Extensive ground experiments under vacuum conditions should be conducted in advance to examine the thermal properties of meteoritic material and to understand the process of vaporization (e.g., shape of vapor plume, vapor deposition process, etc.).

6 Future prospects

Facing the global consequences that are expected from a large NEO impact as well as the problems related to the development and operation of a NEO deflection system, an international collaboration is highly recommended. Possibly, this could be done by the Spaceguard Foundation supported by the United Nations (UN). The UN has shown interest in the topic of NEO impact hazard, by co-sponsoring and organizing the United Nations International Conference on Near-Earth Objects in April 1995. The UN would be an internationally accepted authority that should be equipped with the power to control all related activities, especially the handling of nuclear explosives if selected for a NEO deflection mission.

The probability of a 1 km NEO impact onto the Earth within the next 100 years is about 1 in 3000 (Morrison *et al.* 1994). However, it seems very probable that mankind will start to use NEO resources within the next 100 years, too. Although a large NEO impact onto the Earth is much less probable than a NEO resources exploitation mission in the near future, the issue of protecting the Earth against NEO impacts should have a higher priority. Results from NEO exploration and mitigation systems testing could then be used for resource utilization activities. Searching for NEOs would also provide us with information about potentially interesting NEOs for resources exploitation. NEO resources could provide propellants and raw material possibly needed for future NEO mitigation systems. Therefore, activities for NEO mitigation are the basis for future resource utilisation in our solar system and cannot be regarded as a waste of time and efforts, even if a case of emergency will not appear in the near future.

To prevent future NEO impacts the following activities are proposed.

Phase 1

The discovery of all relevant NEOs has the highest priority because only from the extrapolation of NEO orbital data a future impact can be predicted. Only about 2700 NEOs are known today; the estimated number of NEAs larger than 100 m is about 320 000; and the number of hazardous comets is unknown. Assuming the current level of search activities and detection rate it will take several years to achieve the NASA goal of finding 90% of all NEAs larger than 1 km. This clearly shows that we have to intensify our efforts in NEO detection. The cost for an extended worldwide NEO detection program could be about $20 million per year (Morrison *et al.* 1994).

Phase 2

In situ exploration of NEOs is very important to obtain more information about their composition, mass distribution, shape, internal structure, etc. There are already some asteroid and comet (including NEO) exploration missions completed, on the way, or in preparation. The research goals of future exploration missions should be extended to the preparation of deflection test missions. Typical costs for small asteroid and comet missions are about $100–300 million.

Phase 3

Tests of deflection systems need to be performed to select optimal systems for certain scenarios. These scenarios depend on the target properties, the available time for the deflection mission, the payload capacity of the launchers, etc. These fundamental tests should be performed well in advance because in the case of an emergency there most probably will be no time for further studies and tests. Depending on the deflection technologies to be tested the costs for such missions may be comparable to the costs of NEO exploration missions.

Phase 4

Basically, there are two mitigation strategies: to destroy the NEO, or to divert it. With current technology a sufficient orbit deflection could be achieved only by means of conventional propulsion systems, kinetic impacts, nuclear explosives, or solar concentrators. However, further studies on mitigation systems have to be performed to prepare and optimize the establishment of a NEO deflection system. Building such a system and keeping it ready for launch at all times would surely be too expensive and dangerous (because of the risk of accidents or misuse). Waiting for an emergency without any prior activity would be a waste of precious warning time and could possibly rule out a successful mitigation mission.

While phases 1 and 2 are already ongoing, phases 3 and 4 should be started immediately. The authors see a promising approach in analyzing and developing all the needed technologies now, testing them, keeping them up to date, and manufacturing the mitigation system on demand.

The topic of hazards due to impacts of NEOs onto the Earth may, or may not, appear as popularized and overstated. Although the impact consequences on modern civilization will be enormous even from objects only a few hundreds of meters across, the low statistical frequencies of impacts support the impression that this hazard is without relevance for us today. NEO hazard mitigation activities are of great importance to mankind. This was also mentioned in the "Resolution on the Detection of Asteroids and Comets Potentially Dangerous to Humankind" passed by the European Council (1996), where further and more extensive NEO search and research, follow-up programs, and NEO deflection studies were requested.

One reason why politicians are not very excited about NEO mitigation programs has been pointed out by Steve Kirsch, a private sponsor of NEO search programs: "If we don't get hit, Senators and Representatives will be criticized for wasting taxpayers' money. And if we do get hit, it won't matter since we'll all probably be dead. So politically, it's a stupid decision to vote for this since you can't win either way" (Kirsch 2002).

Mankind has the knowledge and the skills to protect planet Earth from catastrophic impacts of asteroids and comets – for the first time in the history of evolution. It is a scientific fact that the next impact will occur – the only question is when!

7 Conclusions

The risk of NEO impacts on Earth can be ranked as a medium-class risk, defined by its low probability of occurrence and extremely high consequences. The risk management approach in project management suggests performing intensive countermeasures to prevent this risk becoming reality. This procedure will avoid enormous damage and possibly save millions of lives. It has been shown that the absolute and annual costs for NEO detection and mitigation are negligibly low compared to the expected impact damages, e.g., the expected area-specific average damages for a small 150 m NEO impact would be about $60 million. If this NEO should hit an industrialized area the damages could be more than $50 000 billion. This shows that the costs for NEO hazard mitigation are fully justified.

NEO search activities should be increased towards the detection of smaller objects, e.g., 100 to 200 m in diameter. These objects pose a great threat to mankind since they are much numerous, and they impact on Earth more frequently than the 1 km objects.

Various ideas for the deflection and disruption of hazardous asteroids were reviewed and brief examples for the diversion of an asteroid 150 m in extent were given to prove the feasibility of each system. The most promising non-nuclear technology seems to be the solar concentrator. Here, we proposed a simple design that would meet the limited payload capability of current interplanetary launchers. The NEO deflection capabilities of the solar concentrator were demonstrated on PHAs 210 m and 580 m in extent. If sufficient warning time is given orbital displacements of up to 100 Earth radii can be achieved with a 100 m concentrator. An example for the importance of planetary flybys in advance of a collision was also given. Due to minor changes in the flyby geometry, the outcome of the flyby can increase the total efficiency of the applied mitigation technique. Clearly, very precise orbit determination and prediction are required for a successful mission.

Further, we showed the importance of NEO properties for mitigation interaction. While some properties can be obtained from remote sensing, others require expensive *in situ* exploration. There are some small-body missions ongoing or under development, which are very important for the investigation and understanding of these objects. However, there is no current NEO mitigation research project. Building a complete mitigation system and waiting for an NEO on a collision course with Earth would probably be too expensive and too dangerous. Conducting some advanced mitigation studies now could save precious warning time in case of an emergency and the costs for such studies and tests would be relatively modest. The problem of NEO impacts can only be solved by a pro-active approach and not by waiting for an emergency before starting to solve the problem.

Acknowledgments

The authors wish to acknowledge helpful reviews by Dr. Donald K. Yeomans and Dr. David Dearborn. The authors wish to thank Professor Dr. Stefanos Fasoulas for his support of the ongoing NEO mitigation studies at the Dresden University of Technology. Ralph Kahle is grateful to the German Aerospace Centre for the financial support of his doctoral thesis and wishes to thank Dr. Ekkehardt Kuehrt, Dr. Gerhard Hahn, and Dr. Alan Harris for their inspiring supervision.

References

Asphaug, E., Ostro, S. J., Hudson, R. S., *et al.* 1998. Disruption of kilometre-sized asteroids by energetic collisions. *Nature* **393**, 437–440.
Bernasconi, M. C. and Zurbuchen, T. 1995. Lobed solar sails for a small mission to the asteroids. *Acta Astron.* **35**, 645–655.

Canavan, G. H. 1994. Cost and benefit of near-Earth object detection and interception. In *Hazards due to Comets and Asteroids*, ed. T. Gehrels, pp. 1157–1191. Tucson, AZ: University of Arizona Press.

CIA 2001. *CIA World Fact Book*. Available online at http://www.cia.gov/cia/publications/factbook/

Duerfeld, K. 2002. Analyse von Kosten zur Vermeidung von Asteroiden- und Kometeneinschlaegen. Internal report, Dresden University of Technology, ILR-RSN-D-02-10.

Durham, F. P. 1969. The nuclear rocket program at Los Alamos. American Institute of Aeronautics and Astronautics 5th Propulsion Joint Specialist Conference, AIAA Paper no. 69-556.

Dyson, F., Friedman, G., and Valentine, L. 2002. Mass drivers, a robust solution for planetary defense. In *Extended Abstracts from the NASA Workshop on Scientific Requirements for Mitigation of Hazardous Comets and Asteroids*, Arlington, VA, September 3–6, 2002, eds. E. Asphaug and N. Samarasinha, pp. 38–40.

European Council. 1996. Resolution on the detection of asteroids and comets potentially dangerous to humankind. Resolution no. 1080, March 20, 1996, Strasbourg.

Giorgini, J. D., Ostro, S. J., Benner, L. A. M., *et al.* 2002. Asteroid 1950 DA's encounter with Earth in 2880: physical limits of collision probability prediction. *Science* **296**, 132–136.

Gritzner, C. 1996. Analyse alternativer Systeme zur Bahnbeeinflussung erdnaher Asteroiden und Kometen. Ph.D. thesis, Berlin University of Technology, DLR-FB 96-26.

2001. NEO hazard mitigation publication analysis (NEO-MIPA). European Space Agency study by EUROSPACE GmbH and Dresden University of Technology.

Gritzner, C. and Fasoulas, S. 2002. Justification of NEO impact mitigation activities by risk management. *Mem. Soc. Astron. Ital.* **73**, 747–750.

2004. The problem of NEO rotation in planetary defence missions. *Great Bear J., Russia*, **3**, in press.

Hahn, G. 2002. European Asteroid Research Node: database of physical properties of NEOs. Available online at http://earn.dlr.de/

Hammerling, P. and Remo, J. L. 1995. NEO interaction with nuclear radiation. *Acta Astron.* **36**, 337–346.

Harris, A. W., Canavan, G. H., Sagan, C., *et al.* 1994. The deflection dilemma: use vs. misuse of technologies for avoiding interplanetary hazards. In *Hazards due to Comets and Asteroids*, ed. T. Gehrels, pp. 1145–1156. Tucson, AZ: University of Arizona Press.

Holsapple, K. A. 2002. The deflection of menacing rubble pile asteroids. In: *Extended Abstracts from the NASA Workshop on Scientific Requirements for Mitigation of Hazardous Comets and Asteroids*, Arlington, VA, September 3–6, 2002, eds. E. Asphaug and N. Samarasinha, pp. 48–51.

Horeis, H. and Liebig, M. 1985. Strahlenwaffen – Militaerstrategie im Umbruch. Munich: Verlag für Wehrwissenschaften.

Huebner, W. F. and Greenberg, J. M. 2001. Methods for determining material strengths and bulk properties of NEOs. *Adv. Space Res.* **28**, 1129–1137.

Kirsch, S. 2002. Internet homepage. Available online at http://www.kirschfoundation.org/who/reflection_5.html

Kleiman, L. A. 1967. Project Icarus. Massachusetts Institute of Technology Report no. 13.

Letteer, G. A., Jungerman, J. A., and Castaneda, C. M. 1991. Secondary gamma radiation from neutral particle beams. *Sci. and Global Security* **2**, 199–208.

Melosh, H. J. 1995. Non-nuclear strategies for deflecting comets and asteroids. In *Proceedings of the Planetary Defense Workshop*, Lawrence Livermore National Laboratory, May 22–26, 1995, p. 311.

Melosh, H. J. and Nemchinov, I. V. 1993. Solar asteroid diversion. *Nature* **366**, 21–22.

Melosh, H. J., Nemchinov, I. V., and Zetzer, Y. I. 1994. Non-nuclear strategies for deflecting comets and asteroids. In *Hazards due to Comets and Asteroids*, ed. T. Gehrels, pp. 1111–1132. Tucson, AZ: University of Arizona Press.

Mikisha, A. M., Smirnov, M. A., and Smirnov, S. A. 1996. Assessment of possibilities for diversion of Earth-approaching asteroids. In *Abstracts of 2nd International Conference on Space Protection of the Earth*, pp. 90–94 (in Russian).

Morrison, D. A. 1992. The Spaceguard Survey: Report of the NASA International Near-Earth-Object Detection Workshop, Jet Propulsion Laboratory.

Morrison, D. A., Chapman, C. R., and Slovic, P. 1994. The impact hazard. In *Hazards due to Comets and Asteroids*, ed. T. Gehrels, pp. 59–91. Tucson, AZ: University of Arizona Press.

Oberth, H. 1923. *Die Rakete zu den Planetenraeumen*. Munich: Oldenbourg Verlag. 1978. *Der Weltraumspiegel*. Bucharest: Kriterion Verlag.

Phipps, C. R. 1992. Laser deflection of NEOs. In *Proceedings of the Near-Earth-Object Interception Workshop*, Los Alamos, NM, January 14–16, 1992, pp. 256–260. 1995. Lasers can play an important role in the planetary defense. In *Proceedings of the Planetary Defense Workshop*, Lawrence Livermore National Laboratory, May 22–26, 1995, pp. 325–331.

Powell, J., Maise, G., Paniagua, J., *et al.* 1999. The MITEE family of compact, ultra lightweight nuclear thermal propulsion engines for planetary exploration missions. 50th International Astronautical Congress, IAF-99-5.6.03.

Pravec, P. and Harris, A. W. 2000. Fast and slow rotation of asteroids. *Icarus* **148**, 12–20.

Project Orion. 1964. Nuclear pulsed propulsion project. Technical summary report, General Atomics Division, GDC, RTD-TDR-62-3006.

Remo, J. L. 1994. Classifying and modeling NEO material properties and interactions. In *Hazards due to Comets and Asteroids*, ed. T. Gehrels, pp. 551–596. Tucson, AZ: University of Arizona Press. 2000. Energy requirements and payload masses for near-Earth objects hazard mitigation. *Acta Astron.* **47**, 35–50.

Shafer, B. P., Garcia, M. D., Scammon, R. J., *et al.* 1994. The coupling of energy to asteroids and comets. In *Hazards due to Comets and Asteroids*, ed. T. Gehrels, pp. 955–1012. Tucson, AZ: University of Arizona Press.

Spitale, J. N. 2002. Asteroid hazard mitigation using the Yarkovsky effect. *Science* **296**, 77.

Temple, R. 2000. *The Crystal Sun*. London, UK: Century.

Urias, J. M., DeAngelis, I. M., Ahern, D. A., *et al.* 1996. Planetary defence: catastrophic health insurance for planet Earth. Available online at http://www.au.af.mil/au/2025/volume3/chap16/v3c16-1.htm

Voelker, L. 2002. Konzeption von Solar-Spiegelsystemen zur Bahnaenderung von NEOs. Internal report, Dresden University of Technology, ILR-RSN-G-02-01.

Winglee, R. M., Slough, J., Ziemba, T., *et al.* 2000. Mini-magnetospheric plasma propulsion: tapping the energy of the solar wind for spacecraft propulsion. *J. Geophys. Res.* **105**, 21067–21077.

10

Peering inside near-Earth objects with radio tomography

W. Kofman

Laboratoire de Planétologie, Grenoble

A. Safaeinili

Jet Propulsion Laboratory, California Institute of Technology

1 Why radio tomography?

So far, the planetary explorations have focused on gathering information about the atmosphere, the ionosphere, and the surface of the planets. Most of the remote-sensing techniques have focused on observation of planets and small objects in visible and near-visible range using cameras, or GHz-range radars for surface mapping of planets (e.g., Magellan radar for Venus, Shuttle Imaging Radar (SIR A, B, and C), Shuttle Radar Topography Mapping SRTM and TOPEX/Poseidon for Earth). The radio science techniques used to study the gravity field are important for exploring planetary interior. Although these techniques have provided a wealth of information, there are still a large number of questions that cannot be answered unless we probe the sub-surface. Example of questions that radio tomography can answer are: (1) sub-surface stratigraphy on planets, (2) the existence of paleo-channels, (3) the depth of CO_2 and H_2O ice layers on Mars, (4) the existence of liquid water under the surface of Mars, (5) the existence of an ocean on Europa, and (6) looking inside comets, and testing the rubble-pile hypothesis for asteroids. It is well known that low-frequency electromagnetic waves (specifically, high-frequency (HF) and very-high-frequency (VHF) regimes) can penetrate ice and rocks to a depth of hundreds of meters to a few kilometers. Such radars have been used on Earth for sub-surface investigations. Also, airborne radar sounders have been deployed over the past few decades to investigate glaciers and to measure ice-layer thickness (Gudmandsen *et al.* 1975). More recently, due to current speculations of the existence of underground water reservoirs on Mars and the possibility of an ocean beneath Europa's icy surface, there is a growing interest in spaceborne radio sounders. Mars Advanced Radar for Subsurface and Ionospheric Sounding (MARSIS), which is the first of this new generation of radio sounders, arrived at

Mitigation of Hazardous Comets and Asteroids, ed. M. J. S. Belton, T. H. Morgan, N. H. Samarasinha, and D. K. Yeomans. Published by Cambridge University Press. © Cambridge University Press 2004.

Mars in late 2003 for a 2-year mission. MARSIS is designed to sense the planet's interior to a depth of up to 5 km. MARSIS' main objective is to investigate the sub-surface of Mars and search for water to see if it exists in liquid form under the surface. It will also attempt to map and characterize the surface and sub-surface geological structure of Mars, which is hidden under a layer of surface dust or ice.

The ability of low-frequency radio waves (MHz range) to penetrate objects that are opaque in the high-frequency region of the electromagnetic spectrum, makes low-frequency radars prime instruments to inspect and image the interior of small bodies. The first step on this path will be taken by the CONSERT (COmet Nucleus Sounding Experiment by Radio wave Transmission) experiment (Kofman *et al.* 1998) on the European Space Agency's Rosetta mission. The primary goal of CONSERT is to measure average dielectric and attenuation properties of the comet, provide estimates for the degree of homogeneity of the comet and to image its interior at a coarse resolution. The ultimate goal for a radio tomographic instrument is to provide a means to image the interior structure, or more precisely the volumetric dielectric property distribution, of small objects (asteroids and/or comets) that are few kilometers in size with a good resolution. To achieve this goal, a Radio Reflection Tomography (RRT) system has been proposed by Safaeinili *et al.* (2002). The proposed RRT imaging system is analogous to the ultrasonic reflection tomography instruments that are widely used in medical applications and are capable of providing high-resolution images of the human body's interior (see example shown in Fig. 10.1).

In order to obtain a radio tomographic image of a small object, similar to the analogous ultrasonic medical application, it is necessary for the transmitted radio waves to propagate within the object. In addition, the forward (transmitted) and/or backward (reflected) scattering signal from the interior features need to be received with sufficient strength at the receiving antenna. However, unlike the medical ultrasonic application where a physical aperture is realized and is used for real-time imaging, in the radio tomography application, it is not feasible to obtain real-time images of the interior, since that would require a very large real aperture. Although such a large real aperture may not be possible, a synthesized aperture can be achieved by obtaining measurements at different positions around the object. In the mono-static case (e.g., RRT) (Safaeinili *et al.* 2002), the transmitter and receiver are co-located and scanned around the object. In the bi-static case (e.g., CONSERT experiment on Rosetta) (Kofman *et al.* 1998), the measurements are obtained by varying the transmitter/receiver location from a transponder that is located on the object. This is achieved by having the spacecraft orbit the object as it spins in space.

As mentioned above, radio tomography can be achieved in two distinct ways: (1) tomography using transmitted radio waves, and (2) tomography using reflected

up

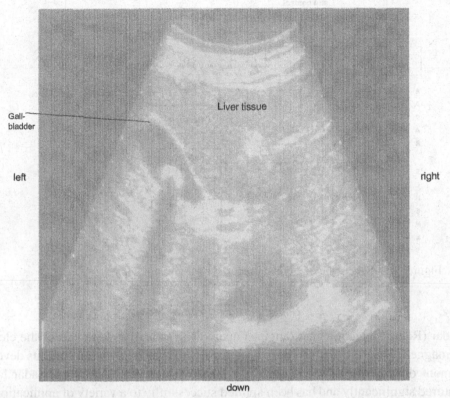

left   right

down

Figure 10.1 An example of a medical ultrasonic image of human tissue. Scan section of the right superior abdomen, going throught the liver and the gall bladder. The liver tissue structure is homogeneous. The gall bladder is transparent in this scan. Within the gall blader, we clearly observe an inhomogeneity with high reflection resulting in a cone of shadow behind it. (Image: CHU Grenoble.)

radio waves. In principle, the two methods are similar in that they rely on the same physical relations; however, in practice they are significantly different. In this chapter, we will attempt to present both techniques and show how each can be implemented. First, we present a brief background on radar theory and the physical and mathematical foundation for both reflection and transmission tomography techniques. Then we address the practical considerations for the implementation of each technique and we conclude with the presentation of a few examples to demonstrate the imaging concept. We finish by indicating some basic parameters of the radar system that we believe will provide an efficient way to explore the interior of the object.

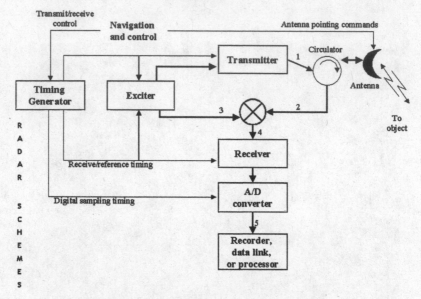

Figure 10.2 Schematic of a typical radar showing major subsystems.

1.1 Background on radar theory

Radar (Radio Detection and Ranging) operates in the radio-wave part of the elec-tromagnetic spectrum and has been used for a variety of application since its devel-opment during World War II. Over the past two decades, spaceborne radar has matured significantly and has been applied successfully to a variety of applications including spaceborne synthetic-aperture radar (SAR) for imaging Earth and other planets surfaces and topography, Specifically for the Earth's surface, SAR has been used for change detection after earthquakes and in volcanic regions, and for future applications for vegetation and soil moisture monitoring.

In this presentation, we will start by introducing a few concepts that are important in order to define a radar.

The main parts of the radar are the transmitter and receiver, as shown in Fig. 10.2. The transmitter sends a signal of a given frequency and bandwidth radiated through the antenna. The receiver through the same or a different antenna senses the reflected, scattered, and/or diffracted signal. In modern radars, the system gen-erates a wide bandwidth signal via digital-to-analog conversion of a sampled wave-form stored in memory. The analog signal is mixed with a carrier frequency (up-conversion) before it is amplified and sent to the transmitter antenna. The receiver amplifies the signal coming from the receiver antenna in the selected ranges, and usually mixes this signal with a digitally generated reference signal to reduce the center frequency of the received signal (down-conversion). The main reason for the signal down-conversion is to reduce the center frequency to be able to process signals which at very high frequency is difficult or impossible and to save memory storage space needed for digitized signal (Henderson and Lewis 1998).

Let us suppose that the radar sends a single pulse of duration T_p. From a point target after processing (pulse compression if necessary) and square-law detection, the peak power of the received signal for narrow-band radar is

$$P_r = \frac{P_T G^2(\theta, \gamma) \lambda^2 K_r}{(4\pi)^3 R^4} \cdot \sigma \qquad (10.1)$$

where K_r is the range compression factor which expresses the peak signal increase due to compression, P_t is the transmitted power, λ the wavelength, G antenna gain, R distance to the target, and σ the radar cross-section of the target.

In order to evaluate the capabilities of the radar to detect and treat signals, one has to compare the signal power, shown in Eq. 10.1 with the noise power that tends to mask the signal. The noise is the radar receiver's output power when no desired radar target is present and is generally described by the noise temperature T_s. The noise power is given by:

$$N_{rec} = K \cdot T_s \cdot B_n \cdot F_n \qquad (10.2)$$

where K is Boltzmann's constant, B_n is receiver noise bandwidth, and F_n is receiver noise factor, which describes how far from the ideal one the given system is, usually of the order of a few decibels.

The important parameter describing the detection capabilities before signal processing is the signal-to-noise ratio (SNR) for a single target and single pulse:

$$SNR = \frac{P_r}{N_{rec}} \qquad (10.3)$$

The equation for SNR indicates that the signal detectability depends on the system parameters (transmitted power, losses, compression, bandwidth) and on the target properties (cross-sections) and environmental factors (noise). This equation does not include the attenuation or multi-path effects that depend on the medium through which the signal propagates. One usually uses multi-pulse radars and this can increase the signal-to-noise ratio in many cases.

The main role of the radar system is to measure the time delay between transmission and reception. The spatial spectrum of the scattered field (also termed the slow-time Doppler effect) that depends on the target or satellite movement can be measured and/or compensated. The main radar parameter determining the range resolution (the radar capabilities to distinguish the return echoes from two adjacent targets) is the pulse length for a simple radar (without modulation) or the compression effect for modulated signals. The compression process is the filtering of the incoming signal by the copy of the transmitted one. This makes the pulse width τ_0 at the output of the filter smaller than the signal width T_p of the input (this is the purpose of this filtering).

Theoretically, $\tau_0 = 1/B$ where B is the bandwidth of the signal. The pulse compression ratio is $K_r = T_p/\tau_o$. Typically one can obtain values between 100 and 10^5.

The range resolution is then given by:

$$S_r = \frac{c_{med}\tau_0}{2} \quad \text{and} \quad S_r = \frac{c_{med}}{2B} \tag{10.4}$$

where c_{med} is wave velocity in the medium where the signal propagates. The range resolution factor is important in the radar system definition.

In imaging radars, the concept of an azimuth filter is important. A systematic movement between the radar and the target leads to systematic phase shifts. If the radar is coherent (i.e., the phase between successive transmitted pulses is known), the movement of the satellite is predictable, and the geometry known, this phase can be calculated and compensated so that the linear integration of the input signals will have the same action as the pulse compression. This is the focusing effect.

This focusing effect can be seen as a construction of the synthetic aperture (SAR). In the real-aperture radar, an antenna of length D illuminates the surface with a half-power one-way beam width

$$\Omega = 0.89\frac{\lambda}{D} \text{ (radians)} \tag{10.5}$$

where λ is the wavelength. Then the spatial resolution at a distance R is:

$$S_a = 0.89\frac{R\lambda}{D} \tag{10.6}$$

In synthetic-aperture radar, the azimuth resolution for a synthetic aperture of length L is:

$$S_a \approx \frac{\lambda R}{2L\sin(\alpha)} \tag{10.7}$$

where α is the angle between the real antenna beam and its velocity vector. This indicates that the aperture synthesis in the azimuth direction (along-track) can essentially increase the radar resolution in that direction.

The transverse resolution (perpendicular to the azimuth or cross-track) is still defined by the real aperture. However in the case of very large lobe, this resolution is determined by the first Fresnel zone defined as

$$S_t = \sqrt{2\lambda R} \tag{10.8}$$

Physically this zone is defined by a region for which the phase differences of reflected signals are less than $\lambda/4$ and this means that these signals add coherently.

1.2 Propagation in a lossy medium

In Eq. 10.1, for the power received from the single point target, we do not take into account the propagation loss. In practice, a signal propagating in comets and asteroids is attenuated due to the object's complex dielectric and magnetic properties. The amplitude and phase of a plane-wave propagating in a medium that is characterized by the dielectric constant ε (phase factor β) and the attenuation factor α is described by

$$E = E_0 e^{-\alpha r} \cos(\omega_0 t - \beta r) \tag{10.9}$$

The complex dielectric constant is $\varepsilon = \varepsilon' - j\varepsilon'' = \varepsilon - j\dfrac{\sigma}{\omega_0}$ ($\sigma =$ conductivity) and the complex permeability is

$$\mu = \mu' - j\mu'' \tag{10.10}$$

where ε_0, μ_0 are to the vacuum dielectric and permeability constants. Usually one uses the reduced dielectric constant and permeability.

$$\varepsilon_r = \varepsilon/\varepsilon_0, \quad \mu_r = \mu/\mu_0$$

The velocity in the medium is given by $c_{med} = (\varepsilon'\mu')^{-1/2}$
One defines the loss tangents:

$$tg\delta_d = \frac{\varepsilon''}{\varepsilon'} = \frac{\sigma}{\omega_0\varepsilon}$$

$$tg\delta_m = \frac{\mu''}{\mu'} \tag{10.11}$$

The attenuation factor is given by

$$\alpha = k \cdot \frac{c}{c_{med}} \sqrt{\frac{1 - tg\delta_d \cdot tg\delta_m}{2} [(1 + tg^2(\delta_d + \delta_m))^{1/2} - 1]}, \quad k = \frac{2\pi}{\lambda}$$

$$\beta = k \cdot \frac{c}{c_{med}} \sqrt{\frac{1 - tg\delta_d \cdot tg\delta_m}{2} [(1 + tg^2(\delta_d + \delta_m))^{1/2} + 1]} \tag{10.12}$$

where c the velocity and λ wavelength in free space.

For a non-magnetic medium with very low losses $tg\delta_d \ll 1$, which is probably the case for asteroids and comets one has

$$\alpha = \frac{\pi}{\lambda} \cdot \frac{c \cdot tg\delta_d}{c_{med}} \tag{10.13}$$

and in the opposite case for $tg\delta_d \gg 1$ $\alpha = \frac{\pi}{\lambda}\frac{c}{c_{med}}\sqrt{2}tg\delta_d$.

From Eq. 10.12, one sees that the attenuation coefficient is directly proportional to the frequency. The penetration depth is given, approximately, by $1/\alpha$ (this characterizes the attenuation by a factor of e^{-1}). In some way, for radar applications, this name can be misleading, because usually the radar capabilities are much larger than the e^{-1} attenuation (it can accept a few tens of decibels of attenuation). The attenuation of the electromagnetic waves in the material medium can be calculated by the following formula in decibels per meter.

$$\frac{8.686 \cdot \pi \cdot \tan \delta_d}{\lambda} \frac{c}{c_{med}} \quad (\text{dB m}^{-1}) \tag{10.14}$$

This will be used to evaluate the penetration depth in the cometary and asteroid materials. Other processes (scattering, multi-path propagation, reflection, etc.) also contribute to the wave attenuation. As we want to use the radar to study interiors using the reflection (diffraction) of signals on the discontinuities we need to define the reflection coefficient:

$$\Gamma_R = \frac{\eta_m - \eta_0}{\eta_m + \eta_u}, \quad \text{where} \quad \eta_m = \frac{\omega_0 \mu}{k_m} = \sqrt{\frac{\mu}{\varepsilon}} \tag{10.15}$$

This factor determines how much of the incoming signal amplitude is reflected. The transmission coefficient determines how much is transmitted. The relation between these two coefficients is given by following equation: $\Gamma_R^2 + \Gamma_T^2 = 1$. These equations allow the estimation of the power of the incoming radar signal and therefore the radar capabilities.

In Fig. 10.3a, we show the attenuation as a function of the loss tangent and for the real part of the dielectric constant varying from 2.0 to 9. We have chosen materials covering the likely composition of asteroids (see Table 10.1). In this figure we indicate the corresponding rocks and minerals. For instance, for granite, the reduced dielectric constant can vary from 5 to 7 and the loss tangent from 0.01 to 0.03. It is clear from Fig. 10.3a that the attenuation will probably be large (for granite between 30 to 60 dB km^{-1}).

By taking into account the porosity of the materials which can reach a large value (typically in the order of 40–50%) and using the Rayleigh mixing formula (Sihvola and Kong 1988), it is possible to calculate the dielectric constant and the attenuation. For porous granite, the attenuation is about 8 to 20 dB km^{-1}, while for basalt, it is 10 to 30 dB km^{-1} (see Fig. 10.3b). If the absorption is on the order of 10 to 30 dB km^{-1}, the radar will be able to investigate objects that are a few kilometers in size. However, if the absorption reaches 60 dB km^{-1} or more, the penetration will be limited to tens or hundreds of meters, at best. These values are valid for a radar operating at a low frequency (e.g., 10 MHz).

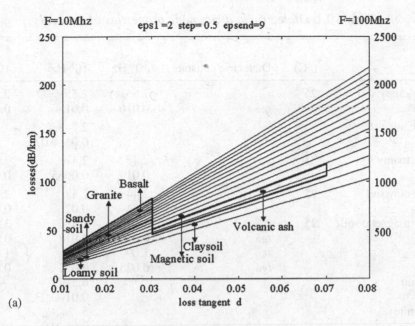

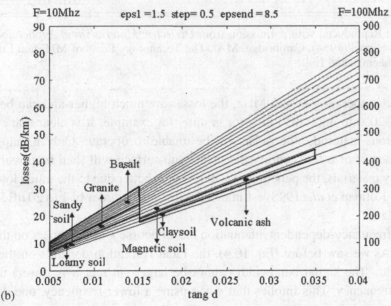

Figure 10.3 (a) Losses in materials as function loss tangent and for different real part dielectric constant varying from 2 to 9. The losses were calculated for 10 and 100 MHz. In the figure we indicate the range of parameter for some materials taken from Table 10.1. (b) Losses for porous materials which were calculated using mixing formula.

Table 10.1 *Dielectric and magnetic properties of minerals*

	T (°C)	Dielectric constant	10^7 Hz	10^8 Hz	10^9 Hz
Dry sandy soil	25	ε	2.55	2.55	2.55
		$tg\delta$	0.016	0.01	0.006
Volcanic ash				2.7–3.4	
				0.03–0.07	
Dry loamy soil	25		2.48	2.47	2.44
			0.014	0.0065	0.0014
Dry clay soil	25		2.44	2.38	2.27
			0.04	0.02	0.015
Dry magnetic soil	25	ε	3.6	3.5	3.5
		$tg\delta$	0.012	0.018	0.022
		μ	1.09	1.09	1.005
		$tg\delta_m$	0.025	0.025	0.099
Basalt				6.5–9.6	
				0.01–0.03	
Granite				5–7	
				0.01–0.02	

Source: Reproduced with permission from *Dielectric Materials and Applications*, by A. von Hippel (1954), Cambridge, MA:, The Technology Press of MIT, and London, UK: Chapman and Hall.

For a radar operating at 450 MHz, the losses are much higher and can be on the order of 2000 to 3000 dB km^{-1} for granite, for example. It is clear that even for very porous materials, the radar will be unable to operate. Only a mapping of the surface or of a few centimeters of the sub-surface will then be possible. For cometary materials, the penetration will be much better due to the much lower loss tangent. Kofman *et al.* (1998) estimated the loss on the order of 1 to 20 dB km^{-1} at 100 MHz.

This frequency-dependent attenuation has of course a large impact on the radar design. As we saw before (Eq. 10.4), the radar resolution depends on the signal bandwidth. The signal bandwidth cannot be larger (in fact it is lower) than the central frequency. This implies that when using a lower frequency, one degrades the space resolution.

This frequency dependence of attenuation makes the diffraction tomography approach probably necessary for asteroid exploration. In fact, in order to use the geometric optics approximation, it is necessary to follow the conditions that the wavelength is much shorter than the characteristic length of the inhomogeneities inside the asteroids. This would imply a high frequency (at least 100 MHz and probably higher) and therefore high attenuation inside the object. The use of high frequency would probably limit the possible imaging to 10 to 100 m depth.

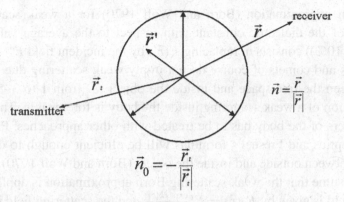

Figure 10.4 Geometry of bi-static measurements.

In some way, it is of course speculative because we do not know the composition of the asteroid, its porosity etc. However, the conservative approach suggests using a low-frequency radar ($\sim$5 to 100 MHz) in order to ensure a good penetration. It is clear that the choice of the low frequency has an influence on the instrument characteristics and essentially on the dimension of the antenna.

The resolution of the radar will also suffer from this choice as we discussed above. The exact choice of the frequency results from a trade-off between science and technical limitations. If it were possible to imagine a mission with two antennas (the high-gain communication antenna used for the high frequency radar purposes, the other for the low frequency), we would propose to have multi-frequency radars.

1.3 Tomographic approach

In order to explain the radar tomographic approach, let us first assume that the scattering of the waves inside the object is due to the changes of the dielectric constant. From wave equations, one obtains that:

$$E^S(\vec{r}) = \frac{1}{4\pi\varepsilon_0}k^2 \iiint (\varepsilon(\vec{r}') - \varepsilon_0)E(\vec{r}', \vec{r}_t)G(\vec{r}, \vec{r}')dV' \qquad (10.16)$$

where $\vec{r}$ is the vector from the origin of the coordinates to the observation point and $\vec{r}'$ is the vector from the origin to the scattering point (see Fig. 10.4), where $G(\vec{r}, \vec{r}')$ is the Green function, which for homogeneous backgrounds in three-dimensional cases is

$$G(\vec{r}, \vec{r}') = \frac{1}{4\pi\varepsilon_0}\frac{\exp(jk|\vec{r} - \vec{r}'|)}{|\vec{r} - \vec{r}'|} \qquad (10.17)$$

$E(\vec{r}')$ is the unknown electric field inside the body and $k = 2\pi/\lambda$ wave number (Herman *et al.* 1987).

The Born approximation (Born and Wolf 1970) for a weak scattering (the fluctuation of the dielectric constant with respect to the average value is weak ($\frac{\Delta \varepsilon}{\varepsilon_{aver}} < 5 - 10\%$)) consists in replacing $E(\vec{r}')$ by the incident field E^{inc}.

Asteroids and comets of course do not imply weak scattering due to the transition between the free space and inside the object (ε_r from 1 to 3–5), however the assumption of a weak scattering inside the body is foreseeable. The reflection on the borders of the body has to be treated with other approaches. Probably the geometric optics and Fresnel's formulas will be efficient enough to describe the transition between outside and inside the object (Born and Wolf 1970).

Let us assume that the weak scattering Born approximation is applicable; then the initial field is given by $E_0(\vec{r}) = E^t \frac{\exp(jk|\vec{r}|)}{|\vec{r}|}$ and the scattering field is:

$$E^s(\vec{r}) = \frac{k^2}{4\pi \varepsilon_0} E^t \iiint (\varepsilon(\vec{r}) - \varepsilon_0) \frac{\exp(j\vec{k}|\vec{r}_t - \vec{r}'|)}{|\vec{r}_t - \vec{r}'|} \cdot \frac{\exp(j\vec{k}|\vec{r} - \vec{r}'|)}{|\vec{r} - \vec{r}'|} d\vec{r}'$$

(10.18)

In order to simplify the above equation, let us assume that the source and the receiver distance is large compared to r' (the radius of the body under investigation). Then it is possible to use the Fraunhofer approximation for the Green's function:

$$|\vec{r} - \vec{r}'| \simeq r - \vec{n} \cdot \vec{r}' \quad \vec{n} = \frac{\vec{r}}{r}$$

$$|\vec{r}_t - \vec{r}'| \simeq r_t + \vec{n}_0 \cdot \vec{r}' \quad \vec{n}_0 = -\frac{\vec{r}_t}{r_t}$$

in exponential (phase factor) and $|\vec{r} - \vec{r}'| \simeq r$ $|\vec{r}_t' - \vec{r}'| \simeq r_t$, for the denominator approximation. Then, within this approximation, the scattered field is given by

$$\widetilde{E}(\vec{r}) \simeq \frac{k^2 E^t \exp(jk(r_t + r))}{4\pi \varepsilon_0 \quad r_t r} \int (\varepsilon(\vec{r}') - \varepsilon_0) \exp(-jk(\vec{n} - \vec{n}_0) \cdot \vec{r}') d\vec{r}' \quad (10.19)$$

This formula means that the observed scattering field is the spatial Fourier transform of the dielectric constant of the scattering body at the spatial frequency $\vec{q} = k(\vec{n} - \vec{n}_0)$ that depends on the difference between scattered and incident wave vectors.

$$\vec{E}_{\vec{q}}(\vec{r}) \approx \int (\varepsilon(r') - \varepsilon_0) \exp(-j\vec{q} \cdot \vec{r}') d\vec{r}' \quad (10.20)$$

By measuring this for various positions of the transmitter and receiver and for large bandwidth signals, one obtains the set of measurements $\vec{E}_{\vec{q}}(\vec{r})$ which can be inverted by the Fourier transform if the sampling in space is dense enough. Now the radar will measure the scattered signal around the body and using this formula, one can in principle, invert the data and deduce the dielectric properties ($\varepsilon(r') - \varepsilon_0$) of the body.

The Born approximation is a "weak scattering approximation" and requires that the total scattered field is small in comparison with the incident field, and that the magnitude of $(\varepsilon(r) - \varepsilon_0)$ and the extent of the body volume is small. This second condition can be relaxed by using the Rytov approximation (1976) (Žernov and Lundborg 1993).

While the Born approximation gives the linear relation between the amplitude of the signal, $\varepsilon(r)$, the Rytov one gives the linear relation between $\varepsilon(r)$ and the phase of the receiving signal.

$$E = E_0(\vec{r}, k) \exp(\Psi(\vec{r}, \omega, t))$$

The obtained equation for the first order approximation in this case is given by:

$$\psi_1 = -k^2 E_0^{-1}(r) \int G(\vec{r}, \vec{r}') \cdot (\varepsilon(r') - \varepsilon_0) E_0(r') dr' \qquad (10.21)$$

This equation and the Born approximation equation have exactly the same structure, which means that mathematically these two problems have the same solution. However, the physical assumptions are not the same and the Rytov approximation is probably more appropriate for the asteroid observations. One of these two approaches may not be valid and in this case, a more general non-linear fitting between the data and some assumed models should be developed (see Chapter 3).

For the large-scale inhomogeneities one uses the geometric optics approximation to solve the wave propagation trough media. In the limit of the high frequency, the Rytov's approximation converges to this classical geometric optics. In this case, the radar tomography reduces to the classical tomography using the back-propagation algorithm.

1.4 Bi-static and mono-static measurement approaches

In general, radar measurements are either bi-static or mono-static. In the bi-static case, the transmitter and receiver are not generally co-located. Mono-static radar, the most common type of radar, is a special case of bi-static radar for which the receiving and transmitting antennas are in the same location and very often there is only one antenna for reception and transmission. The treatment and interpretation of the radar signal for the two cases is quite different and their use depends on the particular application at hand.

In the following, we will focus on both bi-static and mono-static radars and their application in radio tomographic imaging of small (kilometer-size) near-Earth objects. Both bi-static and mono-static radars can be used for tomographic imaging of asteroids and comets using radio transmission or a radio reflection tomography approach. CONSERT (Kofman *et al.* 1998) is an example of radio transmission

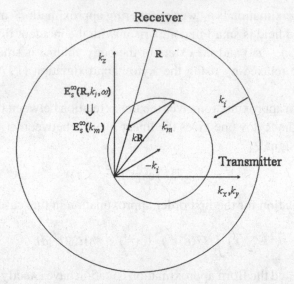

Figure 10.5 The bi-static measurements represented in k-space. The transmitter and receiver k-vectors and scattering vector are indicated in the figure.

tomography (bi-static approach). Recently, a radio reflection tomographic system using a mono-static approach has been proposed for imaging of small bodies (Safaeinili *et al.* 2002).

The choice between mono-static and bi-static approach depends on the scientific objectives of the mission. For example, for cometary materials (i.e., porous ice with dielectric constant of about 2), only 2% of the incoming energy is reflected; this means that 98% of the transmitted energy goes through the object. This high transmission coefficient and mission mass and power were the reasons why the CONSERT instrument on the Rosetta mission was designed as a transmission experiment. The presence of the lander on the Rosetta mission facilitated this choice. The transparency of the object allows the use of higher frequencies (e.g., 100 MHz for CONSERT) providing the capability to use geometric optics principles for imaging of the comet. On the other hand, if the objective of the mission is to perform high-resolution volumetric imaging of an object, possibly with higher attenuation such as asteroids or comets with potentially significant inhomogeneity (i.e., larger internal reflections which correspond to large jumps of composition), then the choice will be a reflection tomography instrument.

In Fig. 10.5 the general bi-static situation is illustrated in the wave number space (k-space) for the far field approximation. Equations 10.19 and 10.20 describe this situation and show that one measure the Fourier component of dielectric constant. The spatial frequency $\vec{k}$, which is the difference between scattered and incident wave vectors, is indicated in the figure.

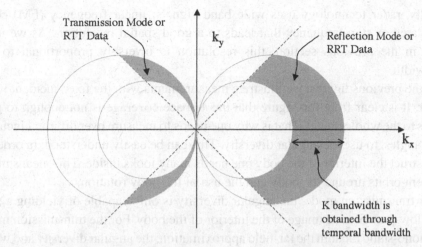

Figure 10.6 Transmission and reflection method compared for a signal with frequency band between ω_1 and ω_2.

By changing the receiver position, the $\vec{k}$ vector is changing and moving on the sphere (called the Ewald sphere, which is in fact the surface on which the dispersion relation relating the frequency and k vector of real propagating waves is fulfilled) depending on the angle between receiver and scattered wave vectors. Therefore, observed by radar scattered field depends only on these two relative positions and after some phase corrections (Eqs. 10.19 and 10.20), one has directly the Fourier transform of dielectric constant, the image of the interior of the object.

In the transmission mode the reception is on opposite side of the incoming incident wave and in the radio reflection case the reception is on the same side. The object is illuminated by the signal with the wave vector $\vec{k}_i$ and the observations are made on the observation plane (surface).

The tomography principle (Fourier diffraction slice theorem) relates the two-dimensional Fourier transform of the measured field on the plane with the Ewald sphere (in the k-space) components of the three-dimensional Fourier transform of the object dielectric constant (Herman *et al.* 1987). The transformed data have to be disposed on the sphere centered on $-\vec{k}_i$ vector (radius of sphere k_i).

The inverse Fourier transform operation maps the spatial spectrum of the object (i.e., k-space) into the volumetric image of the object. The space resolution of the reflection system is better than that of transmission one as soon as the wide band-width signal is used by radar. In Fig. 10.6, we compare these two approaches for the signal of frequency band between ω_1 and ω_2. The transmission mode leads to the narrow coverage in k-space that leads to the poorer spatial resolution as compared with the mono-static case. In other words, the mono-static radar can better detect the fast changes in the dielectric constant than the transmission experiment.

Usually, radar technology uses wide band signals, linear frequency (FM) chirp modulated or step-frequency that leads to a good spatial resolution. As we indicated in the previous section, this resolution is inversely proportional to the bandwidth.

In the previous figures, we illustrate measurements with the fixed incident wave vector. It is clear from this figure that the k-space coverage is not enough to give access to the whole object. That is why one needs to measure over different incident wave angles, to use the angular diversity. This can be easily understood: in order to reconstruct the interior of the body one needs many looks inside. This means many different orbits around the body and the use of the body rotation.

For transmission mode, the angular diversity is only capable of yielding a spatially low-pass filtered image of the interior of the body. For the transmission mode and mono-static radar in the far-field approximation, the angular diversity and wideband signals lead to the image construction by summing signals arriving at the same time, on a given point of the measurement surface. In the far-field conditions the observation space and k-space are "superimposed" which means that the image can be obtained by simpler than non-far-field mathematical processing.

The received signal $\vec{E}^s(\vec{r})$ has to be transformed to the k-space by corresponding to each $\vec{r}$ a given $\vec{k}$ vector, $\vec{E}_k^s(\vec{r})$ and by phase normalization. Equation 10.22 shows how to obtain by Fourier transform the image of the objects. For the mono-static radar, the operations are similar:

$$O(\vec{r}') = \frac{r^2}{2\pi^2} \iiint \frac{E^S(\vec{K})e^{-jkr}}{k^2 F(\omega)} e^{j\vec{k}.\vec{r}'} d^3 K \qquad (10.22)$$

where $F(\omega)$ is a spectrum of the radar signal.

It is clear from the above that without the wide-band signals and angular diversity, one can not rebuild the whole image of the interior.

2 Radio transmission tomography (RTT) approach

In transmission tomography mode, the receiver is placed on the opposite side of the object with respect to the transmitter. Usually the transmission mode is used in medical or non-destructive analyses applications. The X-ray absorption tomography technique also uses the transmission mode. The most frequent technique uses a continuous wave sinusoidal signal and in this case the receiver measures the phase and amplitude of the transmitted signal. Assuming a weak scattering case, the inversion technique is carried out by simply Fourier transforming the three-dimensional spatial spectral image of the object. Here we provide the characteristics and specific design of the CONSERT data collection strategy and inversion.

2.1 CONSERT experiment

CONSERT is the experiment on board the Rosetta mission that will provide information about the deep interior of the comet. In this experiment, an electromagnetic signal is transmitted between the lander, located on the comet surface, and the orbiter. The transmitted signal will be measured as a function of time and as a function of the relative position of the orbiter and the lander for a number of orbits. A simulation of electromagnetic wave propagation through assumed cometary materials showed that it is possible to propagate signals between the lander and the orbiter.

The basic principle for the experiment is straightforward. An electromagnetic wavefront will propagate through the cometary nucleus at a smaller velocity than in free space and loses energy in the process. Both the change in velocity and the energy loss depend on the complex permittivity of the cometary materials. They also depend on the ratio of the wavelength used to the size of any inhomogeneities present.

Thus, any signal that has propagated through the medium contains information concerning this medium. The change in velocity of the electromagnetic wave induced by propagation through the cometary material is calculable from the time taken by the wave to travel, while the loss of energy is deducible from the change in signal amplitude.

If a sufficient number of orbits were available, one would be able to obtain many cuts of the interior of the comet and therefore to build up a tomographic image of the interior. The transmission mode was chosen because 98% of electromagnetic energy at 90 MHz goes through the comet and in this condition the radio reflection tomography will be not very efficient. During the initial planning stages, it was assumed that two landers would be available. Unfortunately, one of them was later cancelled, significantly diminishing the amount of data that we might expect and also restricting our ability to build up a complete image of the cometary interior.

The experiment originally proposed was a one-way transmission of a radio wave from the orbiter to the lander. As it was not possible to get sufficiently stable clocks within the given constraints of mass and power consumption, the initial concept was abandoned and the transponder technique was adopted. The simplest explanation of this technique is to imagine the lander as a simple reflector of the signal coming from the orbiter. The signal is thus measured in the time reference of the orbiter and this enables one to relax the constraints on the stability of clocks. It is technically impossible to use the lander as a simple reflector; however, it is possible to use it as a delayed active reflector. The method is described in Barbin *et al.* (1999).

This evolution to a transponder system makes the experiment scientifically richer. Indeed, in the first version, we planned to measure only the amplitude and the phase of a sinusoidal signal. These measurements would contain the information corresponding mostly to the main path of the signal propagating through the comet. The new version of the experiment, which uses a wide bandwidth signal, will allow the measurement of a signal propagating through main and secondary paths.

The orbiter will send a signal that will be picked up by the lander. As the orbiter moves along its orbit, the path between it and the lander will vary and so pass through differing parts of the comet. In addition, the rotation of the comet nucleus will also change the relative position of the lander and the orbiter. Hence, over several orbits, many different paths will have been obtained. This allows us to think that it will be possible to build a coarse image of the cometary interior, permitting us to distinguish cometesimals if there are any (see Fig. 10.10). Figure 10.7 describes the objectives and principles of CONSERT.

To improve the signal-to-noise ratio, one transmits the periodic signal for the coherent integration of the received signal. Taking into account the relative radial velocity between the orbiter and the lander, and the offset between two clocks, one can accumulate the sounding signal over a 28 ms period. This coherent integration creates a 54 dB increase in the signal-to-noise ratio as compared to a single pulse of the same power. The coherent integration is made in two ways: the first is the transmission of the phase shift coded signal, called pseudo-random code, and the second is the use of the periodic repetition of this coded signal (Barbin *et al.* 1999). In Fig. 10.8 the block description of the CONSERT experiment is shown.

The CONSERT experiment has a transmit/receive antenna on both the orbiter and on the lander. The choice of the center frequency of 90 MHz has a strong impact on antenna size. A fundamental requirement of the antenna is that it must have a broad main antenna lobe centered on the antenna normal. This is to ensure that the whole comet is "illuminated" by the main lobe when the antenna normal is pointing towards the comet. To be independent of the spacecraft orientation and of the possible polarization properties of the cometary materials, we use a circular polarization for the transmission. In order to reduce interaction with the orbiter spacecraft body and other instruments, the antenna is on top of a mast, away from the main structure. Two crossed dipoles oriented 90° to each other and fed 90° out of phase form a circular antenna. The orbiter antenna is formed of two crossed dipoles and two crossed dipole reflectors. On the lander there are severe restrictions on mass and on the location of the antenna. Two monopoles oriented 90° to each other and fed 90° out of phase form a circular polarized antenna. These monopoles form the lander antenna. The mass and average power used of the CONSERT experiment are 2.9 kg and 3 W on the orbiter and 1.8 kg and 2.5 W on the lander.

The coded signal is transmitted from the orbiter through the comet and is received by the lander. This signal is coherently integrated and compressed, which means that the cross-correlation between the model of the coded signal and the received integrated signal is calculated, and the time-delay of the strongest cross-correlation

CONSERT

(Comet Nucleus Sounding Experiment by Radiowave Transmission)

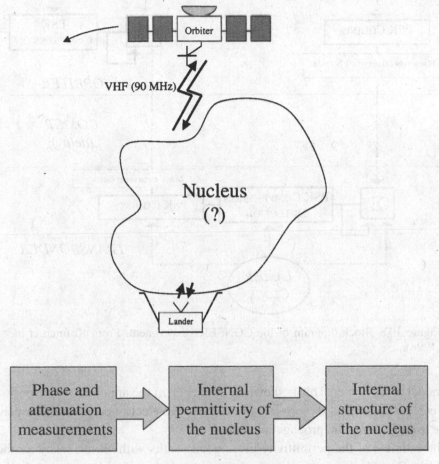

Figure 10.7 Schematic description of the CONSERT experiment and its objectives.

peak is measured by the instrument. Then the coded signal is transmitted from the lander to the orbiter with a delay equal to the one just measured. This is the way the system utilizes the lander as a reflector. The signal is received on the orbiter where it is coherently integrated and stored in the memory, and will be sent to the Earth.

As mentioned before, from the measurements of the time delay, we will be able to deduce the average properties of the cometary materials, starting with their dielectric constant. A report on the simulation results for the two-dimensional case has been published by Herique *et al.* (1999). Figure 10.9 shows a comparison between the dielectric constant obtained from the simulated measurements and the one used in

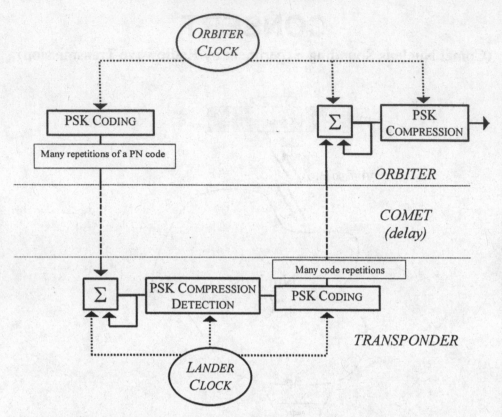

Figure 10.8 Block diagram of the CONSERT experiment. From Kofman *et al.* (1998).

the model of the comet. This is shown as a function of the angular relative position between the lander and the orbiter. The estimated dielectric constant corresponds to the average along the propagation path.

The estimate of the permittivity gives good results without too many artifacts and bias. The calculation is robust and requires a low a priori knowledge. It gives information on the internal large-scale structure of the nucleus and it allows the inversion of a stratified nucleus. We are currently developing a process for small-scale perturbations, which is based on an estimation of the correlation of the arrival time along the orbit. This correlation is interpreted with a model of propagation based on statistics of the perturbations.

Another method for inversion is described in the paper by Barriot *et al.* (1999). This method is closer to the one used in the ray tomography. Figure 10.10 shows the quality of the inversion using Tikhonov's inverse formulas (Tarantola 1987), which is a least-squares non-linear approach. Even the use only of direct rays allows an acceptable inversion. One can clearly distinguish the large structures inside the

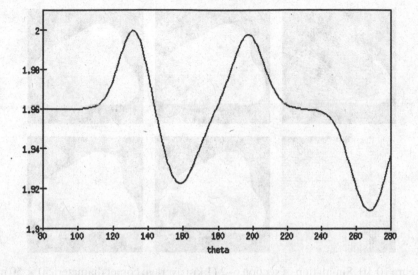

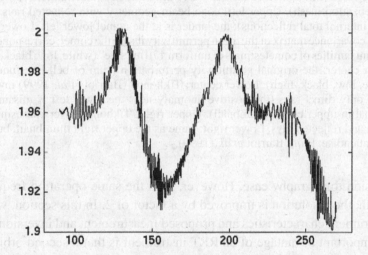

Figure 10.9 (a) Angular permittivity calculated by integration of the analytical model of nucleus. (b) Angular permittivity calculated by formula from the back propagated signal. From Herique *et al*. (1999).

nucleus. In this article it is demonstrated that reliable mapping (i.e., without false images) of the internal permittivity of comet Wirtanen is feasible in the context of the perturbation theory of geometrical optics.

3 Radio reflection tomography (RRT) approach

In this mode, data is collected (received) at the transmitter location. Assuming weak scattering case, the inversion technique is similar to the one described in the

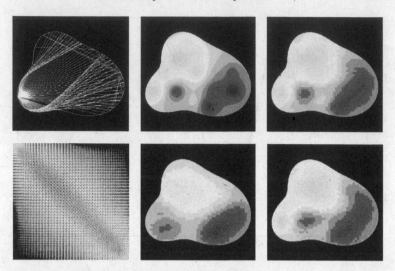

Figure 10.10 Simulation of section 6–2 (1 km average comet diameter, 50 × 50 m discretization mesh). Upper left: cone-beam geometry with reflected rays (up to three internal total reflections); the lander is at the comet lower left. Lower left: a priori covariance matrix of the mean permittivity inside the comet, corresponding to random families of cometesimals of uniform 170 m "size" (white, low; black, high). Upper center: the original permittivity perturbation (sum of bell-like functions) (white, low; black, high). Lower center: Tikhonov (Barriot *et al.* 1999) inversion using only direct rays (the positive anomaly near the lower left is missing; see original in upper center thumbnail). Upper right: Tikhonov inversion using both direct and reflected rays. Lower right: same as the upper right thumbnail, but with 5% data noise. From Barriot *et al.* (1999).

transmission tomography case. However, with the same operation frequency and bandwidth, the resolution is improved by a factor of 2. In this section, we provide RRT instrument characteristics and proposed measurement and inversion schemes.

One important advantage of an RRT instrument is that a second orbiter and/or lander is not necessary. Similar to the RTT, in order to be able to image the interior of the object, radar measurements should be made over a region encompassing the object with sufficiently dense sampling to ensure adequate sampling of the spatial spectral domain of the object. The sampling requirement depends on the radar's operation frequency and the object's angular extent as viewed by the radar. One of the main challenges in implementing an RRT system is obtaining the data with proper sampling. Another critical issue is the determination of the radar position with respect to the object. This is important since radio tomography relies on coherent combination of radar data. Currently, the radar positioning is a major driver in the design of a coherent tomographic imaging system and it also limits the maximum operation frequency (most likely <20 MHz). If the position accuracy required for the coherent processing cannot be met, the imaging will be done by combining

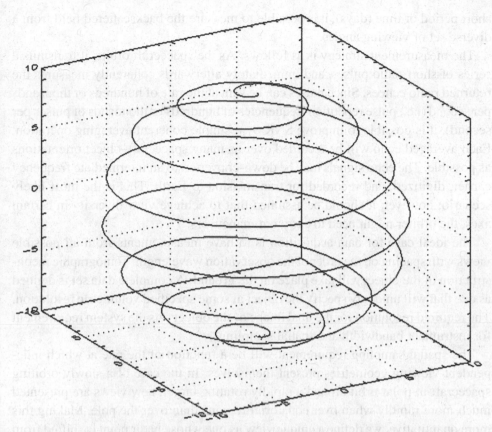

Figure 10.11 Spatial coverage during half an orbit for an object with a rotation period of 4 h and a spacecraft orbit period of 64 hs. The spiral curve is the sub-spacecraft point projected on a spherical surface in the target's rest frame. This demonstrates how a wide range of viewing geometries can be sampled from a polar orbit. (From Safaeinili *et al*. (2002).

the data after ignoring the carrier phase (incoherent imaging scheme). This is not ideal since the resulting image will be low resolution (defined by the signal band-width rather than carrier frequency).

As mentioned above, collection of a complete data set is crucial in producing a high-resolution and artifact-free volumetric image of the object. A promising measurement strategy is to set the spacecraft in a nearly polar orbit (the pole being defined by the target's primary axis of rotation). Small bodies have very weak gravity fields, and the orbital period of the spacecraft around the object (potentially days) will probably be long in comparison to the spin period of the object itself (typically 4 to 12 h). Due to the slow orbital velocity of the spacecraft and the object's rotation, the nadir direction from the spacecraft is a helix spanning from one pole to the other (Fig. 10.11), which samples a wide range of viewing geometries. Over a relatively short period of time (days), it is possible to measure the backscattered field from a diverse set of viewing angles.

The measurement strategy is as follows. As the spacecraft orbits, it transmits a series of short radio pulses, and immediately afterwards coherently measures the returned radio echoes. Since pulses can be sent at the rate of hundreds or thousands per second (i.e., pulse repetition frequencies of hundreds or thoudands of pulses per second), it is possible to improve SNR by a simple coherent averaging operation. Each averaged echo will be recorded over as many spacecraft–object orientations as possible. The echo signals may be down-converted to an intermediate frequency carrier, digitized, and recorded for transmission to Earth. This is the most likely scenario; however, it should be assumed that to achieve view diversity in certain axes, the orbiter might need to carry out maneuvers.

The ideal case for data acquisition is to have measurements from all possible views with spacing on the order of the observation wavelength. Tomographic reconstruction of the object will take place on the ground. A complete data set is defined as one that will uniquely specify the object at some specified volumetric resolution. The required resolution for the final image of the object sets the system requirement for instrument bandwidth and spatial sampling.

The spatial sampling requirement will be a function of the rate at which independent viewing geometries present themselves. In the case of a slowly orbiting spacecraft in polar orbit around a rapidly rotating target, new views are presented much more rapidly when over equatorial regions than over the pole. Making this more quantitative, we define a unique view as one whose nadir point is shifted from the previous view by half the observing wavelength. In this case, the sampling rate is a function of the sub-spacecraft latitude (measured from the spin equator), φ, the radius of the target, a, the observing wavelength, λ, and the angular velocity of the target's rotation, ω:

$$T \leq \frac{\lambda}{a} \frac{1}{2\omega \cos \varphi} \tag{10.23}$$

where T is the time between radio pulses. Using typical values (see Section 4.4) of $\lambda = 60$ m, $a = 2000$ m, $\omega = 2 \times 10^{-4}$ rad s^{-1}, over the equator we find a maximum required rate of one pulse every 75 s. This is easily maintained from both a power and data rate perspective. Near the poles, the required pulse repetition rate becomes much slower.

3.1 Inversion scheme

The data collected by an RRT or RTT system are the amplitudes and phases of each of the radio echoes from the object over the operational bandwidth. The data can be collected either in time in case of a pulsed system or frequency domain in case of a continuous wave (CW) system. These echoes are collected over as many orbits as possible to maximize the number of viewing geometries. The data do not need to be collected at the same distance from the object, so observations can be made as the orbiter reduces its distance to the target.

In this section, we discuss the general theory of how such observations can be inverted to determine the interior structure of the target body. In a later section, we will give examples of the inversion process and illustrate the impact of incomplete coverage on the image quality.

In general, the most accurate scheme for this type of inversion is the iterative approach since in general case the inversion model is non-linear in terms of the scatterer's volumetric properties. Iterations are done over the object's electromagnetic parameters using a conjugate gradient search approach (Safaeinili and Roberts 1995; Lin and Chew 1996) driven by the observed difference between model-predicted radio echoes and the actual measured radio data as given in the following equation:

$$\Delta(\xi) = \sum_x \sum_t \left(\frac{F(t, r; \xi) - D(t, r)}{\sigma(t, r)} \right)^2 \qquad (10.24)$$

In this equation, $F(t, r; \xi)$ is the model-predicted data set as a function of time, t, and spacecraft position vector relative to the target center of coordinate, r, for a target described by the parameter set ξ (e.g., dielectric constant distribution inside the object). $D(t, r)$ is the observed data set.

The parameter set ξ is defined throughout the volume of the object. The summation extends over all data samples, and is weighted by the uncertainty of each measurement, $\sigma(t, r)$. The difference function, or error in inversion, is given by $\Delta(\xi)$ which is a function of the body dielectric property map. The inversion is complete when $\Delta(\xi)$ is minimized and is equivalent to the noise level. The resulting ξ is the desired solution.

So far, due to the heavy computational load of inversion algorithms that use the more exact full wave propagation model, most of the work in the field has focused on the inversion of objects with sizes close to a wavelength. The inversion of larger objects has been carried out by using an approximate forward wave propagation model (e.g., Born–Rytov approximations or distorted Born) that ignores higher-order scattered fields (i.e., interaction between scatterers within the object). These fast approximate techniques have been successfully used in medical ultrasonic imaging, producing accurate and useful three-dimensional tomographic images.

The key to the effectiveness of these algorithms is that multiple-scattered echoes are, generally, much weaker in comparison with the direct scattering ones due to attenuation of the echo at every reflection (note that in most targets of interest, the reflection coefficient is usually much less than 1). Further, the increased propagation path of the higher-order scattering echoes cause additional attenuation. In most cases, the error introduced by ignoring the multiple reflections will manifest itself as part of the systematic noise in the image.

In cases where the largest contrast is at the boundary of the object, the error due to the multiple scattering will cause points outside the physical boundary of the object to have dielectric values different from that of free space. The presence of such an artifact is a good indication of the presence of multiple scattering.

It is clear that to achieve the best inversion result, an iterative approach should be used to update the field inside the object to account for the presence of multiple scattering. Future work is needed to address this issue by taking advantage of simplifications that can be obtained by using appropriate approximations (e.g., distorted Born approximation, and by the application of geometric optics, etc). Otherwise, the computational load of a full wave equation approach will be prohibitive.

The forward model computation of the predicted data set is the central component of the inversion process. The frequency of the transmitted radio waves and the size and electromagnetic properties of the object require a model that includes diffraction effects and possibly multiple scattering. Since all of the data processing and the tomographic inversion are carried out on the ground, it is possible to use a very detailed forward model. The development of a forward model for this problem will ultimately benefit from many years of work on electromagnetic wave propagation and scattering modeling.

Although the inversion scheme is the most important aspect of this problem, it should be emphasized that the data quality and information content in the RRT experiment data does not depend on our current ability to invert such data. As the inversion and computational resources improve, it will be possible to obtain better and more accurate inversion results using the same data, assuming that the data were obtained with high accuracy and sufficient sampling.

3.2 Radio reflection tomography inversion examples

In the following, we use simulated radar data from a set of two-dimensional objects to show the inversion concept. For this demonstration, we ignore internal attenuation and multiple scattering effects. More accurate modeling will be carried out in the future as part of our ongoing research into this technique. The main purpose for this section is to illustrate the capability and limitations of radio reflection tomography assuming that all interactions other than first-order reflections can be ignored.

Safaeinili *et al.* (2002) present a study in which they simulate radar measurements using the physical model described by Eq. 10.19, corresponding to the weak

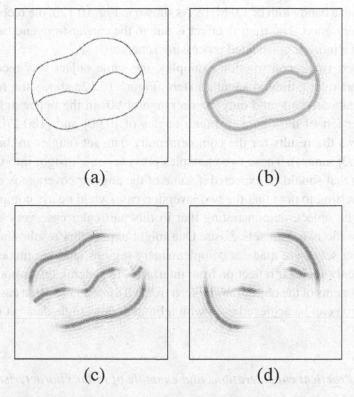

(a)   (b)

(c)   (d)

Figure 10.12 (a) The model used for "Twin-Rock"; the object includes an interface that runs through the length of the object. The object is approximately 4 × 4.5 km and the radar measurements were done at a circular orbit of 10 km from the center of the object. (b) This is the reconstruction of an object with crack in the middle of the object. The reconstruction was performed with data collected over the entire 360 angular range. (c) Object reconstruction using two 90° view-angle (second and the fourth quadrant). The surfaces that are parallel to the observation direction are not present in the reconstruction. (d) Reconstruction of the object with the two complementary 90° view angles relative to the previous example (first and third quadrant). From Safaeinili *et al.* (2002).

scattering case. The test object is a three-dimensional body that lies in a plane with a crack through it. Radar data were simulated at equal angular separations over a circular orbit, in the plane of the object, with a 10 km radius from the center of the object. For each radar position, data were generated over a relatively broad bandwidth around a prescribed center frequency. In practice, such data is obtained by using either a linearly frequency modulated signal or a pulse that contains a few cycles of the carrier frequency.

The object (Twin-Rock), 4–5 km in length, was reconstructed using the wave back-propagation technique with data collected over the full 360° view-angle with angular sampling of 0.5° (Fig. 10.12a). The center frequency of the radar was

2.0 MHz with a bandwidth of 1.0 MHz. As shown in Fig. 10.12b, the reconstruction quality is very good. The ringing effect is due to the carrier frequency and can be removed by a more sophisticated processing scheme.

In the next two reconstruction examples, the same object was reconstructed by using data over a limited angular extent. Figure 10.12c shows the reconstruction when data are collected only over a range of 90° in the upper right and 90° in the lower left of the figure (azimuth angles of [0,90] and [180,270]). Figure 10.12d shows the results for the complementary data set (angles in the [90,180] and [270,360] ranges). These two examples provide some insight into the type of degradation that should be expected if some of the angular coverage is missing.

It is interesting to note that the two inversion cases yield nearly complementary images of the object, demonstrating that in this particular case very little overlap between the two data sets exists. One might expect this result since the two measurement sets were made at complementary regions, and the model assumes essentially only specular reflection from interfaces. In general, this is not true since some components of the object may be seen from all directions. In that case, a partial reconstruction can be achieved even with a limited view-angle data set (Safaeinili *et al.* 2002).

3.3 Practical considerations and example of radar characteristics

A radio reflection tomography instrument would consist of one or more radar systems located on a spacecraft. The operating frequency needs to be in the low to mid high-frequency band (e.g., 1–20 MHz) to provide deep penetration into the asteroid material.

In a mono-static system, such as RRT, the range resolution, Δr, is given by $\Delta r = c/2Bn$ (Eq. 10.4), where c is the speed of light in free space, B is the bandwidth, and $n = \sqrt{\varepsilon_r}$ is the refractive index of the medium. One assumes $\mu_r = 1$.

The resolution in the plane normal to propagation vector (lateral resolution) is given by dimension of the first Fresnel zone $\Delta x \propto \sqrt{\lambda R}$ (Eq. 10.8), where λ is the wavelength and R is the range to the target (e.g., $R = 10$ km for this case). For a resolution on the target of 20–100 m, it is therefore necessary to have a bandwidth of 1 to 5 MHz.

To explore a potential design of an RRT system for small bodies, and taking the above considerations into account, we consider a radar system that operates in the 2 to 20 MHz range, with an adjustable instantaneous bandwidth of 1 or 2 MHz, generated as either a short pulse or chirp transmission. The antenna for this radar can be a 30 m dipole similar to the one used for MARSIS (Picardi *et al.* 1998).

The transmitted pulse illuminates the object and the reflected energy is received at the same location by the same or a different antenna. This received energy is amplified and down-converted to a lower frequency prior to conversion to digital form to minimize data volume and reduce sensitivity to frequency stability. Due to the slow rotation of the object and low operation frequency, there is sufficient time to coherently integrate many pulses (possibly thousands) to increase the signal-to-noise ratio. These data are then stored in the orbiter for transmission to Earth. When the orbiter's position relative to the object has moved sufficiently, this process is repeated until sufficient data have been acquired around the object.

In order to obtain the highest resolution image of the object's interior, the coherency between measurements at different positions and times needs to be preserved. Generally, this is a difficult task since a number of factors, such as position knowledge accuracy and frequency variation due to oscillator drift, degrade the coherency between independent measurements. The requirements are more relaxed at lower frequencies due to a longer wavelength that makes the positioning accuracy less stringent. The required positioning accuracy for a tomography measurement can be estimated as $\lambda/10$, where λ is the wavelength.

The entire data, collected by the RRT system, is down-linked to the ground for analysis and inversion. The data volume for an RRT system will depend on both the operation frequency and the bandwidth. The operation frequency and size of the object define the number of measurement positions around the object (see Eq. 10.23) and the bandwidth defines the sampling frequency and hence impacts the number of samples per echo. Assuming a target radius of ~1 km, a combined 5 MHz system with a bandwidth of 1 MHz as sketched out above would need to collect less than 1 gigabits of data to fully characterize the target interior at full instrument resolution.

Figure 10.13 shows the expected power of coherent and incoherent received signals versus depth or equivalent propagation time. The measurement is simulated for a rough interface modeled by a random Gaussian distribution of heights defined by its height standard deviation (σ_h) and its slope root mean squares (rms). The simulation includes a complete link budget, the antenna gain, and the signal processing (i.e., clutter reduction, coherent integration, and compression). All the field components are estimated as a function of time using different theoretical or empirical approaches.

The surface echo coming from the Fresnel zone is the first received signal corresponding to the star in the figure at the null depth. It is incoming from the nadir direction which corresponds to maximum antenna gain. This power depends on the Fresnel reflection coefficient modified by the roughness attenuation. This one is estimated using the classical Kirchoff approximation (Ishimaru 1978). The surface backscattering coming from the whole antenna pattern is spread in time and corresponds to both curved lines in the figure. The backscatter coefficient is estimated for a given surface statistic using two methods:

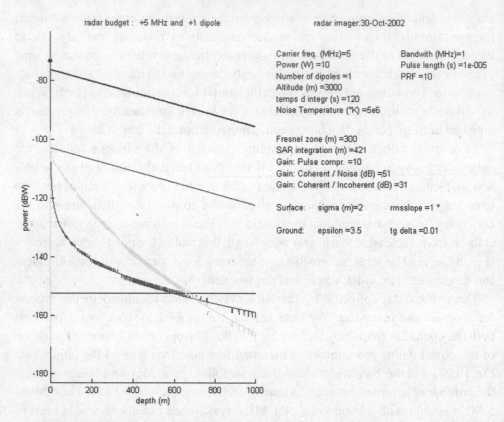

Figure 10.13 Radio reflection tomography (RRT) instrument penetration simulations are shown. The radar parameters are: central frequency of 5 MHz, the signal bandwidth of 1 MHz, and the dipole antenna. The signal was integrated over 1200 pulses; surface clutter and galactic noise are indicated in the figure.

(1) The light gray curve is based on theory (Ishimaru 1978). The backscattering coefficient is then the sum of the small perturbation model and of the Kirchoff first-order model. The small perturbation model is given by an electromagnetic field development at large incidence angles and tends to underestimate the component near nadir. The Kirchoff · first-order model is a classical approximation in the geometrical optics for rough enough surfaces at low incidence angles.

(2) The lower (black) curve is based on an observational law: the backscatter (reflection) coefficient is given by the semi-empirical Hagfors law (Hagfors 1968). The continuous horizontal line corresponds to the galactic noise level after processing.

The deep echoes coming from internal layers or from the bedrock corresponds to the two parallel straight lines. The graphs show the expected power versus depth for two kind of interfaces: a surface transition (−10 dB reflection coefficient) for the bedrock reflection (−10 dB reflection coefficient) and a very low internal transition with

Table 10.2 *Radar characteristics and requirements*

Radiated power	≥10 W
Frequency (limited by position accuracy requirement)	<20 MHz (for tomography purpose)
Pulse repetition frequency	10–100 Hz (due to slow orbital speeds PRF can be low in comparison to planetary sounders)
Bandwidth	>2 MHz (~35 m in comets and asteroids)
Antenna	Dipole (or cross-dipole)

Other requirements:
 – Precise measurement of spacecraft's position in object fixed coordinate system (i.e., $\lambda/10$)
 – Requires viewing the object at a large number of object/TX/RX orientations, for RTT and a large number viewing angles for RRT (large number is defined as sufficient to provide complete coverage of the k-space)
 – In cases where an adequate k-space coverage is not feasible, a precise determination of object's shape will provide good constraint for tomographic inversion
 – Object cannot be metallic (M-type) and needs to have sufficiently low absorption to allow at least a partial penetration of the radio waves

a -37dB reflection coefficient. The estimate of the power includes the roughness effect on interfaces (in transmission for the surface, in reflection from the deep interface), and the dielectric absorption (0.01 loss tangent).

From Fig. 10.13 one can see that the radar of 5 MHz frequency with 1 MHz bandwidth can penetrate more than 1000 m and detect internal transitions. This calculation was made for a relatively smooth surface ($\sigma_h = 2$ m and rms slope of $1°$).

We have shown that the RRT instrument is an efficient method to probe the interior of small bodies. The general characteristics of such radar systems are summarized by the general parameters and requirements in Table 10.2.

4 Final remarks

In this chapter, we have explained the principles of radio tomography and its possible applications for planetary exploration. The tomography technique is now widely in use in everyday life; medicine, airport controls, seismic research, etc. The planetary applications of this technique are the new and future field. The CONSERT experiment is on the Rosetta mission spacecraft and will be launched in 2004. MARSIS is the radar for the investigation of Mars subsurface on the Mars Express mission, and will begin operation in 2004. A proposal for radar studies of Europa, Jupiter's satellite, was prepared, but it is presently on hold. New projects are in preparation for exploration of asteroids in the NEO programs. The development of radar tomography is important because it is probably the most powerful technique for exploring the interior of solar system bodies. We have presented a possible technical solution for low-frequency radar, the choice of frequency being determined by the absorption properties of materials, the radar parameters, and the constraints on the orbits. The processing of the data and the volumetric image reconstruction is challenging but is well within the available computational resources on the ground. Radio tomography in space is in its early stages but it has the potential to contribute to planetary exploration efforts in a significant way.

Acknowledgments

The part of this chapter relating to the RRT system was carried out for the Deep Interior Discovery proposal, submitted to the National Aeronautics and Space Administration in August of 2000. The research described in this chapter was carried out by the Jet Propulsion Laboratory, California Institute of Technology under a contract with the National Aeronautics and Space Administration and by Laboratoire de Planétologie de Grenoble supported by Centre National de la Recherche Scientifique and Centre National d'Etudes Spatiales.

References

Barbin, Y., Kofman, W., Nielsen, E., et al. 1999. The CONSERT instrument for the ROSETTA mission. Adv. Space Res. 24, 1115–1126.

Barriot, J.-P., Kofman, W., Herique, A., et al. 1999. A two-dimensional simulation of the CONSERT experiment (Radio tomography of comet Wirtanen). Adv. Space Res. 24, 1127–1138.

Born, M. and Wolf, E. 1970. Principles of Optics. New York: Pergamon Press.

Gudmandsen, P., Nilsson, E., Pallisgaard, M., et al. 1975. New equipment for radio echo sounding. Antarctic J. US 10, 234–236.

Hagfors, T. 1968. Relations between rough surfaces and their scattering properties as applied to radar astronomy in Radar Astronomy, eds. J. V. Evans and T. Hagfors, pp. 187–216. New York: McGraw-Hill.

Henderson, F. and Lewis, A. 1998. Principles and Applications of Imaging Radar. New York: John Wiley.

Herique, A., Kofman, W., Hagfors, T., et al. 1999. A statistical characterization of comet nucleus: inversion of simulated radio frequency data, Planet. Space Sci. 47, 885–904.

Herman, G., Tuy, H., Langenberg, H., et al. 1987. Basic Methods of Tomography and Inverse Problems. Philadelphia, PA: Adam Hilger.

Ishimaru, A. 1978. Wave Propagation and Scattering in Random Media. New York: Academic Press.

Kofman, W., Barbin, Y., Klinger, J., et al. 1998. Comet nucleus sounding experiment by radiowave transmission. Adv. Space Res. 21, 1589–1598.

Lin, J. H. and Chew, W. C. 1996. Three-dimensional electromagnetic inverse scattering by local shape function method with CGFFT. In Proceedings of the 1996 AP-S International Symposium and URSI Radio Science Meeting, July 21–26, 1996, Baltimore, MD, pp. 2148–2151.

Picardi, G., Plaut, G., Johnson, W., *et al.* 1998. The subsurface sounding radar altimeter in the Mars Express mission. Proposal to the European Space Agency, Infocom, no. N188-23/2/1998.

Safaeinili, A. and Roberts, R. A. 1995. Support minimized inversion of incomplete acoustic scattering data. *J. Acoust. Soc. America* **97**, 414–424.

Safaeinili, A., Gulkis, S., Hofstadter, M., *et al.* 2002. Probing the interior of asteroids and comets using radio reflection tomography. *Meteor. and Planet. Sci.* **37**, 1953–1964.

Sihvola, A. H. and Kong, J. A. 1988. Effective permittivity of dielectric mixtures. *Geosci. Remote Sens., IEEE Trans.* **26**, 420–429.

Tarantola, A. 1987. *Inverse Problem Theory: Methods for Data Fitting and Model Parameters*. New York: Elsevier Science Publications.

von Hippel, A. 1954. *Dielectric Materials and Applications*. Cambridge, MA: Technology Press of MIT.

Zernov, N. and Lundborg, B. 1993. The statistical theory of wave propagation and HF propagation in the ionosphere with local inhomogeneities. IRF Scientific Report, Uppsala, Sweden.

11

Seismological investigation of asteroid and comet interiors

James D. Walker

Southwest Research Institute

Walter F. Huebner

Southwest Research Institute

1 Overview

Understanding the interior structure and composition of asteroids and comets is important for understanding their origin and evolution. In addition to basic science objectives, understanding the interior structure of near-Earth objects (NEOs) will be essential to addressing mitigation techniques should it become apparent that such an object has the potential to impact Earth. NEOs are comprised of asteroids and comets in near-Earth space. Work is progressing to find, catalog, and determine the orbits of NEOs larger than 1 km, but little is known about NEOs' bulk properties, such as strength and structure (Huebner *et al.* 2001; Greenberg and Huebner 2002). Should a Potentially Hazardous Object (PHO) threaten Earth, attention would focus on countermeasures. All conceived countermeasures rely on knowledge of the bulk material properties of NEOs, in particular material strength, structure, mass distribution, and density. It is believed that NEO compositions range from nickel–iron through stony and carbonaceous to ice-and-dust mixtures. Their structure can be monolithic, fractured, assemblages of fragmented rock held together only by self-gravity ("rubble piles"), porous, or fluffy. Types of collision mitigation and countermeasures will vary widely depending on composition and structure.

Much of the lack of knowledge of the interior properties of NEOs is due to the fact that most study has been by remote sensing. Information such as rotation rate and shape can be determined remotely through radar and light curve analysis, but determining the geophysical and geological properties requires something more. Two techniques of imaging the interior are apparent: radio tomography (electromagnetic waves) and seismology (mechanical waves). Both techniques require sensors in the vicinity of or touching the object. Both techniques can provide global

Mitigation of Hazardous Comets and Asteroids, ed. M. J. S. Belton, T. H. Morgan, N. H. Samarasinha, and D. K. Yeomans. Published by Cambridge University Press. © Cambridge University Press 2004.

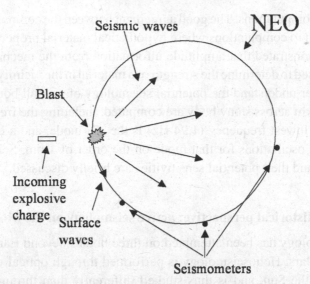

Figure 11.1 Scenario for seismology on an NEO: either an impactor or more likely an explosive is used as a seismic disturbance for distributed seismometers.

information about the object. This global information about the object from radio tomography and seismology is to be contrasted with the local information that is obtained though techniques such as surface and near-surface sampling. For an account of radio tomography see Chapter 10; this chapter focuses on mechanical means (seismology) of determining mechanical properties of the object.

Properties such as density, stiffness, and strength can be investigated through analysis of mechanical (seismic) waves. With the careful placement of a small number of seismometers on the surface, it should be possible to record property-discriminating wave traces (Fig. 11.1). These wave traces will allow a determination of the structure of the NEO. There are two methods for producing the seismic disturbance: impact and explosives. Both methods produce a seismic pulse that is then propagated throughout the body by mechanical waves. At a distance, it is possible to quantify the seismic impulse from both sources, and in some cases they are indistinguishable. Though impactors are a potential seismic source (and were used in the Apollo missions), it is more likely that explosives would be used. To this end, there are various types of charge geometries available, with different strengths and weaknesses as to producing a seismic pulse. The best choice depends on the surface where the charge is to be used.

This chapter reviews the Apollo active seismology program and then describes a quantification of the seismic source applicable to both an explosive and impactor. Cased and bare charges as sources are then compared. Experiments were performed with aluminum and a foamed aluminum to examine the role of charge depth and to compare with computations. The good agreement between the computations and the experiments led to computations where hypothetical material properties were used and it was demonstrated that amplitude information from the mechanical (sound) waves can be used to determine the strength of a material in the vicinity of the source. Finally, to better understand the potential seismology of a small body, the normal modes for a 1-km across stony body are computed, including the frequencies. The mode with the lowest frequency (1.94 Hz) is the $_0S_2$ mode, and a charge of 1 kg would produce oscillations for that mode on the order of 1 nm. Seismometers in micro-gravity and their potential sensitivities are briefly discussed.

2 Historical perspective: active seismology on the Moon

To date, seismology has been attempted on three bodies beyond Earth: the Moon, the Sun, and Mars. Helioseismology is performed through optical observation of the surface of the Sun, and is thus studied differently than through mechanical seismometers that would be used on a comet or asteroid (Brown 1994). As to Mars, the Viking landers carried seismometers, but on Viking 1 it failed to uncage and on Viking 2 it appeared no information from Mars ground motion was obtained (the seismometer was mounted on top of the spacecraft and it appeared there was a strong correlation between wind and seismometer signal due to motion of the lander in the wind).

As to the Moon, seismology on the Apollo missions was very successful. On four of the Apollo flights (12, 14, 15, and 16) passive seismometers were placed on the Moon and three of the flights (14, 16, and 17) included active seismographic experiments (Kovach *et al.* 1971, 1972, 1973; Latham *et al.* 1971, 1972, 1973).

There are two seismic sources, explosives and impactors, and the Apollo mission used both. The impactors employed were the Saturn IV B upper stage rocket body and the lunar module ascent stage. The masses of the former were on the order of 14 000 kg and the latter 2350 kg. Impact velocities were roughly 2.5 km s^{-1} for the Saturn IV B upper stages and 1.7 km s^{-1} for the lunar module ascent state. Based on information from these impacts, it was concluded that at a depth of 6+ km, the lunar material's primary wave sound speed is 5–6 km s^{-1}. By the end of the Apollo program, nine spacecraft impacts had been recorded seismically, ranging in distance from 67 to 1750 km from the seismometers.

In addition to the impacts, three Apollo flights included direct active seismo- logical experiments. Apollo 14 and 16 carried mortar packages that were placed on the moon and contained four grenades. After the astronauts' departure, these grenades were to be launched to distances of 150, 300, 900, and 1500 m and had explosive loads of 695, 775, 1024, and 1261 g. However, there was concern about dust formation interfering with other experiments on Apollo 14 and so the grenades were never launched. On Apollo 16, three of the four grenades were launched, but the resulting data were ambiguous due to interference with the seismic signal generated by the launch of the grenades. The fourth grenade was not launched because of an apparent tilting of the mortar package in the previous launch.

Much more successful was a hand-held thumper utilizing Apollo standard initiators (exploding bridgewires) that the astronauts could place on the lunar surface and produce small disturbances. Three geophones stretched out in a straight line of length 90 m (one geophone at each end and one in the middle) recorded arrival times (less than a half a second since distances were so short). On Apollo 14 the regolith layer was determined to have a sound speed of 104 m s^{-1} and to have a thickness of 8.5 m, and the underlying layer directly below the regolith was determined to have a sound speed of 299 m s^{-1}. These signals provided information to a distance of 40–45 m from the geophones. Beyond that, the measured signal often was too weak to provide a clear arrival time against the background noise. On Apollo 16, the surface-layer sound speed was determine to be 114 m s^{-1} by the thumpers, and this surface-layer sound speed result was combined with information from the grenades and the lunar module ascent to allow a determination of the local surface structure of the Moon in the vicinity of the Apollo 16 Descartes landing site. In particular, the lunar regolith thickness was estimated to be 12.2 m (with sound speed 114 m s^{-1}) and beneath that was a layer with a sound speed of 250 m s^{-1}.

Apollo 17's active seismology experiment involved the placement of explosives by the astronauts during their exploration of the site. A small box was placed on the ground and its location recorded. The box contained explosive and communications equipment to allow the explosive to be detonated after the astronauts had departed for Earth. On three extravehicular activities (EVAs) the astronauts placed eight charges (Table 11.1). An array of four geophones, with a maximal distance between geophones of 50 m, was placed on the lunar surface. Beginning the day after the astronauts' departure, over the next 4 days the explosives were detonated and the traces from the geophones recorded. The data for Apollo 17's Taurus–Littrow landing site are shown in Fig. 11.2 (Kovach *et al.* 1973).

Based on the results of the tests it was estimated that the 250 m s^{-1} layer has a thickness of 248 m and beneath that there is a layer with a sound speed of 1200 m s^{-1}. (If the interface between the two layers is assumed flat, then elementary calculus shows that the thickness of the surface layer is given by

$$T = \frac{x}{2}\sqrt{\frac{c_2 - c_1}{c_2 + c_1}} \qquad (11.1)$$

where x is the distance at which the two sound speed time vs. distance lines cross, and c_1 is the sound speed in the surface layer and c_2 is the sound speed in the lower layer. From Fig. 11.2, $x = 612$ m, hence the layer thickness is 248 m.) It is

Table 11.1 *Explosive charges used by Apollo 17*

Explosive package	Explosive mass (g)	Approximate distance to geophones (m)
EP-8	113	100
EP-4	57	225
EP-3	57	260
EP-2	113	375
EP-7	227	750
EP-6	454	1200
EP-5	1361	2325
EP-1	2722	2700

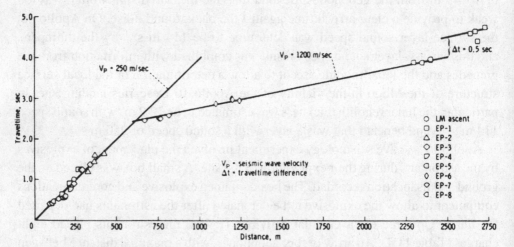

Figure 11.2 Travel times for waves produced by the eight charges deployed by Apollo 17. From Kovach *et al.* (1973).

speculated that the data points in the upper right-hand corner of the plot experienced a greater than expected travel time due to the presence of Camelot crater along the travel path. The impact forming the crater could have forced the lower velocity layer down further, thus slowing travel through the otherwise deeper, faster layer and delaying the arrival of the seismic wave. Based on the Apollo 17 lunar module impact, there is an even deeper layer with roughly 4000 m s^{-1} sound speed.

On Earth, water in the ground dissipates propagating waves, and such dissipation does not occur on the Moon. Hence, it is thought that the passive seismometers left on the Moon could detect a 7 to 10 kg meteoroid impact anywhere on the Moon. Meteoroid impact mass estimates are based on the measured amplitude of the waves vs. the amplitudes of the spacecraft impacts where the masses are known (Latham *et al.* 1972). The smallest meteoroid impact detected is estimated to be on the order of 100 g.

The Apollo experience showed that information about the lunar surface could be obtained from seismological techniques. The best data came from well-characterized explosive charges and impacts, and most information relied on initial arrival times. The inferred wave speeds indicate the cohesiveness: the regolith has a very low sound speed and is therefore unlikely to be cohesive, whereas the lower rock has a higher sound speed and is likely to be cohesive. Cohesiveness is an important issue with regard to asteroid and comet nucleus characterization.

Analysis of both the active and passive seismometry data from the Moon has provided some idea of the material properties of the Moon. For example, once one gets below the regolith and low sound speed zone, inferred values for the lunar upper crust are very roughly: density 3.35 g cm^{-3} (Hood 1986), a longitudinal wave speed $\alpha \sim 5.0$ km s^{-1} (Nakamura 1983) and Poisson's ratio $1/4$ (Goins *et al.* 1981).

3 Comparison of impactors and explosives as seismic sources

The seismology community's approach to quantifying loading sources has been the introduction of the seismic moment. This section will describe how the traditional notion can be extended to the loading condition of an explosion or impact on the surface, and then present computations with explosives and impacts to provide an estimate of the seismic load that can be delivered by an explosive or with an impact.

Essentially, the seismic moment is a force couple tensor that is an interior body force, with the terms of the tensor given by the force multiplied by the distance apart:

$$M = F_x \Delta x \tag{11.2}$$

It is possible to define the moment in the plane tangent to the surface, $M_{rr} + M_{\theta\theta}$, and it is possible to compute the moment with large-scale numerical simulations of either the explosive event or the impact event. In contrast, typically earthquakes occur within the Earth, and the seismic source can be completely described by a moment. However, for an explosive or impact at the surface there is an unbalanced force (not a moment) that also must be included. The momentum downward due to the impact or explosive charge is straightforward to determine from a calculation of the event:

$$p_z = \int \rho v_z dV \tag{11.3}$$

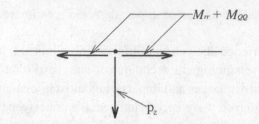

Figure 11.3 Schematic of downward momentum and tangential moment.

where p is the momentum, dV is the volume element, and the integral is taken over a region large enough to include all the material motion due to the seismic source. This integral gives the amount of momentum that is imparted to the body by the explosive or by the impact. Walker (2003) describes how to compute these two values.

Based on theoretical grounds, it is possible to show that the following ratio gives a measure of how spherical the loading is:

$$\frac{M_{rr} + M_{\theta\theta}}{2\alpha p_z} \tag{11.4}$$

For a completely buried spherical charge, this ratio equals 1 (Walker 2003). This ratio is displayed in the computations below to provide some sense of how much of the load is transferred into the sides of the crater vs. down into the crater formed by the impact or explosion. If the ratio is greater than 1, the implication is that more load is going into the lateral direction than is being delivered downward into the body. Almost all loadings and impacts produce such a loading. From a seismic point of view, where the measurements are being made at a sufficient distance from the load, the parameters $M_{rr} + M_{\theta\theta}$ and p_z completely define the load (Fig. 11.3). Thus, these two terms quantify both seismic sources.

To compute the downward momentum and radial moment terms for some specific cases, the hydrocode CTH (McGlaun et al. 1990) was used. For the explosive cases, 100 g of Comp C-4 was modeled with the Jones–Wilkins–Lee (JWL) equation of state. An $L/D = 1$ cylinder of explosive was used, with $D = 4.30$ cm. The detonation was handled by programmed burn, with the detonation always beginning at the top of the explosive, opposite the surface to be loaded. The target material was modeled as 6061-T6 aluminum, using literature values for the Mie–Grüneisen equation of state and the Johnson–Cook constitutive model. The reason for using aluminum in these calculations is that its post-yield behavior is well characterized. For an actual asteroid or comet nucleus loading calculation, a more relevant constitutive model should be used; however, the post-yield behavior of rock materials varies considerably. The view taken here is that loading an aluminum target gives a qualitative

Table 11.2 *Results of explosive computations*

Explosive mass (g)	Depth of placement	$M_{rr} + M_{\theta\theta}$ (dyn cm)	p_z (g cm s^{-1})	$(M_{rr} + M_{\theta\theta})/(2\alpha p_z)$
100	0 (on surface)	1.7×10^{13}	0.95×10^7	1.36
100	$D/2$ (half embedded)	5.3×10^{13}	1.30×10^7	3.1
100	D (fully embedded)	8.6×10^{13}	1.35×10^7	4.8
100	$3D$	1.18×10^{14}	0.79×10^7	11.4
50	$D/2$ (symmetric half space)	1.51×10^{14}	1.18×10^8	0.97

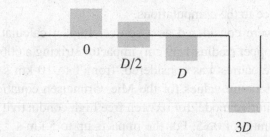

Figure 11.4 Different explosive depths used in computations.

idea of the response of explosives vs. impactors, and experience has demonstrated that CTH provides good results for both impact and explosive-loading problems involving aluminum. Constitutive values for this aluminum were a bulk sound speed (the zero-particle-velocity intercept of the particle velocity-shock velocity Hugoniot curve) $c_0 = 5.35$ km s^{-1} and Poisson's ratio $v = 0.33$, giving a longitudinal elastic wave speed $\alpha = 6.58$ km s^{-1}. In the computations, a mesh of 210 by 350 was used, with square zoning in the vicinity of the explosive with ten cells across the radius of the explosive ($\Delta r = \Delta z = 0.215$ cm). The aluminum target block was 150 cm in radius and 150 cm thick. The explosive was placed at various depths in the aluminum target: from sitting on top, half embedded, fully embedded, and three diameters down (Fig. 11.4). Results are given in Table 11.2. It can be seen that embedding the explosive (so that the top of the explosive is flush with the surface of the target) increases the downward momentum by 40%, but more strikingly it increases the radial moment term by a factor of 5. Thus, there are large gains relative to the seismic load by placing the charge deeper within the target. For the charge placed down $3D$, there is a decrease in downward momentum because now the charge is opening a cavity and pushing up on target material as well as down. Finally, the same calculation was performed where instead of a free surface a rigid surface was used, providing a reflecting boundary. The computation led to a value

Table 11.3 *Results of impact computations*

Copper impactor mass (g)	Impact velocity (km s^{-1})	$M_{rr} + M_{\theta\theta}$ (dyn cm)	p_z (g cm s^{-1})	$(M_{rr} + M_{\theta\theta})/(2\alpha p_z)$
100	1	1.24×10^{13}	1.05×10^7	0.90
100	3	1.13×10^{14}	3.4×10^7	2.5
100	5	3.20×10^{14}	7.3×10^7	3.3
100	7.5	6.73×10^{14}	1.3×10^8	3.9
100	10	1.05×10^{15}	1.9×10^8	4.2

for the ratio of 0.97, very close to the theoretical value of 1 stated above. This result gives added confidence to the computations.

Next, impactors were considered as a source. Impact calculations were performed for a 100-g copper (radius 1.39 cm) impactor striking a 6061-T6 Al target. A range of impact velocities was considered, from 1 to 10 km s^{-1}. The copper was modeled with literature values for the Mie–Grüneisen equation of state and Johnson–Cook constitutive model for oxygen free high conductivity (OFHC) copper with a Poisson's ratio of 0.35. For the impacts up to 5 km s^{-1}, the aluminum target was 150 cm in radius and 150 cm thick. The zoning was 270 by 320, with square zoning in the vicinity of the impactor with ten cells across the radius of the impactor ($\Delta r = \Delta z = 0.13$ cm). For the two higher-velocity impacts, it was necessary to increase the size of the target to achieve convergence to 200 cm in radius by 200 cm thick, and the zoning was increased to 300 by 350, with the same size zoning in the vicinity of the impactor as for the lower velocity impacts. Table 11.3 presents the results of the computations. The moment term is nearly scaling with velocity squared. A least-squares fit through the data points yields

$$M_{rr} + M_{\theta\theta} = 1.29 \times 10^{13}[v/(1 \text{ km s}^{-1})]^{1.95}\text{dyn cm} \qquad (11.5)$$

A least-squares fit through the momentum points yields

$$p_z = 0.97 \times 10^7[v/(1 \text{ km s}^{-1})]^{1.27}\text{g-cm s}^{-1} \qquad (11.6)$$

The superlinear dependence on velocity is due to the added momentum imparted to the surface as debris from the surface and impactor is thrown backwards.

It is seen that a 100 g copper impactor striking the aluminum surface at 1 km s^{-1} gives a seismic signal quite similar to 100 g C-4 charge sitting on the surface. Adjusting the impactor for an exact match, using the curve fits in Eqs. 11.5 and 11.6, a 56 g copper impactor striking at 1.55 km s^{-1} gives the same moment and downward momentum terms as 100 g of C-4 sitting on the surface (both the moment and momentum scale linearly with explosive charge mass and impactor mass). Such an impact would be seismically indistinguishable (from a distance) from the explosive detonation of 100 g of C-4 on the surface. The kinetic energy of the striking round is 6.7×10^{11} erg, while the internal energy of 100 g of C-4 is 5.6×10^{12} erg, based on a JWL detonation energy $\rho E_0 = 5.62 \times 10^{10}$ erg g^{-1}. To match the same moment-to-momentum ratio as seen for the fully embedded explosive (flush with the surface) would require an impact speed above 10 km s^{-1}. However, choosing to look at the 10 km s^{-1} impact and matching the moment, the mass of the striking copper impactor would be 8.2 g. The kinetic energy of the impactor is 4.1×10^{12} ergs, where again the energy in 100 g of C-4 is 5.6×10^{12} ergs. The question of using energies to compare impacts and explosions clearly relies on the coupling of the explosive to the surface: in general, one cannot match the kinetic energy of the impactor with the chemical energy of the explosive and expect the same seismic load. Also, the computed moment does depend strongly on the target material, with softer material resulting in a smaller moment for the same explosive charge (Hannon 1985). The examples here were with aluminum because it is a material that is well characterized. They will be used to provide a qualitative input loading value for the ringing-sphere problem analyzed below.

The cost for either explosives or impactors as seismic sources is the cost of the space mission to move material to the NEO site. The explosives must essentially be soft-landed: impacted at less than 1–2 km s^{-1} (most likely at less than 100 m s^{-1}), so the Δv from Earth is similar to a soft landing. There is no point in impacting an explosive at higher velocities. Due to the soft landing, the explosive location would be precisely known. Knowing the exact location of the source considerably aids in solving the inverse problem of determining wave speeds and their domains within the asteroid or comet body. Impacting projectiles could be at any obtainable velocity: for example, 10 km s^{-1} is likely not to be a problem for an impact due to the relevant orbital mechanics. However, at higher velocities there is more uncertainty as to the impact point and so guidance and navigation of the impactor become of prime importance. From a mission scenario point of view, there also needs to be time to deploy the seismometers and it is desirable to have a large amount of time (hours) between impacts or explosive events so that the asteroid or comet body can ring down, reducing the background noise, making it much easier to identify the seismic motion due to the source. Also, a long time between tests is desired to listen for modal ringing in the body.

4 Cased charges vs. bare charges

Explosives are solids that phase change to high temperature gaseous detonation products at a rapid rate. The detonation wave travels through a solid explosive at a speed of 6–8 km s^{-1}. The gaseous explosive products have extremely high pressures, and drive material surrounding them. Either the charge can be bare, so that the expanding gases drive the asteroid's or comet nucleus's surface material directly to produce the seismic pulse, or the charge can be cased in a material and the load can be transferred through impact of the casing material as well as the gaseous products. It is likely that for the seismic applications the charges would be lightly cased. However, there are scenarios where heavier and/or shaped casing of the charge may be desired. For example, if the surface had a thick layer of porous ice, then it would be necessary to transmit the load deeply into a more (presumably) highly consolidated interior. Of course in this scenario it is also necessary to have seismometers that can detect the motion of the more consolidated layers.

Typically, the casing material on an explosive is a metal such as steel or copper. Explosives are very efficient at driving metals. There are various forms of cased charges, from cylindrical casing around the outside, where the case fragments into pieces that are thrown in all directions, to more directional techniques such as explosively formed projectiles (EFPs) where a single fragment is launched in a predefined direction at velocities on the order of 3 km s^{-1}, to conical shaped charges, where a solid stretching metal jet is formed, with the front tip traveling at (for example) 8 km s^{-1} and the back traveling at 4 km s^{-1} (the jet subsequently particulates and the resulting particles then travel at their respective velocities based on the original distribution). These various configurations could conceivably have differing applications or scenarios for producing a seismic pulse. The downside of using a metal casing is that it is massive – there is a trade-off between weight of explosive and the weight of the casing.

To give a qualitative idea of the results of the trade-off between explosive mass and casing mass, the geometry of a spherically symmetric charge within an evacuated spherical cavity in ice was considered. This geometry is idealized, but it allows a simplified analysis. The explosive chosen for the studies was a plastically bonded RDX referred to as PBX-9010. This explosive is 90% RDX mixed with plasticizers to make it castable. It is a common explosive used in military ordinance, and is considered relatively insensitive. Thus, it seems this explosive (or the slightly better and much more expensive HMX) is a possibility for actual use in a space system. As a baseline mass, 1 kg was assumed. For PBX-9010, this mass corresponds to a sphere of radius 5.1 cm. Keeping the total mass of the charge the same, including a copper case reduces the explosive mass. Table 11.4 shows the relationship between different thicknesses of copper case, the resulting charge mass and copper mass, and the copper launch velocity based on a Gurney formula (the Gurney formula is an experimentally calibrated energy balance)

$$v = \sqrt{2E} \left(\frac{M}{C} + \frac{3}{5} \right)^{-1/2} \tag{11.7}$$

Table 11.4 *Spherical charges with total mass 1 kg*

Copper case thickness (cm)	Explosive radius (cm)	Charge radius (cm)	Explosive mass (g)	Copper mass (g)	Copper fragment velocity (km s^{-1})
0.00	5.11	5.11	1000	0	3.62
0.05	4.87	4.92	866	134	3.22
0.10	4.65	4.75	753	247	2.91
0.15	4.45	4.60	657	343	2.64
0.20	4.25	4.45	576	424	2.42
0.25	4.08	4.33	507	493	2.23
0.30	3.91	4.21	447	553	2.07
0.35	3.75	4.10	396	604	1.92
0.40	3.61	4.01	351	649	1.79
0.45	3.47	3.92	312	688	1.67
0.50	3.34	3.84	279	721	1.57

where M/C is the ratio of the mass of the metal to the mass of the explosive (Walters and Zukas 1989). The $\sqrt{2E}$ term is an experimentally determined velocity, and for PBX-9010 it was taken to be 2.8 km s^{-1}. This launch velocity is the velocity at which the copper fragments will impact the asteroid or comet nucleus material, presuming there is space enough for the acceleration before impact.

However, though this provides information about the momentum transferred by the metal case, it does not provide information about the amount of momentum transferred by the explosive gases themselves. To obtain an idea of the momentum transferred by the coupled action of the copper fragments and explosive gases, CTH computations were performed with a spherical charge as described above inside a spherical cavity in ice, with a cavity radius of 10.2 cm.

Results from three configurations will be presented. In the first, the charge is bare and only the explosive product gases load the inside of the cavity in the ice. In the second and third, the charge is encased in a sphere of copper and both copper fragments and explosive product gases load the inner surface of the cavity. In particular, copper cases of 0.25 cm thickness and 0.50 cm thickness were considered. The PBX-9010 was modeled in CTH with the JWL equation of state, the constants taken from Dobratz and Crawford (1985). The copper for the explosive casing was modeled using a Mie–Grüneisen equation of state and the Johnson–Cook strength model, with the constants taken from the CTH library. The ice was modeled with a Mie–Grüneisen equation of state ($\rho_0 = 0.93$ g cm^{-3}, $c = 3.5$ km s^{-1}, $s = 1$, and $\gamma = 1$) with an elastic-perfectly plastic constitutive response ($v = 0.35, Y = 10$ MPa). Ice strength depends upon temperature, and this strength (which is approximate) corresponds to a temperature of $-20\,^\circ$C. The latter two Mie–Grüneisen numbers are estimates.

Using these models for the materials, both one-dimensional and three-dimensional calculations of the explosive/NEO interaction were performed. The one-dimensional calculations were performed with 0.5 mm zoning and the three-dimensional calculations with 4 mm zoning. Good agreement occurred between one-dimensional and three-dimensional computations with the same 4 mm zone size. This is partially due to CTH's interface tracker, which prevented the asymmetric break-up of the copper jacket material. Of most interest here are impulses delivered to the inside of the spherical ice cavity. For the three cases, the impulses at the inner surface of the ice cavity (integrated over the interior cavity surface) were roughly 5250 kg m s^{-1}, 2000 kg m s^{-1}, and 1200 kg m s^{-1}, for the bare charge, 0.25 cm thick copper cased charge, and the 0.50 cm thick copper cased charge, respectively. Thus, the total transferred momentum decays a little more quickly than 1/(explosive mass), and the amount of the momentum in the copper fragments (using Table 11.4) is 0 kg m s^{-1}, 1100 kg m s^{-1}, and 1130 kg m s^{-1}. Thus, more and more of the momentum is going into the fragments rather than being transferred to the ice directly by the explosive product gases. (As an aside, the computations predict roughly 2.1 km s^{-1} and 1.6 km s^{-1} impact velocities for the copper fragments for the 0.25 cm and 0.50 cm copper cased charge, respectively, values that are in agreement with the Gurney velocity predictions in Table 11.4.)

Looking at total momentum, it might be thought that the best choice is the uncased charge, and this is certainly true if the asteroid or comet body is dense. However, for a low-density material that is in the way of the higher-density material that it is desirable to load, using fragments to transfer the load through the low-density material can be the better choice.

5 Experiments and supporting computations

A number of experiments were performed to explore signal strength based on explosive depth. In addition, initial experiments were performed with a low-density aluminum foam (6% dense) to explore signal strength transmitted through a low density material.

The tests were performed at the Southwest Research Institute ballistics range. The primary test article was a large 6061-T6 aluminum plate, 60.96 cm in diameter and 7.62 cm thick. On the top surface of the plate six strain gages were attached; three each were on two radial lines that were perpendicular to each other. The distances of the gages from the top-surface center of the plate were 10.16 cm (gages 1 and 4), 17.78 cm (gages 2 and 5), and 25.4 cm (gages 3 and 6).

Figure 11.5 Craters from tests. Notice the tiers of crater diameter corresponding to the tests performed at different explosive depths. (The outer ring nearly spanning the photograph demarcates the inner diameter of the steel pipe used to protect the wiring.)

RP-83 detonators were used as the explosive source. The detonator includes an exploding bridgewire, 80 mg of PETN, and then 1.03 g of RDX mixed with a binder. The detonator assembly is sheathed in 0.018 cm of aluminum and in total is 0.71 cm in diameter. A large diameter steel pipe was set on top of the aluminum test plate surrounding the detonator to protect the gage wires from fragmentation damage during the test.

Tests were performed with the intent that the detonator be placed at various depths of penetration into the surface. Thus, the initial tests were performed near the surface, then two tests were performed drilling a 1.27 cm deep hole in the plate, and then two tests were performed drilling a 2.54 cm deep hole in the plate. As each test further opened the crater and produced localized permanent deformation of the plate, an ideal hole exactly the same size as the detonator no longer existed (Fig. 11.5). However, the comparison computations were performed with an assumed ideal hole for the detonator placement with the diameter equal to the diameter of the charge.

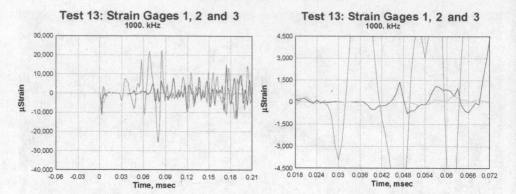

Figure 11.6 Strain gage traces from gages 1, 2, and 3, (right is an enlargement of initial arrival).

The CTH hydrocode was used to perform computations in conjunction with the experiments. Axial symmetry was assumed. Zoning was 280 by 300, with the cell size 0.05 cm square in the vicinity of the charge; as the charge was 0.3175 cm in radius, this corresponded to six cells across the radius. The height of the cylinder of explosive in the computations was 1.67 cm, giving a total charge mass of 1.0 g. The detonation was handled through programmed burn. The JWL equation of state with constants for HMX was used to model the explosive products. The 6061-T6 aluminum plate was modeled with a Mie–Grüneisen equation of state and the Johnson–Cook constitutive model. Strains from the computations were measured at corresponding distances to those in the test article.

The strain gage data collected included the initial arrival time, the first peak's amplitude, and the width in time of the first peak. Strain gage data is shown for one test's gages 1, 2, and 3 in Fig. 11.6. At zero time there is noise from the firing system, but that decays and it is possible to see the arrival of the three pulses. The arrival time was recorded for each test. The peak amplitude of the first arrival at each gage was also recorded. In the enlargement on the right of the figure it is possible to see that the more distant gage traces have lower amplitude. Also, the zero crossing of the initial pulse was recorded, to allow a calculation of the pulse width. Similar results were recorded from the computations.

It should be pointed out that there was considerable difficulty in reducing the noise from the capacitor box and the FS-10 firing unit. In the end, both units were moved 36.5 m away from the test article and the capacitor box was wrapped in copper mesh. The 36.5 m is the maximum distance according to the manufacturer for the cables and firing system being used. The initial time comes from a signal from the FS-10 unit. Aluminum tape was placed on top of the gages and the aluminum plate was grounded to the Nicolet recorder in attempts to reduce the noise. It

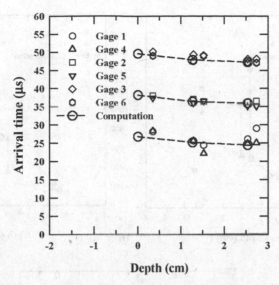

Figure 11.7 Arrival times for the tests and computations for the three gage locations.

appeared that distance was the most important action taken in reducing the noise. The fact that the delay between the firing signal and breakout was 10 μs produced enough time for the decay of the capacitor noise to be sufficient so that the arrival at the first strain gage signal could be seen.

The first data examined were the arrival time information for all the gages. Figure 11.7 shows these data, along with the results from the corresponding computations. The arrival time data are plotted vs. depth of charge placement. With all the data points being used (save from one test where there was a different time zero) – that is, with no distinction being made as to the charge depth – a least-squares fit of gage distance vs. arrival time produced a wave speed in the aluminum of 6.61 km s^{-1} with a time between initial signal and detonation of the detonator of 10.0 μs (the published break-out time is 5.38 μs). The sound speed from constants used in the hydrocode modeling is $\alpha = 6.58$ km s^{-1}, and, as can been seen, the agreement is excellent.

Next, the peaks were examined. Figure 11.8 shows the peaks vs. depth of charge placement for both the experiments and the computations. The deeper the charge is placed in the target, yielding a larger loading moment, the larger the amplitude of the first arriving signal. The peaks from the computations appear low compared to those seen in the test for the nearest gages (10.16 cm). There is good agreement at the gages located at 17.78 cm. Save for one outlier point, there is good agreement between the computations and the data for the 25.4 cm gages.

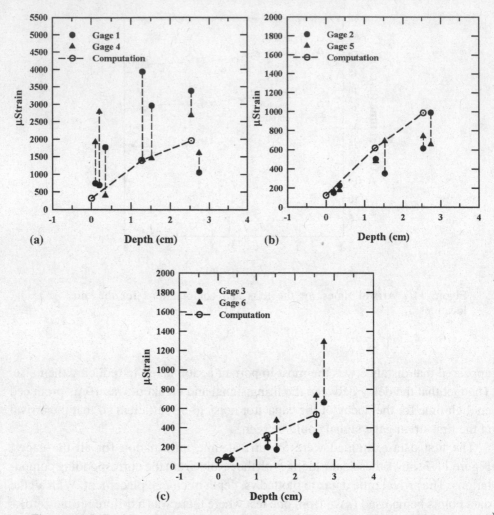

Figure 11.8 Peak strain vs. charge depth at 10.2 cm (a), 17.8 cm (b), and 25.4 cm (c).

In a similar fashion, the width of the first arriving pulse was examined. Figure 11.9 shows the pulse width vs. charge depth for both the experiments and the computations. The results imply that as one moves away from the detonation center, there is little dependence of pulse width on charge depth.

The purpose of these tests was to show that good data could be obtained recording both amplitude and arrival time information. The good agreement of the tests with the computations allows us to now ask questions regarding the effect of adjusting the strength of the aluminum material, and determining whether such an effect could be observed in tests. Figure 11.10 shows the results of decreasing the flow stress by a factor of 2 and increasing it by a factor of 2 on the arriving peak magnitude

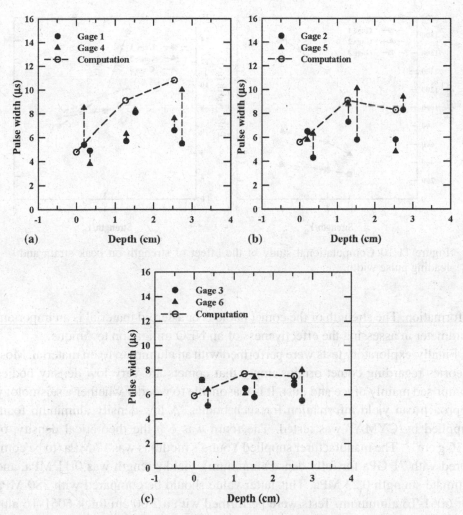

Figure 11.9 Leading pulse width vs. charge depth at 10.2 cm (a), 17.8 cm (b), and 25.4 cm (c).

and pulse width. Adjusting the flow stress was achieved by multiplying the initial yield term (*A*) and the work hardening leading coefficient (*B*) in the Johnson–Cook constitutive model. As can be seen, the strength of the material does affect the observed magnitudes and pulse widths. Essentially, the physics is this: when the charge goes off, the stronger material allows passage of higher-amplitude waves into the material, and so larger amplitudes are observed. However, due to the greater strength, a smaller crater is opened, venting out the explosive product gases more quickly, leading to shorter pulse widths. These computations show it is possible to obtain strength of the material in the vicinity of the charge from amplitude

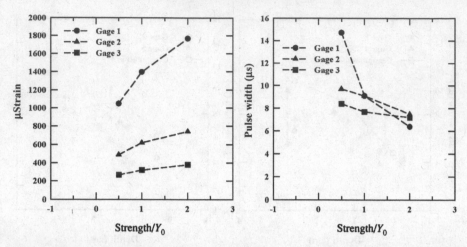

Figure 11.10 Computational study of the effect of strength on peak strain and leading pulse width.

information. The strength of the comet nucleus or asteroid material is an important parameter in assessing the effectiveness of an NEO mitigation technique.

Finally, exploratory tests were performed with an aluminum foam material. Most theories regarding comet origins imply that comets are very low density bodies comprised mainly of ice and dust. It is reasonable to wonder whether a seismology approach can yield information for such bodies. A low-density aluminum foam supplied by CYMAT was tested. The foam was 6% the theoretical density, or 0.16 g cm^{-3}. The manufacturer supplied Young's modulus was 17 MPa (to be compared with 71 GPa for fully dense aluminum), yield strength was 0.11 MPa, and ultimate strength 0.23 MPa. This latter value should be compared with 380 MPa for 6061-T6 aluminum. Tests were performed with a 2.60 cm thick 6061-T6 aluminum plate that was 15.24 cm square sitting on top of the aluminum foam block. The foam block was 6.67 cm thick and 20.3×30.5 cm. This block was then placed on top of the aluminum plate described above, save that the plate was turned upside down so that the gages were on the bottom (the reason for turning the plate over was that the crater lips from the previous tests prevented the aluminum foam block from sitting flat on the upper surface). In the top 6061-T6 Al block the detonator was placed at a 1.27 cm depth. Two tests were performed in this fashion with nearly identical results. The arrival time at the gages was 213 μs and 215 μs for the respective tests. The amplitude of the signal was 12 to 25 microstrain, depending on the gage. The initial arrival was very spread out in time. There was a small permanent indention on the upper surface of the foam due to the bottom of the top aluminum plate deforming because of the explosive load of the detonator. The crater formed in the top plate had an inner diameter of 14 mm with a depth of 19 mm (the bottom of the crater was pointed).

Computations were also performed. The foam was modeled as an elastic-plastic solid homogeneous material. The density, elastic, and strength constants supplied by the manufacturer were used to give a Mie–Grüneisen equation of $\rho = 0.16$ g cm^{-3}, $c_0 = 265$ m s^{-1}, $s = 1$ (a guess), $\gamma = 2$ (same as dense aluminum), and $c_v = 0.06 \times 10^{11}$ erg g^{-1} eV (fully dense Al value of 1.02×10^{11} erg g^{-1} eV would probably have been a better choice, but it does not greatly influence the results). The strength was given by the Johnson–Cook model for 6061-T6 aluminum with $A = 0.11$ MPa and $B = 0.12$ MPa to reflect the provided strength parameters. The Poisson's ratio was assumed to be 0.33 (other values, such as 0, led to calculations that terminated early due to numerical difficulties). The crater diameter in the upper aluminum plate (where the detonator was placed 1.27 cm down in a fully dense 6061-T6 aluminum plate) was 16 mm in the computation, as compared with the 14 mm seen in the test.

In the aluminum foam, more permanent deformation was seen in the computation than was seen in the experiment; however, the arrival time in the computation was 206 µs, very similar to that observed experimentally. The amplitudes at the gage locations were 6 microstrain, somewhat smaller that that observed in the tests.

An additional computation was performed with the aluminum foam block replaced by fully dense 6061-T6 Al. The arrival time at the gage 1 location in this instance was 29 µs and the peak amplitude, was 523 microstrain. Thus, the foam reduced the amplitude of the signal by a factor of 100. This result implies that there is considerable question as to how much signal could be sensed if the asteroid or comet body was very low density. Also, though not an issue in these tests, the moment is greatly reduced for explosions in weak, low-density materials, hence the earlier discussion about using cased charges rather than bare charges.

In conclusion, the testing program showed good agreement between the experiments and computations on arrival time, peak amplitude, and initial pulse width. This agreement allowed an exploration of the role of strength of the aluminum material. It was shown computationally that the peak amplitude depended on the aluminum strength, and therefore it is possible to obtain strength information from the amplitude of the seismic signal in the characterization of an asteroid or comet nucleus. It was also shown that very low density materials transmit correspondingly low seismic signals, thus making it difficult to do seismology on a body composed of very low density materials.

6 Oscillations of a sphere

One of the most intriguing ideas for the seismic analysis of a small body is to use seismology to excite the normal modes, and then examine the frequencies. Though it has been demonstrated mathematically that one cannot "hear the shape of a drum," i.e., perform the complete inversion problem given the eigenfrequencies, still, given the observed outer geometry, determining the vibrational frequencies of a cohesive body will provide considerable information about its interior structure and properties. In order to develop a sense of the frequencies and motions involved, the simplest case will be discussed, namely that of the oscillations of a sphere (though admittedly few asteroids are thought to have a nice spherical shape). This problem has classical origins, being first solved by Lamb (Dahlen and Trump 1998; Takeuchi and Saito 1972), and presented here are the basics with the calculation of the frequencies for a 1-km diameter sphere with reasonable properties. This approach ties in naturally with a knowledge of the downward momentum p_z and the moment term $M_{rr} + M_{\theta\theta}$ described above. For the most part, the discussion follows the notation of Dahlen and Trump (1998). They present part of the solution where gravity is involved; for this application, gravity is ignored due to the small size of the body. For more general shapes typical of observed asteroids, numerical procedures are required to determine the frequencies. Also, this discussion does not address the effect of asymmetries and rotation, which split the frequencies and complicate the spectrum.

6.1 Determination of the normal modes

Spherical symmetry will be assumed, where the explosive is detonated on the z axis, defined by $\theta = 0$. The equations for elasticity are

$$(\lambda + 2\mu)\frac{\partial e}{\partial r} - \frac{2\mu}{r \sin(\theta)}\frac{\partial}{\partial r}(w_\varphi \sin(\theta)) = \rho\frac{\partial^2 u_r}{\partial t^2} \tag{11.8}$$

$$\frac{(\lambda + 2\mu)}{r}\frac{\partial e}{\partial \theta} + \frac{2\mu}{r \sin(\theta)}\frac{\partial}{\partial r}(r w_\varphi \sin(\theta)) = \rho\frac{\partial^2 u_\theta}{\partial t^2} \tag{11.9}$$

where λ and μ are the Lamé constants (μ is the shear modulus), u_r and u_θ are the displacements, ρ is the density, and

$$e = \frac{1}{r^2 \sin(\theta)}\left[\frac{\partial}{\partial r}(r^2 u_r \sin(\theta)) + \frac{\partial}{\partial \theta}(r u_\theta \sin(\theta))\right] \tag{11.10}$$

$$w_\varphi = \frac{1}{2}\left[\frac{1}{r}\frac{\partial}{\partial r}(r u_\theta) - \frac{1}{r}\frac{\partial u_r}{\partial \theta}\right] \tag{11.11}$$

e is the dilatational portion of the deformation and w_φ is a rotation matrix component. To find the normal vibrational modes of the sphere, a certain form of the solution is assumed, in terms of spherical harmonics:

$$\vec{u} = \hat{r}U(r)Y_{l0}(\theta) + \hat{\theta}\frac{1}{k}V(r)\frac{\partial}{\partial \theta}Y_{l0}(\theta) \tag{11.12}$$

Here, the Y_{l0} terms are the real spherical harmonics with $m = 0$ since it has been assumed there is no φ dependence (φ is the angle in the x–y plane). The $1/k$ term in the θ direction makes the angular terms orthonormal when integrated over the surface of the sphere. The functions U and V depend only on the radius. l is the degree of the spherical harmonic, and $k = \sqrt{l(l+1)}$. In particular,

$$Y_{l0}(\theta) = \left(\frac{2l+1}{4\pi}\right)^{1/2} \frac{1}{2^l l!} \left(\frac{1}{\sin(\theta)} \frac{d}{d\theta}\right)^l (\sin(\theta))^{2l}$$

$$= \left(\frac{2l+1}{4\pi}\right)^{1/2} P_l(\cos(\theta)) \tag{11.13}$$

These (up to the normalization coefficient) are the Legendre polynomials P_l of degree l. They satisfy the Helmholtz equation:

$$\frac{\partial^2 Y_{l0}}{\partial \theta^2} + \cot(\theta)\frac{\partial Y_{l0}}{\partial \theta} + k^2 Y_{l0} = 0 \tag{11.14}$$

Inserting this assumed solution form into the elastic equations, one obtains

$$e = \left[\dot{U} + \frac{2U}{r} - \frac{kV}{r}\right] Y_{l0} \tag{11.15}$$

$$w_\varphi - \frac{1}{2}\left[\frac{\dot{V}}{r} + \frac{V}{kr} - \frac{U}{r}\right] \frac{\partial Y_{l0}}{\partial \theta} \tag{11.16}$$

To obtain Eq. 11.15, Eq. 11.14 was used. The elastic equations become

$$\alpha^2 \left[\ddot{U} + \frac{2\dot{U}}{r} - \frac{2U}{r^2} - \frac{k\dot{V}}{r} + \frac{kV}{r^2}\right] + \beta^2 \left[\frac{k\dot{V}}{r} + \frac{kV}{r^2} - \frac{k^2 U}{r^2}\right] + \omega^2 U = 0 \tag{11.17}$$

$$\alpha^2 \left[\frac{\dot{U}}{r} + \frac{2U}{r^2} - \frac{kV}{r^2}\right] + \beta^2 \left[\frac{\ddot{V}}{k} + \frac{2\dot{V}}{kr} - \frac{\dot{U}}{r}\right] + \omega^2 \frac{V}{k} = 0 \tag{11.18}$$

Here it has been assumed that the temporal dependence is of the form $\sin(\omega t)$ or $\cos(\omega t)$. The longitudinal (P wave) and shear (S wave) wave speeds are

$$\alpha = \sqrt{\frac{\lambda + 2\mu}{\rho}} = \sqrt{\frac{K + 4\mu/3}{\rho}}, \quad \beta = \sqrt{\frac{\mu}{\rho}} \tag{11.19}$$

respectively, where K is the bulk modulus. Typically, the longitudinal wave speed is on the order of twice the shear wave speed, and the relationship between the two wave speeds is given by

$$\alpha = \sqrt{2\frac{1-v}{1-2v}}\beta \tag{11.20}$$

where ν is Poisson's ratio, which is typically between 1/4 and 1/3. For lunar materials, Poisson's ratio is typically 1/4, so the ratio of the speeds is 1.73. Equations 11.17 and 11.18 are to be solved simultaneously subject to the constraint that U and V are well behaved at the origin and that the surface traction is zero at the outer surface of the sphere. Assuming the rotation is small at the surface, the surface tractions are approximated by the normal stress and shears perpendicular to the undeformed sphere to give

$$\sigma_{rr}|_{r=a} = (\lambda e + 2\mu\varepsilon_{rr})|_{r=a} = 0 \tag{11.21}$$

$$\sigma_{r\theta}|_{r=a} = 2\mu\varepsilon_{r\theta}|_{r=a} = 0 \tag{11.22}$$

(The other shear condition, $\sigma_{r\varphi} = 0$, is satisfied everywhere due to the symmetry assumptions.) The required strains are given by

$$\varepsilon_{rr} = \frac{\partial u_r}{\partial r}, \quad \varepsilon_{r\theta} = \frac{1}{2}\left[\frac{1}{r}\frac{\partial u_r}{\partial \theta} + \frac{\partial u_\theta}{\partial r} - \frac{u_\theta}{r}\right] \tag{11.23}$$

The stresses can then be written as (using Eq. 11.12)

$$\sigma_{rr} = R(r)Y_{10}(\theta) \tag{11.24}$$

$$\sigma_{r\theta} = S(r)\frac{\partial}{\partial \theta}Y_{10}(\theta) \tag{11.25}$$

where

$$R(r) = (\lambda + 2\mu)\dot{U} + \lambda\frac{2U}{r} - \lambda\frac{kV}{r} \tag{11.26}$$

$$S(r) = \mu\left[\frac{U}{r} + \frac{\dot{V}}{k} - \frac{V}{kr}\right] \tag{11.27}$$

The boundary conditions become $R(a) = 0$ and $S(a) = 0$.

With this background, it is now possible to calculate the normal modes. There are three cases to consider.

Case 1: $\omega = 0$. This is the special case of no oscillations. Equations 11.17 and 11.18 admit power law solutions of the form $U(r) = Ar^b$, $V(r) = Br^b$, and it can be shown there are four solutions with $b = -l-2, -l, l-1, l+1$. However, there is only one solution that can satisfy the boundary conditions and that is for the special case $l = 1, b = 0$. For this solution, $U = 1$ and $V = \sqrt{2}$ combines with the angular terms

$$Y_{10} = \sqrt{\frac{3}{4\pi}}\cos(\theta) \tag{11.28}$$

$$\frac{1}{k}\frac{\partial}{\partial \theta}Y_{10}(\theta) = \frac{1}{\sqrt{2}}\sqrt{\frac{3}{4\pi}}\cos(\theta) \tag{11.29}$$

to give (from Eq. 11.12)

$$\vec{u} = \hat{r}\sqrt{\frac{3}{4\pi}}\cos(\theta) - \hat{\theta}\sqrt{\frac{3}{4\pi}}\sin(\theta) = \sqrt{\frac{3}{4\pi}}\hat{z} \qquad (11.30)$$

which corresponds to a translation along the z-axis – that is, this solution will correspond to the total momentum imparted along the axial direction to the body by the explosion or impact that produces the seismic pulse.

Case 2: $l = 0$. The angular term is

$$Y_{00} = \frac{1}{\sqrt{4\pi}} \qquad (11.31)$$

Due to the lack of θ dependence, Eq. 11.18 is automatically satisfied ($V = 0$, $k = 0$), and Eq. 11.17 becomes

$$\ddot{U} + \frac{2\dot{U}}{r} + \left\{\left(\frac{\omega}{\alpha}\right)^2 - \frac{2}{r^2}\right\}U = 0 \qquad (11.32)$$

This equation is solved by a spherical Bessel function of order one, where

$$U(r) = j_1(\gamma r) = \frac{\sin(\gamma r)}{(\gamma r)^2} - \frac{\cos(\gamma r)}{\gamma r} \qquad (11.33)$$

$$\gamma = \frac{\omega}{\alpha} \qquad (11.34)$$

and ω satisfies the equation (from Eq. 11.26)

$$(\lambda + 2\mu)j_0(\gamma a) = 4\mu\frac{1}{\gamma a}j_1(\gamma a) \qquad (11.35)$$

which reduces to (putting in the actual spherical Bessel function terms)

$$\frac{\lambda + 2\mu}{4\mu}\sin(\gamma a) - \frac{\sin(\gamma a)}{(\gamma a)^2} + \frac{\cos(\gamma a)}{\gamma a} = 0 \qquad (11.36)$$

This equation was then solved with a search and bracket technique, followed by an application of Newton's method to determine the roots. These roots are denoted $_n\omega_0$ with the integer index n going from 0 to plus infinity.

Case 3: $l > 0$. This is the general case. There are four solutions to Eqs. 11.17 and 11.18, but only two are well behaved at the origin. The two linearly independent solutions are

$$(1)\ \gamma = \frac{\omega}{\alpha},\qquad
\begin{aligned}
U &= \frac{l}{r} j_l(\gamma r) - \gamma j_{l+1}(\gamma r) \\[4pt]
V &= \frac{k}{r} j_l(\gamma r)
\end{aligned}$$

$$(2)\ \gamma = \frac{\omega}{\beta},\qquad
\begin{aligned}
U &= \frac{k}{r} j_l(\gamma r) \\[4pt]
V &= \frac{l+1}{r} j_l(\gamma r) - \gamma j_{l+1}(\gamma r)
\end{aligned}
\tag{11.37}$$

Showing that these satisfy the equations requires use of the fact that the spherical Bessel functions satisfy

$$\frac{d^2}{dz^2} j_l(z) + \frac{2}{z}\frac{d}{dz} j_l(z) + \left(1 - \frac{k^2}{z^2}\right) j_l(z) = 0$$

$$\frac{l+1}{z} j_l(z) + \frac{d}{dz} j_l(z) = j_{l-1}(z) \tag{11.38}$$

$$\frac{l}{z} j_l(z) - \frac{d}{dz} j_l(z) = j_{l+1}(z)$$

(Abramowitz and Stegun 1970). Here, $z = \gamma r$, and the more useful forms of these expressions where prime denotes the derivative with respect to z (not r) are

$$\gamma^2 j_l''(\gamma r) + \frac{2\gamma}{r} j_l'(\gamma r) + \left(\gamma^2 - \frac{k^2}{r^2}\right) j_l(\gamma r) = 0$$

$$\frac{l+1}{\gamma r} j_l(\gamma r) + j_l'(\gamma r) = j_{l-1}(\gamma r) \tag{11.39}$$

$$\frac{l}{\gamma r} j_l(\gamma r) - j_l'(\gamma r) = j_{l+1}(\gamma r)$$

The two solutions of Eqs. 11.37 do not satisfy the boundary conditions, but a linear combination of them yields the unique solution that does. To determine the coefficients on the final solution, it is necessary to calculate the stress terms R and S for both these solutions:

$$(1)\ \gamma = \frac{\omega}{\alpha},\qquad
\begin{aligned}
R(r,\gamma) &= \left[-(\lambda + 2\mu)\gamma^2 + \frac{2\mu l(l-1)}{r^2} \right] j_l(\gamma r) + \frac{4\gamma\mu}{r} j_{l+1}(\gamma r) \\[6pt]
S(r,\gamma) &= \frac{2(l-1)}{r^2} j_l(\gamma r) - \frac{2\gamma}{r} j_{l+1}(\gamma r)
\end{aligned}
\tag{11.40}$$

$$(2)\ \gamma = \frac{\omega}{\beta},\qquad
\begin{aligned}
R(r,\gamma) &= \frac{2\mu k(l-1)}{r^2} j_l(\gamma r) - \frac{2\mu k \gamma}{r} j_{l+1}(\gamma r) \\[6pt]
S(r,\gamma) &= \left[\frac{2(l-1)(l+1)}{kr^2} - \frac{\gamma^2}{k} \right] j_l(\gamma r) + \frac{2\gamma}{kr} j_{l+1}(\gamma r)
\end{aligned}$$

The reason the effort has been exerted to write these expressions in terms of spherical Bessel functions, rather than a sum of spherical Bessel functions and their first derivatives, is that it will be necessary to compute derivatives to use Newton's method to determine the roots. What is required to satisfy the boundary conditions $R(a) = 0$ and $S(a) = 0$ is that these two boundary stress terms be linearly dependent. Thus, the equation being solved to find the normal mode frequencies is

$$f(\omega) = \begin{vmatrix} R_1(a, \omega/\alpha) & R_2(a, \omega/\beta) \\ S_1(a, \omega/\alpha) & S_2(a, \omega/\beta) \end{vmatrix}$$

$$= R_1\left(a, \frac{\omega}{\alpha}\right) S_2\left(a, \frac{\omega}{\beta}\right) - R_2\left(a, \frac{\omega}{\beta}\right) S_1\left(a, \frac{\omega}{\alpha}\right) = 0 \quad (11.41)$$

Solving this involves a bracket and search, with the final calculation of the $_n\omega_l$ term through Newton's method. The derivative for the Newton step iteration is

$$\frac{df}{d\omega} = \frac{1}{\alpha}\frac{dR_1}{d\gamma}S_2 + \frac{1}{\beta}R_1\frac{dS_2}{d\gamma} - \frac{1}{\beta}\frac{dR_2}{d\gamma}S_1 - \frac{1}{\alpha}R_2\frac{dS_1}{d\gamma} \quad (11.42)$$

where

$$\frac{dR_1}{d\gamma} = -2(\lambda + 2\mu)\gamma j_l + \left[-(\lambda + 2\mu)\gamma^2 + \frac{2\mu l(l-1)}{r^2}\right]rj_l'$$

$$+ \frac{4\mu}{r}j_{l+1} + 4\mu\gamma j_{l+1}'$$

$$\frac{dS_1}{d\gamma} = \frac{2\mu}{r}[(l-1)j_l' - j_{l+1} - \gamma r j_{l+1}']$$

$$\frac{dR_2}{d\gamma} = \frac{2\mu k}{r}[(l-1)j_l' - j_{l+1} - \gamma r j_{l+1}'] \quad (11.43)$$

$$\frac{dS_2}{d\gamma} = \frac{\mu}{k}\left\{-2\gamma j_l + \left[\frac{2(l-1)(l+1)}{r^2} - \gamma^2\right]rj_l' + \frac{2}{r}j_{l+1} + 2\gamma j_{l+1}'\right\}$$

In each of these equations, γ is the appropriate value for case 1 or 2 and $r = a$. The spherical Bessel functions and their derivatives are evaluated at $z = \gamma a$.

Once the frequency is found, it is then possible to write down the solution of the equations that satisfied the boundary condition as

$$U(r) = \frac{R_2}{\sqrt{R_1^2 + R_2^2}}U_1(r) - \frac{R_1}{\sqrt{R_1^2 + R_2^2}}U_2(r)$$

$$V(r) = \frac{R_2}{\sqrt{R_1^2 + R_2^2}}V_1(r) - \frac{R_1}{\sqrt{R_1^2 + R_2^2}}V_2(r)$$

$$\quad (11.44)$$

This solution satisfies $U(a) = 0$ and $V(a) = 0$, thus satisfying the boundary conditions. Thus, the normal mode corresponding to the computed frequency has been determined.

For a general solution, the normal modes can then be summed using as initial conditions the downward momentum and perpendicular moment values defined above. Then, for a spherically shaped asteroid or comet nucleus, it is possible to compute the motion and acceleration at potential seismometer locations due to the signal produced by the seismic source. The normal modes can be written as

$$_n\vec{S}_l = \hat{r}_n U_l(r) Y_{l0}(\theta) + \hat{\theta} \frac{1}{k} _n V_l(r) \frac{\partial}{\partial\theta} Y_{l0}(\theta) \qquad (11.45)$$

where the $l \geq 0$ is the index for the spherical harmonic and n is the index based on the ordering of the vibration frequencies $_n\omega_l$ where it will be assumed that the lowest index is $n = 0$. The various modes are assumed normalized (they are orthogonal by construction), so that

$$\int \rho(_n\vec{S}_l \cdot _n\vec{S}_l) \, dV = 1 \qquad (11.46)$$

or in particular

$$\int\limits_0^a \rho(_nU_l^2 + _nV_l^2) r^2 \, dr = 1 \qquad (11.47)$$

due to the normalizing assumption on the angular terms described above. Given this normalization, the general solution can be written

$$\vec{S}(r, \theta, t) = \frac{p_z}{(4/3)\pi a^3 \rho} \hat{z} t$$

$$+ \sum_{n=0}^{\infty} \sum_{l=0}^{\infty} _n\vec{S}_l(r, \theta)[_n A_l(1 - \cos(_n\omega_l t)) + _n B_l \sin(_n\omega_l t)] \qquad (11.48)$$

The first term reflects the total momentum imparted to the body, resulting from the Case 1 solution, and the next term is the sum over all the modes; p_z is assumed negative for a downward (toward the asteroid's center) impulse. Notice that for each l there are an infinite number of frequencies. The solution in Eq. 11.48 assumes there are no inelastic losses. With some work it can be shown that the coefficients resulting from the moment/momentum loading at the $r = a$, $\theta = 0$ surface point is

$$_n A_l = \frac{1}{_n\omega_l^2} \sqrt{\frac{2l+1}{4\pi}} \frac{1}{a} \left(_n U_l(a) - \frac{1}{2} k_n V_l(a) \right)(M_{\theta\theta} + M_{\varphi\varphi}) \qquad (11.49)$$

$$_n B_l = \frac{1}{_n\omega_l} \sqrt{\frac{2l+1}{4\pi}} _n U_l(a) p_z \qquad (11.50)$$

Table 11.5 *Normal modes with frequencies ($_n\omega_l/2\pi$) less than 8 Hz, with composition in terms of functions, the loading coefficients, and a surface displacement estimate*

Mode	Frequency (Hz)	R_1	R_2	$_nA_l/(M_{\theta\theta}+M_{\varphi\varphi})$ (s^2 g $^{1/2}$ cm^{-1}) $\times 10^{-15}$	$_nB_l/p_z$ (s g$^{-1/2}$) $\times 10^{-9}$	$\sqrt{_nU_l^2(a)+_nV_l^2(a)}$ (g$^{-1/2}$) $\times 10^{-7}$
$_0S_0$	3.26			1.20	1.23	0.90
$_1S_0$	7.71			−0.18	−0.44	0.75
$_0S_1$	2.52	0.6407	−0.7678	−4.05	−0.97	1.07
$_1S_1$	4.98	0.7751	0.6318	−1.01	−1.10	0.83
$_2S_1$	5.69	0.6065	−0.7951	−0.02	−0.60	0.73
$_3S_1$	7.87	0.1643	0.9864	0.23	4.62	0.74
$_0S_2$	1.94	0.6444	−0.7647	−2.49	−3.86	0.83
$_1S_2$	3.58	0.7590	−0.6511	−4.13	−0.36	1.25
$_2S_2$	6.12	0.5650	−0.8251	1.16	0.84	0.86
$_3S_2$	7.19	0.8182	−0.5749	−0.18	−0.94	0.74
$_0S_3$	2.88	0.8043	−0.5943	−3.31	−3.60	0.88
$_1S_3$	4.74	0.8208	−0.5713	−3.69	0.17	1.31
$_2S_3$	7.13	0.5076	0.8616	−1.36	−0.69	0.92
$_0S_4$	3.68	0.8976	−0.4408	−3.74	−3.47	0.96
$_1S_4$	5.93	0.8451	−0.5346	−3.25	0.52	1.31
$_0S_5$	4.43	0.9488	−0.3160	−3.97	−3.39	1.04
$_1S_5$	7.08	0.8465	0.5324	−2.86	0.75	1.28
$_0S_6$	5.16	0.9750	−0.2220	−4.11	−3.32	1.11
$_0S_7$	5.88	0.9880	−0.1543	−4.20	−3.27	1.18
$_0S_8$	6.58	0.9943	−0.1066	−4.26	−3.23	1.24
$_0S_9$	7.28	0.9973	−0.0735	−4.30	−3.20	1.30
$_0S_{10}$	7.98	0.9987	−0.0506	−4.33	−3.17	1.35

Table 11.5 shows the normal modes with frequency less than 8 Hz for a sphere with material properties $\rho = 3$ g cm^{-3}, $\alpha = 4000$ m s^{-1}, and $\nu = 1/4$. These numbers are suggested by those seen near the lunar surface, and correspond to a Young's modulus $E = 40$ GPa and a shear wave speed of $\beta = 2310$ m s^{-1}. The radius of the sphere is $a = 500$ m. Thus, the frequencies of oscillation are for a body 1 km across with possible properties for a stony asteroid. The main purpose of this table is to give both a qualitative idea and a quantitative estimate of the seismic behavior that would be exhibited by such a body. The unusual units are due to the normalization in Eq. 11.47.

The lowest frequency mode is $_0S_2$, with a frequency of 1.94 Hz and a period of 0.515 s. This mode of oscillation is seen in the Earth with a period is 53.9 min, though

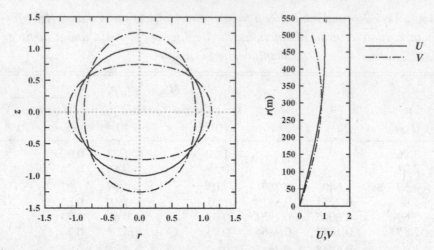

Figure 11.11 Surface (left) and radial (right) oscillation behavior of $_0S_2$.

the Earth structure is stratified and much more complicated (the radial solutions are found numerically). Figure 11.11 shows pictorially how the $_0S_2$ mode oscillates. This mode would be excited on the spherical body by explosives or an impact and, based on the magnitude of the $_nA_l$ and $_nB_l$ terms, it should be observable. The numbers in the table can be used to get a rough idea of the behavior of the mode. If a 1-kg charge were placed on the surface, then based on Table 11.4 a moment of $M_{\theta\theta} + M_{\varphi\varphi} \sim 2 \times 10^{14}$ dyn cm^{-1} and a downward momentum of $p_z \sim -1 \times 10^8$ g cm s^{-1} could be expected. Then $_0A_2 \sim 0.5\, g^{1/2}$ cm^{-1} and $_0B_2 \sim 0.4\, g^{1/2}$cm^{-1}, and with a magnitude of the normal mode estimate of $2\sqrt{_nU_l^2(a) + _nV_l^2(a)} \sim 2 \times 10^{-7}1/g^{1/2}$ an estimate for the displacement at the surface due to the 1-kg charge loading is 1×10^{-7} cm, or 1 nm. Velocity and acceleration estimates are obtained by multiplying by $_n\omega_l$ and $_n\omega_l^2$, and so if only the $_0S_2$ mode was involved the velocity and acceleration would be on the order of 10 nm s^{-1} and 100 nm s^{-2} (or 10^{-8} gs), respectively. Of course, an actual explosive load excites many modes, so the displacement and accelerations are more complicated and are larger. However, this estimate gives some idea of what needs to be measured to observe the normal modes.

A question related to detection of the normal modes is the length of time they oscillate. Such an estimate can be obtained from the Q value. The magnitude of a mode decays roughly as

$$\exp\left(-\frac{1}{2}\frac{\omega t}{Q(\omega)}\right) \tag{11.51}$$

For the Moon, values of Q near the surface are estimated to be in the range of 4000 to 7000, implying little dissipation (Hood 1986). In particular, if a Q value of 5000 is assumed, then the half-life of the 1.94 Hz $_0S_2$ oscillation is $Q\ln(4)/_0\omega_2 = 570$ s, or 9.5 min. Based on this time estimate, being able to run the seismometers for 30 min after a test would be beneficial for identifying normal modes. Higher frequency modes (i.e., all the other modes) decay more quickly.

7 Seismometers

Two questions arise with regards to seismometers for an asteroid or comet. First: how sensitive must it be; and second: how is it attached to the surface of the comet or asteroid.

Recently Kovach and Chyba (2001) have published a chart of seismometer sensitivities for extra-terrestrial applications. Essentially, seismometers exist (and are light enough for space missions) that can detect the small accelerations (10^{-9} g) and ground motions of 1 nm that are relevant to the seismology problem on a small body with a small source, as described above. It is of interest that moonquakes were in the vicinity of 5×10^{-9} g and that on Earth the background noise is on the order of 2×10^{-8} g at periods on the order of 1 s. (Thus, readings on an asteroid or comet seismology mission would be of surface motions less than Earth's background noise. Also of concern is the fact that comet outgassing will produce seismic noise whose magnitude may interfere with an active seismic investigation.) As Kovach and Chyba (2001) describe, seismometers have been built for space applications with reasonably small masses (<0.5 kg) and volumes with the required sensitivities.

Of more concern is how to attach the seismometer to the asteroid or comet surface so that the mechanical measurements can be made. To date, the attachment has always been with gravity; however, for a spherical body 1 km in diameter as examined above, the gravity is 0.042 cm s^{-2} or 4.3×10^{-5} g. It is likely that this force is insufficient to maintain good frictional contact with the surface material to allow measurement of the ground motion. Currently, work is ongoing at various institutions including that of the authors to examine attachment methods, such as spikes embedded into the surface or fluids released on the surface that harden and thus produce a footprint for the seismometer that is attached to the ground.

8 Conclusions

Work has been presented addressing some of the issues related to a seismology experiment on an asteroid or comet nucleus. The intent of a seismological investigation of such a body is to determine the material properties and the internal structure. Knowledge of the material properties and internal structure is required if it was determined that an asteroid or comet was on a collision course with

Earth, since proposed mitigation techniques require such information to assess their effectiveness.

It is clear that seismology will answer many questions about internal structure and material properties, especially of asteroids and bodies that are fairly dense and consolidated. A comet nucleus of low density would be difficult to study through seismology, as was shown experimentally with tests involving aluminum foam, and may be better explored through radio tomography. Also, a body that is not cohesive will not transmit large seismic signals, thus limiting the amount of information on internal structure that could be obtained: however, it could be learned through seismology that such a body was a loose-aggregate rubble pile rather than a body comprised of several large pieces separated by fractures. Not discussed in detail here, though clearly of interest, are two additional topics: seismic tomography, which requires many sources and many seismometers and thus does not seem likely for a near-future space mission, could provide a detailed reconstruction of the interior of a body; and passive seismology, where the sensors listen to the natural noises of the body, mostly driven by the thermal cycle caused by rotation of the body with respect to the Sun as well as the possibility of micrometeoroid impact. Given an active seismic experiment on a future mission, it is expected that an extended passive seismic experiment could yield still more data on the interior of the object.

It is likely that the seismic source for an active seismic experiment will be an explosive charge, and the loading from such a charge was quantitatively characterized. The modal frequencies of a small spherical body were presented to allow an estimation of ground motions for a seismic event and to determine the frequency as an item of interest in its own right. Such normal mode frequencies could be determined through a seismological experiment, and could be used to determine information about the interior structure of the object. Given the current state of the art in seismology and the currently level of interest in NEOs, it seems that the time is ripe for seismology to be included in a mission to small bodies of the solar system.

Acknowledgments

The authors wish to thank Ray Burgamy and Jerry Nixon for their assistance with the experiments.

References

Abramowitz, M. and Stegun, I. A. 1970. *Handbook of Mathematical Functions*.
       Washington, DC: National Bureau of Standards, US Government Printing Office.
Brown, T. 1994. Asteroseismology, *Ann. Rev. Astron. Astrophys.* **32**, 37–82.

Dahlen, F. A. and Trump, J. 1998. *Theoretical Global Seismology*. Princeton, NJ: Princeton University Press.

Dobratz, B. M. and Crawford, P. C. 1985. *LLNL Explosives Handbook*, UCRL-52997. Livermore, CA: Lawrence Livermore National Laboratory.

Goins, N. R., Dainty, A. M., and Toksöz, M. N. 1981. Lunar seismology: the internal structure of the Moon. *J. Geophys. Res.* **86**, 5061–5074.

Greenberg, J. M. and Huebner, W. F. 2002. Summary of the Workshop on Geophysical and Geological Properties of NEOs: "Know Your Enemy." *Int. Seminars Nuclear War Planet. Emergencies* **26**, 419–432.

Hannon, W. J. 1985. Seismic verification of a Comprehensive Test Ban. *Science* **227**, 251–257.

Hood, L. L. 1986. Geophysical constraints on the Moon's interior. In *Origins of the Moon*, eds. W. K. Hartman, R. J. Phillips and C. J. Taylor, pp. 361–410. Houston, TX: Lunar and Planetary Institute.

Huebner, W. F., Cellino, A., Cheng, A. F., and Greenberg, J. M. 2001. NEOs: physical properties. *Int. Seminars Nuclear War Planet. Emergencies* **25**, 309–340.

Kovach, R. L. and Chyba, C. F. 2001. Seismic detectability of a subsurface ocean on Europa. *Icarus* **150**, 279–287.

Kovach, R. L., Watkins, J. S., and Landers, T. 1971. Active seismic experiment. In *Apollo 14 Preliminary Science Report*, prep. NASA Manned Spacecraft Center, pp. 163–174. Washington, DC: US Government Printing Office.

Kovach, R. L., Watkins, J. S. and Talwani, P. 1972. Active seismic experiment. In *Apollo 16 Preliminary Science Report*, prep. NASA Manned Spacecraft Center, pp. 10-1–10-14. Washington, DC: US Government Printing Office.

Kovach, R. L., Watkins, J. S., and Talwani, P. (1973). Lunar seismic profiling experiment. In *Apollo 17 Preliminary Science Report*, prep. Johnson Space Center, NASA, pp. 10-1–10-12. Washington, DC: US Government Printing Office.

Kovach, R. L. and Watkins, J. S. 1973. Apollo 17 seismic profiling: probing the lunar crust, *Science* **180**, 1063–1064.

Latham, G. V., Ewing, M., Press, F., *et al.* 1971. Passive seismic experiment. In *Apollo 14 Preliminary Science Report*, prep. NASA Manned Spacecraft Center, pp. 133–161. Washington, DC: US Government Printing Office.

Latham, G. V., Ewing, M., Press, F., *et al.* 1972. Passive seismic experiment. In *Apollo 16 Preliminary Science Report*, prep. NASA Manned Spacecraft Center, pp. 9-1–9-29. Washington, DC: US Government Printing Office.

Latham, G. V., Ewing, M., Press, F., *et al.* 1973. Passive seismic experiment. In *Apollo 17 Preliminary Science Report*, prep. Johnson Space Center, NASA, pp. 11-1–11-9. Washington, DC: US Government Printing Office.

McGlaun, J. M., Thompson, S. L., and Elrick, M. G. 1990. CTH: a three-dimensional shock wave physics code. *Int. J. Impact Engng* **10**, 351–360.

Nakamura, Y. 1983. Seismic velocity structure of the lunar mantle. *J. Geophys. Res.* **88**, 677–686.

Takeuchi, H. and Saito, M. 1972. Seismic surface waves. In *Methods in Computational Physics*, vol. 11, ed. B. A. Bolt, pp. 217–295. New York: Academic Press.

Walker, J. D. 2003. Loading sources for seismological investigation of asteroids and comets. *Int. J. Impact Engng.* **29**, 757–769.

Walters, W. P. and Zukas, J. A. 1989. *Fundamentals of Shaped Charges*. New York: John Wiley.

12

Lander and penetrator science for near-Earth object mitigation studies

A. J. Ball

The Open University, Milton Keynes

P. Lognonné

Institut de Physique du Globe de Paris

K. Seiferlin

Westfälische Wilhelms-Universität Münster[a]

M. Pätzold

Universität zu Köln

T. Spohn

Westfälische Wilhelms-Universität Münster[b]

1 Introduction

Some investigations of the surface or sub-surface of near-Earth objects (NEOs) that are needed to support mitigation demand contact with the surface. The main examples are:

- Seismological methods, requiring both sources and receivers, to examine the internal structure of NEOs and look for cracks and voids that may influence the mitigation strategy and its effects.
- Surface and sub-surface mechanical properties measurements, to determine the material's response to drilling, digging, hammering, impacts, explosive detonations, etc. The type of measurements performed would depend on the mechanical interaction involved in the mitigation strategy being pursued (e.g., whether low or high strain rate).
- Measurements of sub-surface thermal properties and volatile content, with a view to using non-gravitational forces (outgassing) for mitigation.
- Emplacement of a radio beacon to help refine predictions of a NEO's future orbit.
- Radio transmission tomography, to examine the interiors of NEOs.

There are of course many other potential investigations requiring surface contact that appear to be rather less important *for mitigation*, being motivated wholly by

---

[a] Now at Universität Bern.   [b] Now at DLR Berlin.

Mitigation of Hazardous Comets and Asteroids, ed. M. J. S. Belton, T. H. Morgan, N. H. Samarasinha, and D. K. Yeomans. Published by Cambridge University Press. © Cambridge University Press 2004.

science or space resources studies. For example, mitigation studies would seem to require compositional information no more detailed than that that can be determined remotely (e.g., by X-ray and infrared (IR) spectroscopy) and by comparison with meteorite analogs. An object's response to a mitigation technique is determined more directly by a set of key physical properties – particularly mechanical, thermal, and structural. For long-duration, low-force mitigation techniques, these properties are generally needed for the material at (or close to) the surface (see Chapter 5), while for impulsive techniques they are needed for the interior (see Asphaug *et al.* 2002; and Chapter 4, this volume). Sub-surface sample acquisition (e.g., for return to Earth or analysis by on-board geochemical instrumentation) and contact or close-up compositional measurements thus fall outside the scope of this paper. Other examples of measurements that would seem to be motivated exclusively by science include microscopy (various techniques), sub-surface imaging, and some other *in situ* physical measurements such as magnetic susceptibility, properties of the plasma environment and dust impact detection.

However, as concepts for mitigation develop, it is possible that additional requirements for surface or sub-surface investigation will emerge. It is also possible that scientific, space resources and mitigation-related investigations will be combined in a single mission.

Of the many concepts for "surface missions," this chapter focuses on "payload delivery" penetrators, soft landers, and surface and sub-surface mobility, with particular reference to the mitigation-related investigations listed above. Mission objectives may of course be best achieved by other means that are technically simpler to implement but are nevertheless still "surface missions," to some extent at least. Indeed, many such techniques have been demonstrated by previous missions or described in proposals:

- Destructive impact of spacecraft for study of the resulting crater or ejecta, or to act as a large seismic source: Apollo (Latham *et al.* 1970), Lunar Prospector (Goldstein *et al.* 2001), Deep Impact (A'Hearn *et al.* 2000), BepiColombo (Lognonné *et al.* 2003), Clementine II (Hope *et al.* 1997), Don Quijote (see Section 2).
- Passive projectiles, for the same reasons as above or to act as target markers or reflectors: Hayabusa (Yano *et al.* 2000; Sawai *et al.* 2001), Aladdin (Cheng *et al.* 2000).
- Active projectiles, either launched from the spacecraft or from the surface by mortars, to act as seismic sources and/or for study of the resulting crater or ejecta: Apollo (Kovach and Watkins 1973a, 1973b), Deep Interior (Asphaug *et al.* 2001).
- "Touch-and-go" measurements, e.g., for surface sampling: (Hayabusa (Yano *et al.* 2000), HERA (Sears *et al.*, 2002; Chapter 15, this volume).
- Tethering / anchoring to secure a spacecraft at a particular location: Phobos DAS (Sagdeev *et al.* 1988; TsUP 1988), Rosetta Lander (now Philae) (Thiel *et al.* 2001), ST-4 Champollion (Steltzner and Nasif 2000).

- End-of-mission landings for descent imagery and possible extended mission operations on the surface: NEAR–Shoemaker (Dunham *et al.* 2002).
- Hovering at very low altitude (<100 m), for remote investigation of the surface (active or passive); low surface gravity means that minor body surface missions do not necessarily require landing! Phobos 1 and 2 (Sagdeev *et al.* 1988; TsUP 1988), Hayabusa (Uo *et al.* 2001).

Yano *et al.* (2002) discussed a range of *in situ* and remote techniques for structural investigations of NEOs, from micrometer to kilometer scales.

2 Seismological methods

The investigation of NEO interiors by seismological methods will be an important technique for determination of internal structure, being complementary to both radio tomography and gravity field measurements (Huebner and Greenberg 2001). Radio tomography (e.g., Safaeinili *et al.* 2002 and Chapter 10, this volume) is more effective for determining the macroscopic structure of loose or porous material, while seismic studies are more suitable for examining variations in mechanical properties of consolidated material. Gravity field measurements from orbit, together with the object's shape, can provide an indication of any large-scale internal variations in bulk density.

2.1 Impact seismology

A basic seismic investigation would employ a set of sensors (deployed by landers and/or penetrators) to record surface accelerations produced by a large impact at high velocity (e.g., >100 kg at $\sim$10 km s^{-1}). Such an impact can also be used to impart momentum to the NEO and therefore change its velocity. The only seismically recorded examples of such impacts are those performed on the Moon by the impacts of Saturn V upper stages and lunar modules. More recently, such an active seismic experiment has been suggested for the BepiColombo mission to Mercury (Lognonné *et al.* 2003). Signals from the impact of the Apollo 17 upper stage are shown in Fig. 12.1. The momentum of this impact was about 37 000 kg km s^{-1} (14 487 kg at 2.55 km s^{-1}), compared to about 4000 kg km s^{-1} for the impactor spacecraft (Hidalgo) in the Don Quijote ESA study (400 kg at 10 km s^{-1}), 3774 kg km s^{-1} for the Deep Impact impactor (370 kg at 10.2 km s^{-1}), and 203–342 kg km s^{-1} for the three "science probe" impactors of the Clementine II proposal (19.4 kg at 10.5–17.6 km s^{-1}). On the Moon, the impact was detected at distances of 150–1000 km. Such powerful impacts will therefore be detected, even on very large asteroids, and may even produce very large accelerations and strong motions at shorter distances of a few kilometers.

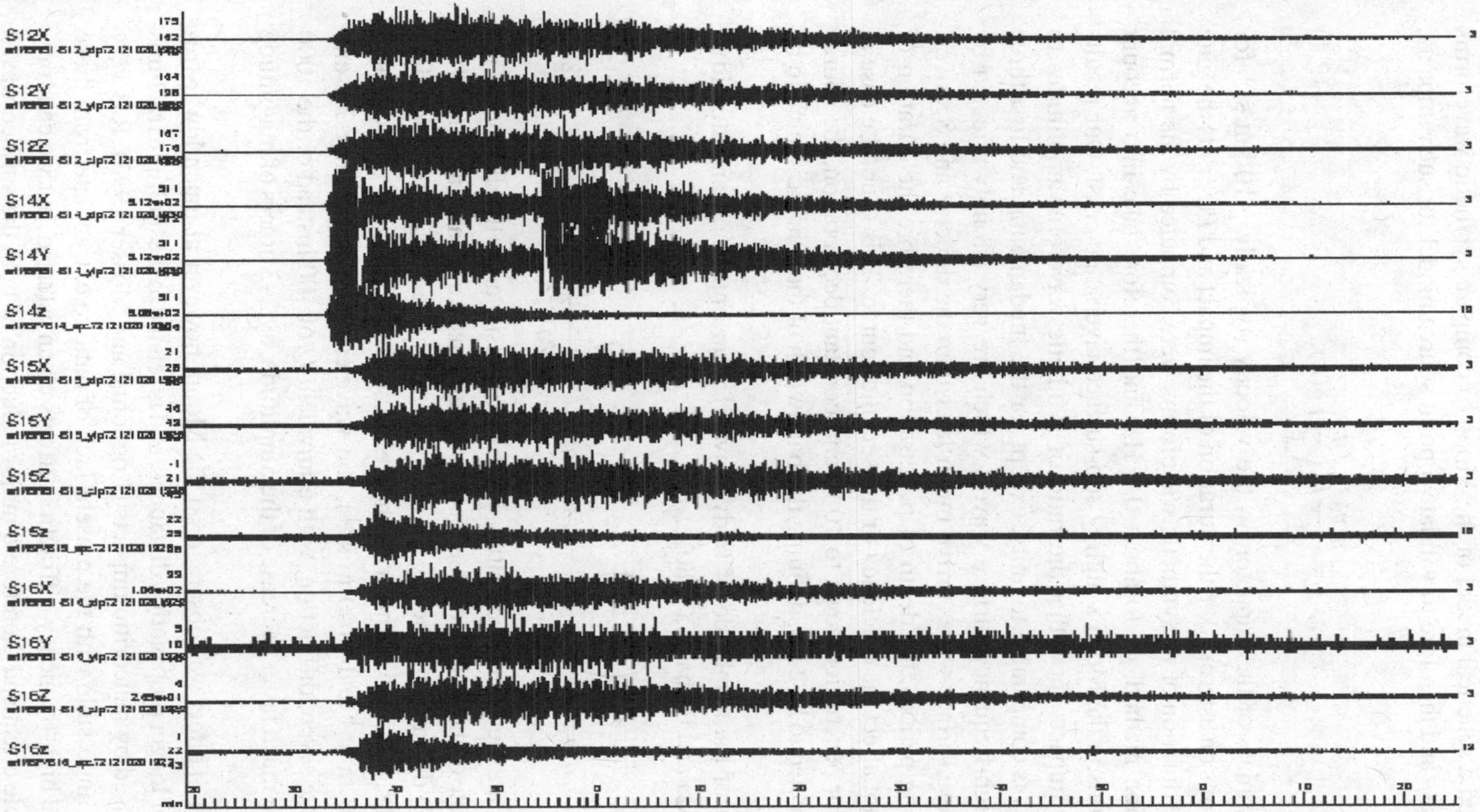

Figure 12.1 Seismic records, from the Apollo seismic network, of the impact of the Apollo 17 Saturn V upper stage (Saturn IV B) on the Moon on December 10, 1972 at 4.21° S, 12.31° W (at distances of 338, 157, 1032, and 850 km from the Apollo 12, 14, 15, and 16 stations, respectively). X, Y, Z are for the long-period seismometers, z is for the short-period seismometer. Sensitivity is 4×10^{-10} m s^{-2} per bit for the long period (at 0.45 Hz) and 1.3×10^{-7} m s^{-2} per bit for the short period (at 8 Hz). Amplitudes at the Apollo 14 station, 157 km from impact, reach about 10^{-5} m s^{-2} with an amplitude mainly related to S waves trapped in the regolith. The first P arrival is typically 10 times smaller.

We have used a spherical model of the Moon to compare synthetic waveforms with seismic data and therefore to validate a point source model for such impacts, expressed by

$$F(t, x) = \frac{mv_0}{\tau} P\left(\frac{t}{\tau}\right) \delta(x),$$
(12.1)

where m is the mass of the impactor, v_0 the velocity (typically $\sim$10 km s^{-1} for impactor missions on asteroids), τ the duration of the impact and $P(t/\tau)$ is the gate function with a duration of τ. Synthetic waveforms were computed by the normal mode summation method, up to about 0.5 Hz, and therefore take into account the reverberations in the Moon's high Q sub-surface layers and crust. Our results from the lunar impacts show that the ratio of synthetic waveform amplitudes to data amplitudes is comparable to unity, except for the fundamental and first three overtones of high-frequency surface waves, which are very strongly affected by diffraction. These surface waves arrive much later than the direct P and S waves, however. At long periods, the depth of penetration and therefore the duration of the source do not affect the efficiency of the seismic source. This is not the case at short period. For penetration down to 10 m and constant deceleration, the source duration is expected to be 2 ms, defining the bandwidth of the seismic source to be about 500 Hz.

An estimate of the amplitude of body waves for an impact on an asteroid is expressed by Aki and Richards (1980):

$$a_p = \frac{F\left(t - \dfrac{r}{c_p}\right)}{4\pi\rho c_p^2 r} \quad \text{and} \quad a_s = \frac{F\left(t - \dfrac{r}{c_s}\right)}{4\pi\rho c_s^2 r},$$
(12.2)

where c_p and c_s are the P and S velocities, ρ the bulk density and r the distance. This is an overestimate of the amplitude due to: (1) conversions at any interface inside the asteroid, and (2) diffraction either by the surface or by lateral variations, including separated blocks. At 500 Hz and with the velocities of moderately consolidated materials (3670 m s^{-1} and 2100 m s^{-1}), the wavelengths are typically 7.4 and 4.2 m. If we have severe diffraction, with equivalent Q of 100 instead of the 1000 value, we have a much faster decrease of the amplitude, by 2–3 orders of magnitude (Fig. 12.2).

This diffraction effect was observed on the Moon, however at much lower frequencies and for longer propagation distances, and is related to the multiple fractures of the crust caused by meteoritic impacts (Lognonné and Mosser 1993). It is very difficult, if not impossible with the present lack of data, to extrapolate from these lunar data to the high-frequency diffraction effects on an asteroid, but we can foresee a comparable effect and therefore a major reduction of both the surface waves

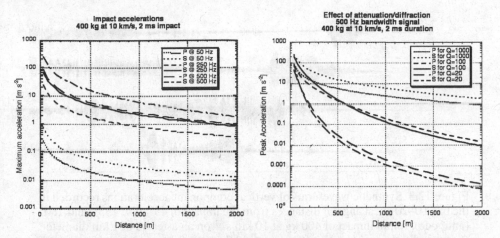

Figure 12.2 Amplitude vs. distance estimates of the P and S waves for the nominal impact and for different frequency bandwidth (left) and diffraction effects (right).

guided by the low velocity sub-surface and of the body waves. For body waves, the scattering reduces the peak-to-peak amplitudes by producing conversions of the wave (mainly P to SV, P to SH) and spreads this energy in time. It was shown that the scattering is able to transfer energy from P to S waves up to an equipartition given by

$$F_p = \frac{v_s^3}{v_p^3} \frac{E_s}{2},$$ (12.3)

where E_p, E_s are the energy of P and S waves, and v_p, v_s the velocities of P and S waves (Aki 1992). Scattering is therefore mainly effective on the P waves, which correspond to most of the signal generated by an explosive source. For S waves, however, the presence of vacuum between building-blocks of the asteroid might produce reverberation inside each block and will therefore reduce the transmission of S waves through several blocks.

All these effects will therefore reduce strongly the amplitudes of waves. The effect can be as strong as a reduction by 10 of the body wave energy, and stronger for the amplitudes due to the length of the coda.

However, the surface gravity of kilometer-sized asteroids is less than 10^{-3} m s^{-2}, and therefore the level of acceleration for a 400 kg impact at several tens of kilometers per second will probably exceed the local gravity. We can therefore expect the generation of dust clouds all around the asteroid after such an impact. Such ejection of material will need to be taken into account in the definition of the orbit of any satellite orbiting the target asteroid during the impact (see Chapter 14). The recording of such an impact by landers will also need strong motion accelerometers, in order to prevent the saturation of seismometers, which have a much higher

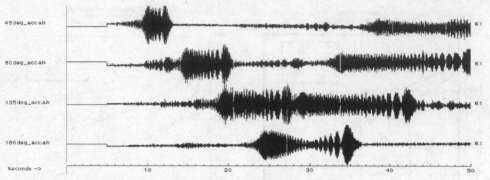

Figure 12.3 Synthetic waveforms of vertical component accelerations recorded in the band 0–20 Hz at angular distances from the impact of 45°, 90°, 135°, and 180° (antipode), for an impact of 400 kg at 10 km s^{-1} on an asteroid of 1 km diameter. Peak amplitudes reach 3.3×10^{-2} m s^{-2}, 2.2×10^{-2} m s^{-2}, 2.1×10^{-2} m s^{-2}, and 2.1×10^{-1} m s^{-2}, respectively. The P wave is 10 to 100 times smaller than the peak amplitude, which corresponds to surface waves. At the antipode, the P wave is of the order of 10^{-3} m s^{-2}. The peak amplitude of the body waves, up to the corner frequency, increases as f^{3} and is therefore 15 times smaller at 20 Hz than at 50 Hz.

resolution and therefore a relatively low clip-level. Moreover, anchoring and/or seismic coupling will be mandatory.

2.2 Asteroid surface waves and normal modes

A large impact will also excite the free oscillations of the asteroid, which will ring after impact, probably for several hours. Seismic waves will also be focused at the antipode of the impact, producing large accelerations. Figure 12.3 shows the expected signals in the frequency band 0–20 Hz, for a simple spherically symmetric asteroid model with a diameter of 500 m. Synthetic waveforms are computed with normal mode summations (Lognonné and Clévédé 2002). The first 2 m are a regolith layer with velocities of 100 m s^{-1} and 58 m s^{-1} respectively for P and S. A 48 m thick layer with velocities of 300 m s^{-1} and 176 m s^{-1} is then taken beneath. A more consolidated material is then found up to a radius of 100 m (500 m s^{-1}, 290 m s^{-1}). The core of the asteroid has velocities of 3670 m s^{-1} and 2100 m s^{-1} corresponding to a concrete model with a density of about 2000 kg m^{-3}. Peak amplitudes of surface waves reach amplitudes 10 times higher than the body waves and even 100 times higher at the antipode. The extrapolation of the surface waves amplitudes at higher frequencies is however impossible to predict and will be very strongly dependent on the shape of the asteroid and regolith layer and of the lateral variations produced by impact craters. However, we might have comparable amplification up to 100 Hz for the surface waves, especially at the antipode, leading to accelerations of more than 10 m s^{-2} in the 0–100 Hz band. Spectral analysis of these records will provide the free frequencies of the asteroid normal mode, strongly sensitive to the mean variation with depth of the shear modulus of the asteroid material but also to the compression modulus and large-scale lateral variations in elastic properties. The spectrum will however be strongly perturbed by the splitting of singlets associated with the shape of the surface and possibly to internal lateral variations.

2.3 Seismic tomography

This more advanced seismic technique has been developed for terrestrial applications (Hanson *et al.* 2000) and is analogous to what might be performed to help examine the interior of a NEO in three dimensions. At least one commercially available three-dimensional tomographics technique (3dT) can probe sub-surface structures and look up to 150 m ahead of tunnel faces. Transducers for the transmission and detection of seismic signals are placed at a number of locations across the surface or within a series of boreholes. Data received from the detectors (large quantities by current interplanetary spacecraft standards) is then inverted using "ray paths" (state-of-the-art software) and converted into three-dimensional images. Low seismic velocities represent cavities or low-density material, while high velocities represent strong, dense material. However, such a technique is difficult to apply to a NEO due to the large number of seismic sources and receivers required, and the ringing effects, which might be quite large.

If controlled seismic sources are used (i.e., sources with known positions and activation times), the critical parameter will be the number of source–receiver paths and not the number of stations as in a network devoted to the detection of natural events. However, if one also wants to use non-controlled sources (e.g., meteorite impacts), three operational seismic stations would be required, ideally several more. The ideal seismic source, in order to reach some comparison with 3dT surveys, must be a mobile device, able to be activated at hundreds of points on the asteroid, for example by generating vibrations or momentum pulses in a spike, and able to move on to a new location.

Seismometers on an asteroid might be able to detect very low amplitude signals due to the lack of micro-seismic noise, as demonstrated by the detection, at several kilometers, of the motion of the Apollo lunar rover on the Moon (Nakamura 1976). The DLR/Russian mole (see Section 7) is able to produce a 0.08 Nm shock energy with its 0.4 kg mass. It is therefore equivalent to an impulse of 0.25 kg m s^{-1}, as compared to the $4 \times 10^6 \text{ kg m s}^{-1}$ impulse of the Hidalgo spacecraft. By scaling the amplitudes found in Fig. 12.2, we obtain a signal amplitude at 1 km for P waves of about $7 \times 10^{-8} \text{ m s}^{-2}$ in a frequency bandwidth of 250 Hz, about three times smaller than the noise level of a micro-seismometer with noise power spectral density of 10^{-8} m s^{-2} Hz$^{-1/2}$ (Lognonné *et al.* 2000). With a sufficient stacking period and a modest increase in the mole's shock impulse capability, it will therefore be possible to get a sufficient signal-to-noise ratio for travel time determination. Such a device might however be difficult to design for repeated operation at many locations.

We might therefore use other possible seismic sources, such as the impact of other stations of the network as they arrive, dedicated impactor spacecraft (e.g., the Clementine II science probes), or end-of-life explosive detonation of other stations. Existing penetrator designs (Vesta/Mars-Aster, Lunar-A, DS-2 Mars Microprobes) would all generate impacts with quite significant impulses (2400, 3705, and 684 kg m s^{-1} respectively), producing signals detectable with large (e.g., 100 or greater) signal to instrument noise ratio at 1 km.

However, the most effective strategy will be to use an orbiting satellite able to shoot small projectiles with explosive heads onto the asteroid in a controlled fashion once the network is installed. Such projectiles can be very light, with yields of 10 to 100 g of explosive, and with a 10 kg mass on board the spacecraft allocated to such a system, we could expect several tens of sources, which will allow more than 100 rays and a reasonable tomographic resolution. An oversizing of the explosive source by a factor 10 will be needed in order to ensure the diffracted P wave amplitudes have a high enough signal-to-noise ratio.

The Don Quijote mission is based on such a strategy, with five seismic measurement points performed by five penetrators equipped with seismometers and an orbiter payload able to release several explosive sources directly from orbit once all the penetrators are deployed. Since these seismic sources are distributed across the object's surface to provide ray paths through its deep interior, boreholes are not necessary, although the seismic stations do need to penetrate a layer of regolith to be sufficiently coupled to the material, especially due to the frequency range of useful bandwidth.

The expected seismic noise on an asteroid might be associated with the thermal cracks related to temperature changes during sunset and sunrise. Such noise was indeed observed on the Moon. These micro-moonquakes or thermal moonquakes increased the level of seismic background noise significantly (Duennebier and Sutton 1974). Such events were assumed to be cracks and slumps near the surface. Due to the small size of an asteroid, such events will probably be detected globally and we can therefore expect a rather high noise level on an asteroid, comparable to that of the Moon at sunset and sunrise, and probably comparable to that of micro-seismometers.

Inversion of the seismic data would require knowledge of the object's shape (for normal modes analysis with a meter-scale resolution) and the location of each station (for body wave analysis, with a decimeter-scale resolution), as well as precise timing information on the source impulses. Only station tilt and not station orientation information would be needed for travel time analysis.

In the case of mitigation missions using an impactor to transfer momentum, the tomography might be performed first before and then after the impact, in order to detect changes in the internal structure. (See Chapter 11 for additional discussion of seismic investigations of NEOs.)

3 Mechanical properties

For many mitigation scenarios the response of the NEO's material to the technique employed depends on both the material's small-scale mechanical properties and, perhaps more critically, its larger-scale structure. This is the case at least for impulsive techniques, which involve mechanical failure of the material at high strain rate. Low force techniques depend, if at all, on the lower strain rate mechanical properties of only the near-surface material.

Some basic constraints on mechanical properties (on scales from global to microscopic and locations from the surface to the deep interior) may be gained from interpretation of remote measurements (from Earth-based telescopes as well as spacecraft), modeling, laboratory simulations, analysis of fireballs and meteoritic materials, and from seismological methods. While instruments dedicated to mechanical measurements – generally employing some kind of destructive technique – are conceivable, such constraints may already be sufficient for the purposes of mitigation. A robust mitigation technique should in any case be able to cope with some uncertainty and variation in mechanical properties. If other constraints are insufficient, the best approach – rather than devising "ideal" experiments to measure specific parameters – might be to perform operations similar to those envisaged for a threatening NEO, if necessary on a reduced scale, and monitor critical "output" parameters such as those of cratering, ejecta dynamics, and momentum transfer. Predictions can then be made for comparable cases based on relatively simple empirical or scaling arguments. The Deep Impact mission adopts a variant of this approach for a comet nucleus; the Don Quijote concept would extend this to an asteroid and incorporate more detailed measurements of the object both before and after the impact event. This practical approach to the characterization of an NEO's mechanical response to potential mitigation techniques has the advantage of avoiding uncertainties arising from the choice of a predictive "bottom–up" model and the assumptions made.

Parameters of relevance include both dynamic and static mechanical properties, since the range of proposed mitigation techniques includes both low and high strain rate operations. Examples of low strain rate operations include landing, drilling, digging, and hammering, in which case the parameters of interest are from soil and rock mechanics, e.g., shear and compressive strength, cohesion, angle of internal friction, bulk density, porosity, grain size distribution, and microstructural texture. Impacts and explosive detonations result in high strain rates, in which case parameters more appropriate to large-scale shock physics are of interest, e.g. Young's modulus, Poisson ratio, shock Hugoniots, and Grüneisen parameter as well as porosity (see Chapter 6). Modeling the shock physics of such impacts or explosions is important for determining how effectively these high-energy mitigation techniques disrupt and/or impart momentum to the target NEO. Such simulations do require validation with real experiments, however.

Soil mechanics properties may vary widely over the surface of an NEO (as suggested by *NEAR–Shoemaker's* imagery of Eros), and between different NEOs, so a thorough assessment of such properties would require visits to multiple locations on multiple objects. In addition, as much contextual information as possible should be obtained at the same locations. This would help us understand what features affect the mechanical properties at a particular location – otherwise one is restricted to a purely statistical approach. Note, however, that simply "scratching the surface" and measuring mechanical properties of the uppermost material might be rather uninformative for our understanding of the deeper ($\sim$1 m or more) layers of practical relevance to mitigation.

Linking mechanical properties to other features observable remotely (e.g., thermal, optical, radar, and topography) via appropriate models of the material's microstructure and surface processes – and obtaining ground truth verification of these links – seems worth pursuing if remotely sensed maps of predicted mechanical properties are required, e.g., to select an appropriate landing site. This suggests a requirement for a series of instrumented mobile devices to make a standard set of mechanical and other measurements at a series of locations across the surface. "Hoppers" such as those built for the Phobos 2 and Hayabusa missions – PrOP-F (Kemurdzhian *et al.* 1988, 1993) and MINERVA (Yoshimitsu *et al.* 2001, 2003), respectively – seem particularly suited to this approach. Indeed, PrOP-F carried a penetrometer and other sensors to carry out a sequence of mechanical, thermal, and electromagnetic measurements at each landing site (Kemurdzhian *et al.* 1989).

Some useful soil mechanics information can be gained by careful measurement of any dynamic operations such as landing, impact penetration, anchoring, drilling, hammering, and "touch-and-go" procedures. Sensors to measure force, torque, acceleration, velocity, or displacement can be used to extract information on strength properties, texture, and layering. An example is the ANC-M accelerometer of the MUPUS experiment mounted in the harpoon anchor of Philae (Kömle *et al.* 2001). Interpretation of such data can be difficult, however, since the operation monitored is unlikely to be optimized for scientific measurements (e.g., the geometry may be complex). Another issue for impact accelerometry measurements is the high sampling rate needed to achieve good spatial resolution at high impact speeds.

Self-inserting "moles" have the potential to measure a profile of mechanical (Pinna *et al.* 2001) and other properties as they advance through the host material; however, like many techniques the properties of interest can be disturbed by the instrumentation, requiring modeling and careful experiment design and data analysis. An instrumented mole has been suggested for measurement of the heat flow and regolith properties on Mercury (Spohn *et al.* 2001).

A further point worth highlighting is that mechanical measurements performed in the low surface gravity environment of NEOs may generate reactive forces or torques that must be balanced (e.g., by anchoring) to avoid ejecting the vehicle from the surface. Drills need to be anchored while operating, while moles require only an initial insertion force, until they start to grip by friction with the surrounding material.

4 Sub-surface thermal properties and volatile content

In order to predict the effect of heating on the material of an NEO (e.g., by concentrated sunlight; see Chapter 9) and the prospects for producing useful non-gravitational forces from outgassing, it is necessary to explore its thermal properties. For example, how effectively could heat be transported from the surface to deeper layers where volatiles (e.g., water ice) may be present? In addition to the volatile-related aspect, thermal properties of NEOs are very important because they may serve as tracers of mechanical properties such as the cohesion and bulk density of the material. Surface thermal properties may easily be measured using remote-sensing techniques, and may even be studied in the laboratory, well in advance of any mitigation activities. Sub-surface thermal measurements require *in situ* techniques, however. The volatile-related aspects are discussed in the first part of this section, while the relation to the mechanical properties follows in the second part.

Whether or not an NEO can (still) contain volatiles (at the surface or buried deep inside) depends strongly on the thermal history of the object, and on its present orbit: depending on its formation region in the solar system, it may or may not have accumulated substantial amounts of volatiles at the time of formation. Concerning the preservation of these primordial volatiles, NEOs fall into two groups: one group is on orbits where the average surface temperature is above $\sim$200 K (note that in vacuum water ice sublimates at around 200 K), the remaining NEOs may get cooler during part of their orbits. Depending on the albedo and surface emissivity, the mean heliocentric distance of a body with an average temperature below 200 K must be larger than about 3 AU. Since an NEO will come as close to the Sun as about 1 AU or closer at least once per orbit (by definition), only NEOs with a rather high eccentricity fall into the second group ($q = 1$ AU, $Q > 3$ AU).

NEOs of both groups have a thermally active layer at and below the surface, where the changing intensity of solar flux will allow the regolith to cool around aphelion, and will warm the regolith well above 200 K as soon as the body is closer to the Sun than about 3 AU. A measure for the layer thickness that undergoes seasonal temperature variations is the thermal skin depth z:

$$z = \sqrt{\frac{\Pi\lambda}{\pi\rho c}},\qquad(12.4)$$

where Π is the orbital period in seconds, λ is thermal conductivity, ρ is the bulk density, and c the specific heat capacity.

For dry regolith and an orbital period of 2 years, z is about 0.25 m. At this depth, the seasonal temperature amplitude is $1/e$ of that at the surface. At about 4.5 z the amplitude is damped to about 1%. Depending on the average temperature and the temperature amplitude (all three depend on the orbit), it may take a few z to reach layers that are not heated above 200 K at least once per orbit. Primordial volatiles may – if at all – be preserved only below this desiccated layer.

Well below the thermally active layers close to the surface, the temperature will eventually reach the *average* temperature the NEO "sees" at its surface. NEOs on orbits with a mean heliocentric distance of about 3 AU or less will eventually loose all volatiles, because the *average* surface temperature is above 200 K, which is too high to keep even water ice stable. In a deep layer that still contains primordial water ice the evaporation process will begin as soon as the temperature reaches approximately 200 K. The specific timescale τ for heat to reach these layers is given by

$$\tau = \frac{d^2\rho c}{\lambda}\qquad(12.5)$$

where d is the depth of the layer, ρ is the bulk density, c is the specific heat capacity, and λ is the thermal conductivity. Though the evaporation process may take some time, the overall timescale will be of a similar order of magnitude. A highly porous and poorly conducting body would keep its volatiles longer than more compact bodies. In order to estimate the longest possible survival times, we assume that layers above d are already dry (if layers above d still contain some volatiles, the thermal conductivity would be even larger; assuming dry regolith gives a more conservative estimate). The thermal conductivity of porous dry regolith is as low as about 0.005 W m^{-1} K^{-1} (a more detailed discussion follows below). With a density of 1500 kg m^{-3} (silicate dust with 50% porosity) and a c of 500 J kg^{-1} K^{-1}, we obtain times of 4.75 years for $d = 1$ m, 475 years for 10 m, 47 500 years for

100 m, and 4.75 million years for 1 km. Layers shallower than d are either already dry (stable end state) or are currently losing volatiles by evaporation (short transitional state). If an NEO has been on an orbit with a mean heliocentric distance <3 AU for geological timescales, it should be entirely dry. All NEOs in our second group fall into this category.

The possible detection of water ice in permanent shadows of craters at the hermean and lunar poles is not in conflict with this conclusion, due to the stability of these planets' rotation axes perpendicular to their orbits, and because both have substantial gravity to hold water molecules (possibly supplied by impacting comet nuclei) long enough to allow them to condense in the cold traps in the permanently shadowed areas. NEOs with the rotational orientation and stability to provide cold traps are presumably very rare. Even these cold spots could contain ice trapped before the NEO's parent body broke up only if the parent body was sufficiently large to prohibit the immediate escape of water vapour, and if the pole of the NEO is identical with the pole of the parent body. We can thus safely exclude the existence of substantial amounts of water or other volatiles in those NEOs with moderately eccentric orbits ($e < 0.5$, or $q = 1$ and $Q < 3$), except those few whose orbits have evolved to reach the warmth of the inner solar system for the first time only recently. Those NEOs on orbits with even higher eccentricity are dry, at least close to the surface.

Important exceptions beyond our group 2 (NEOs with an average heliocentric distance >3 AU) are of course cometary nuclei. Because the relative velocity between the Earth and an impactor of this kind would be significantly larger than for an NEA, the potential damage would also be significantly more serious compared to a NEO with the same mass. This difference would also make mitigation much more difficult – and *in situ* studies even more so.

The material of "dry" NEOs would clearly need heating to a very much higher temperature than that needed for "icy" NEOs for useful outgassing to be initiated. Sintering (and ultimately melting) would modify the material and its thermal properties before this stage is reached, so the worth of *in situ* thermal measurements on dry NEOs using landers or penetrators seems doubtful – remote measurements, laboratory experiments, information from meteorites, and modeling would seem to be sufficient. In contrast to conceivable techniques where heat is required to reach sub-surface layers to liberate volatiles, low thermal conductivity is preferred for "dry" NEOs so that deposited heat can remain at the surface to heat (and ultimately vaporize) the material rather than "leaking" into the interior.

Knowing the thermal properties of a porous body is extremely useful because thermal and mechanical properties, especially thermal conductivity and cohesion, are strongly related to each other (e.g., Seiferlin *et al.* 1995; Keller and Spohn 2002). Cohesion is certainly one of the key parameters that control the behavior of a body when treated with explosives, but cannot be measured directly by remote means. The thermal conductivity of the surface, however, can easily be derived with good confidence from thermal inertia measurements, which can be done with good accuracy with available remote sensing technology. The thermal inertia results from the TES IR spectrometer on Mars Global Surveyor (Jakosky *et al.* 2000; Mellon *et al.* 2000; Jakosky and Mellon 2001) and the THEMIS IR imaging spectrometer on Mars Odyssey are impressive examples of this technique.

A very substantial and relevant knowledge of the thermal properties (conductivity, diffusivity, specific heat capacity, and density) may be gained in the laboratory, well in advance of any mitigation or scientific NEO mission. Based on our present knowledge of asteroids we can assume that the surface layers are porous, or maybe the entire body. Many asteroids have a density that is too low for a compact silicate body, and all asteroids imaged so far are covered by deep layers of dust or regolith. The thermal conductivity of porous media depends only weakly on the composition, a little more strongly on porosity, but very strongly on the microstructure (Seiferlin *et al.* 1995; Keller and Spohn 2002). Without knowing the exact composition of the regolith covering an NEO, studies on loose silicate samples in vacuum can provide relevant data. In the absence of gas that might contribute to the heat transport, as it is the case on Mars (Seiferlin *et al.* 1996; Presley and Christensen 1997a, 1997b, 1997c) and comets (Seiferlin *et al.* 1996), loose powders are very weak conductors of heat. The thermal conductivity may be as low as 0.005 W m^{-1} K^{-1}.

The situation changes significantly when the silicate grains are cemented or indurated. On Mars, water ice, CO_2 ice, salts, sediments, and products of mineral erosion may act as cementing agents. In comets, the loose ice grains will eventually form a solidified network of sintered grains. On asteroids, however, none of these processes should play a major role.

Assuming *in situ* thermal measurements are needed for mitigation studies, there are of course techniques available that can be applied with only minor modifications. The measurement of sub-surface thermal properties – temperature profile and thermal conductivity (or, more easily, diffusivity) – can be achieved at a particular location by means of a string of heatable thermal sensors such as the PEN thermal probe of the MUPUS experiment on the Philae (Seiferlin *et al.* 2001), or the mole- and tether-based sensors proposed for heat flow measurement on the surface of Mercury (Spohn *et al.* 2001). The MUPUS PEN probe is a 10 mm diameter composite tube incorporating a 32 cm long Kapton sheet, onto which 16 titanium heatable thermal sensors have been laser-sputtered. The thermal probe will be deployed to a distance of ~1 m from the lander (away from its shadow) and hammered into the cometary surface by means of an electromechanical hammering mechanism. The sensors are used to measure the temperature profile; however, the probe itself unavoidably creates a thermal "short circuit" that must be corrected for by modeling (Hagermann and Spohn 1999). Heating individual sensors and monitoring the resultant rise in temperature over time enables the local thermal diffusivity to be measured. Measurements of other physical properties including bulk density and penetration resistance are important in the analysis of such thermal data. SPICE, a thermal probe similar to the MUPUS probe, but much smaller and simpler, is currently being developed for the Netlander mission, a network of four Mars landers. A simple thermal probe was also carried by Mars Polar Lander. Thermal experiments can also be incorporated into payload delivery penetrators, such as those of the Lunar-A mission.

Thermal measurements can also be used to deduce the presence of volatiles, via their effect on the thermal conductivity and energy balance. It is also possible to distinguish between volatile species (e.g., between water ice and CO_2) by virtue of their different sublimation temperatures.

5 Radio beacon for radio science and orbit refinement

If an NEO is predicted to undergo a close planetary encounter that could divert it onto a collision course with the Earth, its predicted future orbit will be extremely sensitive to the parameters of the close approach. To assess whether or not the NEO will pass through the "keyhole" putting it on a collision course thus requires extremely accurate knowledge of its current orbit, including measurement of non-gravitational perturbations such as the Yarkovsky effect. One way to achieve this is to fix a radio beacon to the NEO to transmit a stable radio signal that can be used for range and Doppler tracking by ground stations, in the same way as for interplanetary spacecraft.

Delivery of a long-lived station to the surface and anchoring it there to transmit a tracking signal over an extended period of time would thus be required. A similar experiment was planned for the fixed landers of the Phobos 1 and 2 missions (Sagdeev *et al*. 1988; TsUP 1988).

Consider the case of a "transponder only" lander just to fulfill the beacon objective. A state-of-the-art spacecraft transponder receiving and transmitting X-band radio signals in up- and down-link mode, respectively, would require 5 W of radio frequency (RF) power, an omnidirectional low-gain antenna (LGA) on the lander and a 35 m ground station on Earth. This would satisfy a sufficient signal-to-noise ratio (SNR) at any geocentric distance up to 2 AU including small amounts of housekeeping data. One can assume velocity accuracies of 0.1 mm s^{-1} from the Doppler measurements and ranging accuracies of the order of 100 m, which will allow precise orbit determination. Restrictions apply at times of small solar elongations when the NEO is close to the Sun as seen from the Earth or at times of large geocentric distances. However, over extended periods of time, these less precise measurements will average out, and precise determination of the orbit and possible perturbing forces acting on the NEO will nevertheless be feasible. These accuracies stated above are certainly far beyond the required ones to determine whether a NEO will eventually collide with Earth.

Adding instruments to the lander payload suite would require a much higher transmitting RF power in order to be able to receive a sufficient radio carrier SNR on ground. Typically, the required DC power for the radio subsystem is three to five times the RF power. It is then also necessary to supply a medium-gain antenna (MGA) with steering and pointing capabilities leading to a higher complexity of the system.

At ultra high frequency (UHF), a significant (several tens of meters) penetration of the radio waves is expected in dry material. UHF occultation might therefore be monitored in order to measure the complex permittivity of the asteroid's sub-surface. See Chapter 10 for discussion of radio tomography of NEOs, which may involve the use of one or more surface stations, e.g., CONSERT on Philae (Barbin *et al.* 1999).

6 Payload delivery penetrators

Payload delivery penetrators are bullet-shaped vehicles designed to penetrate a surface and emplace experiments at some depth. The basic technology for these has existed for several decades (e.g., Simmons 1977; Murphy *et al.* 1981; Bogdanov *et al.* 1988); however, only in the mid-1990s did proposals for their use in solar system exploration begin to be adopted for actual flight.

Impact speeds range from about 60 to 300 m s^{-1}, depending on factors such as the desired depth, the mass and geometry of the penetrator, the expected surface mechanical properties, the shock resistance of internal components, and constraints imposed by the entry and descent from orbit or interplanetary trajectory. Additional impact damping may be included in the form of crushable material (e.g., honeycomb or solid rocket motor casing), sacrificial "cavitator" spikes protruding ahead of the penetrator's tip (e.g., Luna-Glob high-speed penetrators, with speeds exceeding 1.5 km s^{-1}) and gas-filled cavities (e.g., the Mars 96 penetrators).

Masses have ranged from the tiny DS-2 Mars Microprobes at 2.5 kg each (excluding aeroshell) to 45 kg each for the Mars 96 penetrators (4.5 kg payload).

Penetrators may consist of a single unit, or a slender forebody and a wider aft-body linked by an umbilical tether, the two parts separating during penetration to leave the aftbody at the surface. Expected forebody penetration depths have ranged from ~0.5 m for the Mars Microprobes (impacting at ~190 m s^{-1}) up to 4–6 m for Mars 96, with 1–3 m expected for the single-body Lunar-A penetrators

(13 kg each, 140 mm diameter, impacting at $\sim$285 m s^{-1}). Power is usually provided by primary batteries or radioisotope thermoelectric generators (RTGs). The DS-2 Mars Microprobes' nominal lifetime was only a few hours, while the Lunar-A penetrators are expected to have enough power for about a year. Transmission of data back to Earth is usually by means of an omnidirectional antenna and a relay spacecraft.

Experiments flown on (or proposed for) penetrators include the following:

- Accelerometry/gravimetry/tiltmeter
- Thermal sensors (temperature profile, thermal conductivity/diffusivity, heat flow)
- Imaging
- Magnetometer
- Permittivity/conductivity sensors
- Seismometer
- Spectrometers (gamma ray, neutron, alpha/proton/X-ray, X-ray fluorescence, etc.)
- Sample collection for evolved gas analyser/mass spectrometry/spectroscopic analysis
- Penetrators with combined sampling and pyrotechnic return
- Explosive charge
- Meteorological sensors (not applicable to atmosphereless bodies of course!).

Many of the penetrators flown or proposed have technological features that may be applicable to NEO missions. These include propulsion for braking or acceleration, attitude control, low-temperature shock-resistant components, miniaturization, and experiments for seismology, thermal and mechanical properties, and water ice detection. Sadly, neither the Mars 96 penetrators nor the DS-2 Mars Microprobes completed their missions – Lunar-A now has the task of demonstrating penetrator technology on another world for the first time. Table 12.1 list gives key references for penetrator missions and proposals.

7 Soft landers and mobility

The first soft landers for asteroid-like worlds were those of the Phobos 1 and 2 missions. Both spacecraft carried a fixed lander (the DAS, or Long-Term Autonomous Station) (Sagdeev *et al.* 1988; TsUP 1988), while Phobos 2 also carried a mobile lander or "hopper" (PrOP-F: Kemurdzhian *et al.* (1988, 1993)). Sadly, both missions were lost before either type of lander was deployed.

The payload of the Phobos DAS comprised TV cameras, the ALPHA-X alpha-proton-X-ray spectrometer, the LIBRATION optical Sun sensor, a seismometer, the RAZREZ harpoon anchor penetrometer, and the celestial mechanics experiment mentioned earlier. The lander would have anchored itself to the surface and (having waited for the dust to clear) deployed solar arrays, and was expected to survive for several months.

Table 12.1 *Penetrator missions*

Mission	Reference
Mars 96 Penetrators	Surkov and Kremnev 1998
DS-2 Mars Microprobes	Smrekar *et al.* 1999; Smrekar *et al.* 2001
Lunar-A Penetrators	Mizutani *et al.* 2001
Vesta/Mars–Aster Penetrators	European Space Agency 1988; Surkov 1997
CRAF/Comet Nucleus Penetrator	Boynton and Reinert 1995
Luna-Glob High-Speed Penetrators	Galimov *et al.* 1999; Veldanov *et al.* 1999
Luna-Glob Large Penetrators/Polar Station	Surkov *et al.* 1999; Surkov *et al.* 2001
BepiColombo Mercury Surface Element – Penetrator Option	Pichkhadze *et al.* 2002
BepiColombo Hard Lander/Penetrator Option	European Space Agency 2000
Polar Night lunar penetrators	Lucey 2002

PrOP-F was 0.5 m in diameter, and had a mass of 50 kg including a 7 kg payload. Measurements were to be made at several locations within ~1 km of the initial landing site, spending 20 min at each site. The near-spherical lander had four "whiskers" to turn itself back onto its circular base after each landing, in order to bring its sensors and foot-like hopping mechanism into contact with the surface. Power from the batteries was expected to last at least 4 h. PrOP-F's payload was as follows: ARS-FP X-ray fluorescence spectrometer, fluxgate magnetometer, Kappameter magnetic permeability/susceptibility sensor, gravimeter, temperature sensors, conductometer, and mechanical sensors (including a penetrometer).

The Japanese mission Hayabusa, as well as hovering close to the surface of its target asteroid and gently touching the surface to acquire samples, also carries a "hopper" called MINERVA, much smaller than PrOP-F with a mass of just 0.55 kg (Yoshimitsu *et al.* 2001, 2003). Its payload comprises three charge-coupled device (CCD) cameras, thermal sensors, mechanical sensors, and Sun sensors. It will draw power from solar cells and has an internal motor-driven rotary hopping mechanism that is able to provide some degree of directional and range control for the hopping. Also planned for Hayabusa but later canceled was NASA's MUSES-CN nanorover (Jones 2000), which has wheels for either roving or hopping. Its planned payload comprised a multi-band camera, a near-IR point reflectance spectrometer, an APX spectrometer, and a laser ranging system.

The current Deep Interior proposal (Asphaug *et al.* 2003) includes a number of surface science "pods" (Scheeres *et al.* 2003) for basic measurements of the surface environment. Each pod also carries an explosive charge to perform end-of-life cratering experiments.

Richter (1998) addresses the problem of mobility on minor bodies, and Scheeres (Chapter 14, this volume) discusses hopping trajectories. The final position reached by a vehicle on a hopping trajectory across an irregular small body is very sensitive to the vehicle's initial velocity and details of the topography. Although this makes active navigation extremely difficult, it might however be useful to help distribute a network of stations across the surface from a single starting-point. New concepts for minor body surface mobility include the three-legged "rock crawler" from the Institute of Space and Astronautical Science (ISAS) (Maruki *et al.*, in press.

The idea of self-inserting "moles" to deploy experiments underground or acquire samples has been around for many years in various forms (e.g., Scott *et al.* 1968), but only recently has a mole been adopted for actual flight, in this case for subsurface sample acquisition on Mars (Kochan *et al.* 2001; Richter *et al.* 2001). Other types of mole have been proposed, for example the Sub-Surface Explorer (SSX) (Wilcox 2002) and the Inchworm Deep Drilling System concept (Rafeek *et al.* 2001). The mole-based experiment for heat flow and regolith properties measurements on Mercury (Spohn *et al.* 2001) illustrates some of the potential for integrating a suite of sensors into a mole for *in situ* experiments.

The cometary mission Rosetta (Schwehm, in press) includes a 97.4 kg lander with a 26.7 kg payload designed to carry out a wide range of investigations on the nucleus of comet 67/P Churyumov–Gerasimenko, including imaging, sample analysis, and physical properties measurements (Biele 2002; Biele *et al.* 2002). On touchdown Philae will anchor itself to the surface and start its first scientific sequence, lasting 120 h. A long-term mission will then follow, lasting 3 months, beyond which the lander will eventually fail, probably by overheating as the comet approaches perihelion. Primary batteries, a solar array, and secondary batteries provide power. The ten payload experiments include several that might be adaptable for use in mitigation studies on NEOs. These include MUPUS (thermal and mechanical properties of surface and sub-surface material), CONSERT (radio tomography of the nucleus interior), SESAME (including CASSE sensors for measurements of the acoustic properties of the near-surface layers (Kochan *et al.*, 2000)) and SD^2 (sampling drill and distribution system).

Technology developed for NASA's canceled ST-4 Champollion mission might also be adapted for use in mitigation studies on NEOs, including the Comet Physical Properties Package (CPPP), the pyrotechnically driven telescopic anchoring spike (Steltzner and Nasif 2000), and the sampling drill (Rafeek *et al.* 2000).

8 Discussion

NEO mitigation studies by surface missions are likely to include seismic and radio tomography, mechanical properties measurements, and radio science, and possibly also measurements of thermal properties and volatile content depending on the nature of the object. Such experiments should be performed on a set of example objects reflecting the diversity of the NEO population. There is a choice of surface mission architectures to achieve such goals, namely:

- Single fixed surface station – good for supporting complex sub-surface operations and detailed composition measurements by on-board sample analysis
- Mobile surface station – good for coverage of surface properties at varied locations and achieving multiple paths through the interior for radio tomography and (only if anchoring and the seismic source are repeatable) seismic measurements
- Network of surface stations – good for radio and seismic tomography as well as coverage of surface properties at varied locations.

Of these, the scientific requirements of surface-based NEO mitigation studies seem best served by a network of fixed surface stations, possibly complemented by a mobile surface element for improved coverage of particular properties, and/or the capability for sub-surface mobility (perhaps a penetrator forebody that is also a mole?). Seismological methods represent the strongest driver in the definition of such a mission scenario.

Fixed soft landers may need anchoring, of course – indeed this is essential on fast-rotating NEOs, except near the poles. Surface mobility would appear difficult for such bodies – perhaps this is an area for future investigation? A requirement for anchoring immediately implies a need for some form of attitude control, to ensure correct deployment of the anchoring system.

The likely requirements for anchoring, good seismic coupling, and sub-surface measurements of mechanical and (maybe) thermal properties and volatile content all point towards penetrators rather than soft landers as the appropriate vehicles for surface-based NEO mitigation studies.

The low escape velocity of NEOs means that a penetrator's impact speed has to be achieved either by on-board propulsion – resulting in extra mass and complexity – or with the arrival speed of the carrier spacecraft. A high-speed flyby scenario would seem to be ruled out, given the resulting short duration window for data relay and other measurements, as well as the possible need to perform initial mapping of the NEO to enable impact site selection. A separate spacecraft sent to rendezvous with the NEO could perform these functions, however. On the other hand, if the required penetrator impact speeds are not much larger than 60 m s^{-1}, the carrier could arrive at such a relative speed, deploy its salvo of penetrators, and then fire its thrusters to achieve rendezvous.

Particular practical considerations for penetrators include the availability of ground test facilities capable of accommodating the appropriate impact speed, projectile mass and diameter, and representative target properties. Simulation tools are clearly important, given the cost of impact tests. All system and payload components must have sufficient shock resistance, and shear forces must be considered as well as axial loads, an issue related to the provision of attitude control of the penetrator.

In summary, investigations related to NEO mitigation will require surface and sub-surface instrumentation for a number of important measurements that are impossible to make by other means. Many of the component technologies exist already but integrated technological development and mission studies are currently needed in a number of areas. The main development priorities are:

1. To integrate the technologies for impact seismology and seismic tomography with the required penetrator, soft lander, surface and sub-surface mobility options.
2. To refine techniques for high strain rate interior mechanical properties measurements.
3. Flight demonstration of penetrators and soft landers for airless minor bodies, perhaps also addressing the difficulties of landers, penetrators and mobility for "fast rotators."

These lander and penetrator priorities should of course be addressed in parallel with those for spacecraft investigation of NEOs by other means, such as radio tomography, gravity field measurements, and the other types of "surface mission" listed in the introduction.

Acknowledgments

A. J. Ball wishes to thank the organizers of the workshop for a travel grant, and the Particle Physics and Astronomy Research Council for rolling grant funding. The work of P. Lognonné was funded by the European Space Agency in the frame of the Don Quijote assessment study.

References

A'Hearn, M., Delamere, A., and Frazier, W. 2000. The Deep Impact mission: opening a new chapter in cometary science. Presented at the *51st International Astronautical Congress*, Rio de Janeiro, October 2–6, 2000, Paper no. IAA-00-IAA.11.2.04.

Aki, K. 1992. Scattering conversions P to S versus S to P. *Bull. Seism. Soc. Am.* **82**, 1969–1972.

Aki, K. and Richards, P. G. 1980. *Quantitative Seismology*. San Francisco, CA: W. H. Freeman.

Asphaug, E., Belton, M. J. S., and Kakuda, R. Y. 2001. Geophysical exploration of asteroids: the Deep Interior mission concept. *Lunar Planet. Sci. Conf.* **32** Abstract no. 1867.

Asphaug, E., Ryan, E. V., and Zuber, M. T. 2002. Asteroid interiors. In *Asteroids III*, eds. W. F. Bottke, A. Cellino, P. Paolicchi, and R. P. Binzel, pp. 463–484. Tucson, AZ: University of Arizona Press.

Asphaug, E., Belton, M. J. S., Cangahuala, A., *et al.* 2003. Exploring asteroid interiors: the Deep Interior mission concept. *Lunar Planet. Sci. Conf.* **34**, Abstract no. 1906.

Barbin, Y., Kofman, W., Nielsen, E., *et al.* 1999. The CONSERT instrument for the Rosetta mission. *Adv. Space Res.* **24**, 1115–1126.

Biele, J. 2002. The experiments onboard the Rosetta lander. *Earth, Moon and Planets* **90**, 445–458.

Biele, J., Ulamec, S., Feuerbacher, B., *et al.* 2002. Current status and scientific capabilities of the Rosetta lander payload. *Adv. Space Res.* **29**, 1199–1208.

Bogdanov, A. V., Nikolaev, A. V., Serbin, V. I., *et al.* 1988. Method for analysing terrestrial planets. *Kosmich. Issled.* **26**, 591–603. (In Russian.) Translation in *Cosmic Res.* **26**, 505–515 (NB: the translation omits the first two authors!)

Boynton, W. V. and Reinert, R. P. 1995. The cryo-penetrator: an approach to exploration of icy bodies in the solar system. *Acta Astronaut.* **35** (suppl.), 59–68.

Cheng, A. F., Barnouin-Jha, O. S., and Pieters, C. M. 2000. Aladdin: sample collection from the moons of Mars. *Concepts and Approaches for Mars Exploration*, Houston, July 18–20, 2000, Abstract no. 6108.

Duennebier, F. and Sutton, G. H. 1974. Thermal moonquakes. *J. Geophys. Res.* **79**, 4351–4364.

Dunham, D. W., Farquhar, R. W., McAdams, J. V., *et al.* 2002. Implementation of the first asteroid landing. *Icarus* **159**, 433–438.

European Space Agency. 1988. *Vesta: A Mission to the Small Bodies of the Solar System.* ESA SCI(88)6. Noordwijk, The Netherlands: ESA.

European Space Agency. 2000. *BepiColombo: an Interdisciplinary Mission to the Planet Mercury.* ESA SCI(2000)1. Noordwijk, The Netherlands: ESA.

Galimov, E. M., Kulikov, S. D., Kremnev, R. S., *et al.* 1999. The Russian lunar exploration project. *Astronomich. Vestnik.* **33** (5), 374–385. (In Russian.) Translation in: *Solar System Res.* **33** (5), 327–337.

Goldstein, D. B., Austin, J. V., Barker, E. S., *et al.* 2001. Short-time exosphere evolution following an impulsive vapor release on the Moon. *J. Geophys. Res.* **106**, 32841–32845.

Hagermann, A. and Spohn, T. 1999. A method to invert MUPUS temperature recordings for the subsurface temperature field of P/Wirtanen. *Adv. Space Res.* **23**, 1333–1336.

Hanson, D. R., Haramy, K. Y., and Neil, D. M. 2000. Seismic tomography applied to site characterization. In *Use of Geophysical Methods in Construction, Geotechnical Special Publication* **108**, 65–79.

Hope, A. S., Kaufman, B., Dasenbrock, R., *et al.* 1997. A Clementine II mission to the asteroids. In *Dynamics and Astrometry of Natural and Artificial Celestial Bodies, Proceedings of IAU Colloquium 165*, eds. I. M. Wytrzyszczak, J. H. Lieske, and R. A. Feldman, pp. 183–190.

Huebner, W. F. and Greenberg, J. M. 2001. Methods for determining material strengths and bulk properties of NEOs. *Adv. Space Res.* **28**, 1129–1137.

Jakosky, B. M. and Mellon, M. T. 2001. High-resolution thermal inertia mapping of Mars: sites of exobiological interest. *J. Geophys. Res.* **106**, 23887–23908.

Jakosky, B. M., Mellon, M. T., Kieffer, H. H., *et al.* 2000. The thermal inertia of Mars from the Mars Global Surveyor thermal emission spectrometer. *J. Geophys. Res.* **105**, 9643–9652.

Jones, R. M. 2000. The MUSES CN rover and asteroid exploration mission. *Proc. 22nd International Symposium on Space Technology and Science*, Morioka, Japan, May 28 – June 4, 2000, pp. 2403–2410.

Keller, T. and Spohn, T. 2002. Theoretical aspects and interpretation of thermal measurements concerning the subsurface investigation of a cometary nucleus. *Planet. Space Sci.* **50**, 929–937.

Kemurdzhian, A. L., Bogomolov, A. F., Brodskii, P. N., *et al.* 1988. Study of Phobos' surface with a movable robot. In *Phobos: Scientific and Methodological Aspects of the Phobos Study*, pp. 357–367. Moscow: Space Research Institute, USSR Academy of Sciences.

Kemurdzhian, A. L., Brodskii, P. N., Gromov, V. V., *et al.* 1989. A roving vehicle for studying the surface of Phobos (PROP). In *Instrumentation and Methods for Space Exploration*, ed. V. M. Balebanov, pp. 141–148. Moscow: Nauka. (In Russian.)

Kemurdzhian, A. L., Gromov, V. V., Kazhukalo, I. F., *et al.* 1993. *Planet Rovers*, 2nd edn. Moscow: Mashinostroenie. (In Russian.)

Kochan, H., Feibig, W., Konopka, U., *et al.* 2000. CASSE: the Rosetta lander comet acoustic surface sounding experiment – status of some aspects, the technical realisation and laboratory simulations. *Planet. Space Sci.* **48**, 385–399.

Kochan, H., Hamacher, H., Richter, L., *et al.* 2001. The mobile penetrometer (mole): a tool for planetary sub-surface investigations. In *Penetrometry in the Solar System*, eds. N. I. Kömle, G. Kargl, A. J. Ball, and R. D. Lorenz, pp. 213–243. Vienna, Austria: Austrian Academy of Sciences Press.

Kömle, N. I., Ball, A. J., Kargl, G., *et al.* 2001. Impact penetrometry on a comet nucleus: interpretation of laboratory data using penetration models. *Planet. Space Sci.* **49**, 575–598.

Kovach, R. L. and Watkins, J. S. 1973a. Apollo 17 seismic profiling: probing the lunar crust. *Science* **180**, 1063–1064.

Kovach, R. L. and Watkins, J. S. 1973b. The structure of the lunar crust at the Apollo 17 site. *Geochim. Cosmochim. Acta.* **37** (suppl. 4), 2549–2560.

Latham, G. V., Ewing, M., Dorman, J., *et al.* 1970. Seismic data from man-made impacts on the Moon. *Science* **170**, 620–626.

Lognonné, P. and Clévédé, E. 2002. Normal modes of the Earth and planets. In *International Handbook of Earthquake and Engineering Seismology*, eds. W. Lee, H. Kanamori, and P. Jennings, pp. 125–147. London: Academic Press.

Lognonné, P. and Mosser, B. 1993. Planetary seismology. *Surv. Geophys.* **14**, 239–302.

Lognonné, P., Giardini, D., Bancrdt, B., *et al.* 2000. The NetLander very broad band seismometer. *Planet. Space Sci.* **48**, 1289–1302.

Lognonné, P., Gudkova, T., Giardini, D., *et al.* 2003. An active seismic experiment on BepiColombo. *Geophys. Res. Abstracts* **5**, 10285.

Lucey, P. G. 2002. Polar Night: a mission to the lunar poles. *The Moon Beyond 2002: Next Steps in Lunar Science and Exploration*, Taos, September 12–14, 2002, Abstract no. 3067.

Maruki, T., Yoshida, K., and Yano, H. 2004. A novel strategy for asteroid exploration with a surface robot. *Adv. Space Res.* (in press)

Mellon, M. T., Jakosky, B. M., Kieffer, H. H., *et al.* 2000. High-resolution thermal inertia mapping from the Mars Global Surveyor thermal emission spectrometer. *Icarus* **148**, 437–455.

Mizutani, H., Fujimura, A., Hayakawa, M., *et al.* 2001. Lunar-A penetrator: its science and instruments. In *Penetrometry in the Solar System*, eds. N. I. Kömle, G. Kargl, A. J. Ball, and R. D. Lorenz, pp. 125–136. Vienna, Austria: Austrian Academy of Sciences Press.

Murphy, J. P., Reynolds, R. T., Blanchard, M. B., *et al.* 1981. *Surface Penetrators for Planetary Exploration: Science Rationale and Development Program.* NASA TM-81251 Moffett Field, CA: Ames Research Center.

Nakamura, Y. 1976. Seismic energy transmission in the lunar surface zone determined from signals generated by movement of lunar rovers. *Bull. Seismol. Soc. Am.* **66**, 593–606.

Pichkhadze, K., Vorontsov, V., Rogovsky, G., *et al.* 2002. Technical proposal on forming the expedition with landing onto the surface of Mercury. *Geophys. Res. Abstracts* **4**, EGS02-A-06764.

Pinna, S., Angrilli, F., Kochan, H., *et al.* 2001. Development of the mobile penetrometer (mole) as sampling tool for the Beagle 2 lander on Mars Express 2003. *Adv. Space Res.* **28**, 1231–1236.

Presley, M. A. and Christensen, P. R. 1997a. Thermal conductivity measurements of particulate Materials. 1. A review. *J. Geophys. Res.* **102**, 6535–6550.

Presley, M. A. and Christensen, P. R. 1997b. Thermal conductivity measurements of particulate materials. 2. Results. *J. Geophys. Res.* **102**, 6551–6566.

Presley, M. A. and Christensen, P. R. 1997c. The effect of bulk density and particle size sorting on the thermal conductivity of particulate materials under Martian atmospheric pressures. *J. Geophys. Res.* **102**, 9221–9229.

Rafeek, S., Myrick, T. M., Gorevan, S. P., *et al.* 2000. Sample acquisition systems for sampling the surface down to 10 meters below the surface for Mars exploration. *Concepts and Approaches for Mars Exploration*, Houston, July 18–20, 2000, Abstract no. 6239.

Rafeek, S., Gorevan, S. P., Bartlett, P. W., *et al.* 2001. The Inchworm Deep Drilling system for kilometer-scale subsurface exploration of Europa (IDDS). *Innovative Approaches to Outer Planetary Exploration 2001–2020*, Houston, February 21–22, 2001, Abstract no. 4085.

Richter, L. 1998. Principles for robotic mobility on minor solar system bodies. *Robot. Auton. Systems* **23**, 117–124.

Richter, L., Coste, P., Gromov, V., *et al.* 2001. Development of the "Planetary Underground Tool" subsurface soil sampler for the Mars Express "Beagle 2" lander. *Adv. Space Res.* **28**, 1225–1230.

Safaeinili, A., Gulkis, S., Hofstadter, M. D., *et al.* 2002. Probing the interior of asteroids and comets using radio reflection tomography. *Meteor. and Planet. Sci.* **37**, 1953–1963.

Sagdeev, R. Z., Balebanov, V. M., and Zakharov, A. V. 1988. The Phobos project: scientific objectives and experimental methods. *Sov. Sci. Rev. E: Astrophys. Space Phys. Rev.* **6**, 1–60.

Sawai, S., Kawaguchi, J., Scheeres, D., *et al.* 2001. Development of a target marker for landing on asteroids. *J. Spacecraft and Rockets* **38**, 601–608.

Scheeres, D. J., Asphaug, E. I., Colwell, J., *et al.* 2003. Asteroid surface science with pods. *Lunar Planet. Sci. Conf.* **34**, Abstract no. 1444.

Schwehm, G. (ed.) (in press). ESA Special Publication on the Rosetta mission. ESA SP-1165. Noordwijk, The Netherlands: ESA.

Scott, R. F., Howard, E. A., and Hotz, G. M. 1968. *Burrowing Apparatus*. US Patent 3 375 885.

Sears, D. W. G., Franzen, M., Bartlett, P. W., *et al.* 2002. The Hera mission: laboratory and microgravity tests of the Honeybee Robotics touch-and-go-sampler. *Lunar Planet. Sci. Conf.* **33**, Abstract no. 1583.

Seiferlin, K., Spohn, T., and Benkhoff, J. 1995. Cometary ice texture and the thermal evolution of comets. *Adv. Space Res.* **15**, 35–38.

Seiferlin, K., Kömle, N. I., Kargl, G., *et al.* 1996. Line heat-source measurements of the thermal conductivity of porous H_2O ice, CO_2 ice and mineral powders under space conditions. *Planet. Space Sci.* **44**, 691–704.

Seiferlin, K., Hagermann, A., Banaszkiewicz, M., *et al.* 2001. Using penetrators as thermal probes: the MUPUS case. In *Penetrometry in the Solar System*, eds. N. I.

Kömle, G. Kargl, A. J. Ball, and R. D. Lorenz, pp. 161–184. Vienna, Austria: Austrian Academy of Sciences Press.

Simmons, G. J. 1977. Surface penetrators: a promising new type of planetary lander. *J. Brit. Interplanet. Soc.* **30**, 243–256.

Smrekar, S., Catling, D., Lorenz, R., *et al.* 1999. Deep Space 2: the Mars Microprobe mission. *J. Geophys. Res.* **104**, 27013–27030.

Smrekar, S., Lorenz, R. D. and Urquhart, M. 2001. The Deep-Space-2 penetrator design and its use for accelerometry and estimation of thermal conductivity. In *Penetrometry in the Solar System*, eds. N. I. Kömle, G. Kargl, A. J. Ball, and R. D. Lorenz, pp. 109–123. Vienna, Austria: Austrian Academy of Sciences Press.

Spohn, T., Ball, A. J., Seiferlin, K., *et al.* 2001. A heat flow and physical properties package for the surface of Mercury. *Planet. Space Sci.* **49**, 1571–1577.

Steltzner, A. D. and Nasif, A. K. 2000. Anchoring technology for *in situ* exploration of small bodies. *Proc. IEEE Aerospace Conference* **7**, 507–518.

Surkov, Y. A. 1997. *Exploration of Terrestrial Planets from Spacecraft: Instrumentation, Investigation, Interpretation*, 2nd. edn. Chichester, UK: Wiley-Praxis.

Surkov, Y. A. and Kremnev, R. S. 1998. Mars-96 mission: Mars exploration with the use of penetrators. *Planet. Space Sci.* **46**, 1689–1696.

Surkov, Y. A., Moskaleva, L. P., Shcheglov, O. P., 1999. Lander and scientific equipment for exploring of volatiles on the Moon. *Planet. Space Sci.* **47**, 1051–1060.

Surkov, Y. A., Kremnev, R. S., Pichkhadze, K. M., *et al.* 2001. Penetrators for exploring solar system bodies. In *Penetrometry in the Solar System*, eds. N. I. Kömle, G. Kargl, A. J. Ball, and R. D. Lorenz, pp. 185–196. Vienna, Austria: Austrian Academy of Sciences Press.

Thiel, M., Stöcker, J., Rohe, C., *et al.* 2001. The Rosetta lander anchoring harpoon: subsystem and scientific instrument. In *Penetrometry in the Solar System*, eds. N. I. Kömle, G. Kargl, A. J. Ball, and R. D. Lorenz, pp. 137–149. Vienna, Austria: Austrian Academy of Sciences Press.

TsUP. 1988. *Phobos*. Moscow: TsUP (Spaceflight Control Centre)/Informelektro. (In Russian.)

Uo, M., Baba, K., Kubota, T., *et al.* 2001. Navigation, guidance and control of asteroid sample return spacecraft: MUSES-C. *NEC Res. Development* **42**, 188–192.

Veldanov, V. A., Smirnov, V. E., and Khavroshkin, O. B. 1999. Lunar penetrator: reducing overload and penetration control. *Astronomich. Vestnik.* **33** (5), 490–494. (In Russian.) Translation in: *Solar System Res.* **33** (5), 432–436.

Wilcox, B. H. 2002. *Method and Apparatus for Subsurface Exploration.* US Patent 6 488 105.

Yano, H., Fujiwara, A., Hasegawa, S., *et al.* 2000. MUSES-C's impact sampling device for small asteroid surfaces. *Near-Earth Asteroid Sample Return Workshop*, Houston, December 11–12, 2000, Abstract no. 8021.

Yano, H., Terazono, J., Akiyama, H., *et al.* 2002. Techniques for the structural investigation of aggregate bodies. *Workshop on Scientific Requirements for Mitigation of Hazardous Comets and Asteroids*, Arlington, September 3–6, 2002, Abstract no. 29.

Yoshimitsu, T., Kubota, T., Nakatani, I., *et al.* 2001. Robotic lander MINERVA, its mobility and surface exploration. *Adv. Astronaut. Sci.* **108**, 491–501

Yoshimitsu, T., Kubota, T., Nakatani, I., *et al.* 2003. Micro-hopping robot for asteroid exploration. *Acta Astronaut.* **52**, 441–446.

13

Optimal interception and deflection of Earth-approaching asteroids using low-thrust electric propulsion

Bruce A. Conway

University of Illinois at Urbana–Champaign

1 Introduction

The spectacular collision of the Shoemaker-Levy 9 comet with Jupiter in July 1994 was a dramatic reminder of the fact that the Earth has and will continue to experience such catastrophic events. While the frequency of such massive collisions is very low, smaller objects collide with the Earth regularly and do damage that would be intolerable in any populated region. As an example, the Tunguska (Siberia) event of 1908 is estimated to have involved a 60-m object exploding at a height of 8 km and produced devastation over an area almost the same as that devastated by the eruption of Mt. St. Helens (Morrison *et al.* 1994). The famous 1-km Meteor Crater in Arizona was made by the impact of an even smaller body only 30 m in diameter (Adushkin and Nemchinov 1994). Human casualties due to direct meteorite strikes are rare but known (Yau 1994). The greater danger is due to the fact that the time between large impacts, such as the Tunguska impact which released tens of megatons of TNT equivalent energy, is significant compared to a human lifetime and there is a small chance that any impact will be in a populated area. The relative scarcity of such areas on the Earth may not offer the protection one might think as recent calculations suggest that larger bodies might do more damage if they didn't hit land; predicting that an impact anywhere in the Atlantic Ocean by a 400-m asteroid would devastate the (well-populated) coasts on both sides of the ocean with tsunamis over 60 m high (Hills *et al.* 1994). A consensus is developing that while the probability for collision is low the potential for destruction is immense and thus some resources should be devoted to threat detection and possible interdiction.

The population of Earth-crossing objects is significant and continuously increasing by virtue of discovery. As of January 2003, 2161 Earth-approaching asteroids

Mitigation of Hazardous Comets and Asteroids, ed. M. J. S. Belton, T. H. Morgan, N. H. Samarasinha, and D. K. Yeomans. Published by Cambridge University Press. © Cambridge University Press 2004.

are known, the largest being 1627 Ivar with a dimension of ~8 km and mass of ~10^{15} kg. (For comparison the K/T or Cretaceous/Tertiary boundary impact of 65 million years ago which left a ~200 km crater, releasing an estimated 10^8 MT equivalent energy in the Gulf of Mexico, would have been caused by a body ~10 km in size.) Of these 2161 asteroids. 449 are considered to be Potentially Hazardous Objects (PHOs) because of their size and the distance to which they approach the Earth. The census of Earth approaching asteroids is believed complete only for objects of Ivar's size; however the rate of discovery has improved dramatically over the last decade and in the near future it is likely that nearly all of the potentially hazardous objects will be identified.

Of course it is the prevention of a catastrophic impact with which this work is concerned. The dangerous object must be intercepted, at the earliest possible time, and then deflected or destroyed. One strategy may be to detonate a large nuclear weapon near the surface of the asteroid or comet, vaporizing part of the surface and yielding an impulsive velocity change because of the momentum imparted to the ejected mass (Ahrens and Harris 1994; Solem and Snell 1994). In the first part of this chapter we consider optimization of the trajectory taken by the interceptor spacecraft to the asteroid. Low-thrust, high specific impulse propulsion is used because of the significant advantages it provides in propulsive mass required for a given mission. However, a low-thrust departure from Earth would require many revolutions of the Earth, which would consume a lot of time. It seems reasonable then to use an impulsive velocity change for the initial departure, followed by continuous low-thrust propulsion (Conway 1997).

The analysis is necessarily three-dimensional as many of the Earth-crossing asteroids have significant inclinations. However, it simplifies the analysis to assume that before the application of the departure impulse the intercepting vehicle is in a low-Earth circular orbit, which lies in the ecliptic plane. The magnitude of this departure impulse is something that might in real life depend on the target and/or on the capability of the launch vehicle. For this work the departure impulse chosen is less than or equal in magnitude to that required for the most commonly employed interplanetary trajectory, i.e., the impulse required for escape from the Earth onto a Hohmann transfer ellipse to Mars, which yields a hyperbolic excess velocity relative to Earth ($v_{\infty/E}$) of 2.94 km s^{-1}. The optimizer may choose the point in the initial low-Earth orbit from which to begin the departure. It may also choose the direction in which it is applied via in-plane and out-of-plane thrust pointing angles. The Earth is assumed to be in a circular orbit about the sun and its true longitude on the departure date is found using the ephemeris program MICA from the US Naval Observatory. Specific impulse for the low-thrust motor is chosen from a range of values (2000–4000 s) representative of current technology.

2 Method

The variational equations used are those for the equinoctial elements. This avoids singularity of the elements for circular or equatorial orbits. The equinoctial elements and their relationship to the conventional elements are (Battin 1987):

$$a = a$$
$$P_1 = e \sin \tilde{\omega}$$
$$P_2 = e \cos \tilde{\omega} \qquad (13.1)$$
$$Q_1 = \tan \frac{i}{2} \sin \Omega$$
$$Q_2 = \tan \frac{i}{2} \cos \Omega$$
$$\ell = \tilde{\omega} + M = \Omega + \omega + M$$

The conventional elements are easily recovered using

$$e = \sqrt{P_1^2 + P_2^2}, \ \Omega = \tan^{-1}\left(\frac{Q_1}{Q_2}\right), \quad \text{and} \quad \tan\frac{i}{2} = \sqrt{Q_1^2 + Q_2^2}. \qquad (13.2)$$

In these variables the variational equations become:

$$\frac{da}{dt} = \frac{2a^2}{h}\left[(P_2 \sin L - P_1 \cos L)R + \frac{p}{r}T\right]$$

$$\frac{dP_1}{dt} = \frac{r}{h}\left\{\frac{p}{r}\cos L R + \left[P_1 + \left(1 + \frac{p}{r}\right)\sin L\right]T - P_2(Q_1 \cos L - Q_2 \sin L)N\right\}$$

$$\frac{dP_2}{dt} = \frac{r}{h}\left\{\frac{p}{r}\sin L R + \left[P_2 + \left(1 + \frac{p}{r}\right)\cos L\right]T - P_1(-Q_1 \cos L + Q_2 \sin L)N\right\}$$

$$\frac{dQ_1}{dt} = \frac{r}{2h}\left(1 + Q_1^2 + Q_2^2\right)\sin L N \qquad (13.3)$$

$$\frac{dQ_2}{dt} = \frac{r}{2h}\left(1 + Q_1^2 + Q_2^2\right)\cos L N$$

$$\frac{d\ell}{dt} = n - \left[\frac{a}{a+b}\left(\frac{p}{h}\right)(P_1 \sin L + P_2 \cos L) + \frac{2b}{a}\right]R$$
$$\qquad - \left[\frac{a}{a+b}\left(\frac{r}{h} + \frac{p}{h}\right)(P_1 \cos L - P_2 \sin L)\right]T$$
$$\qquad - \left(\frac{r}{h}\right)(Q_1 \cos L - Q_2 \sin L)N$$

where:

$$T = \text{tangential component of thrust acceleration} = F \cos \beta \cos \gamma$$
$$R = \text{radial component of thrust acceleration} = F \sin \beta \cos \gamma$$
$$N = \text{normal component of thrust acceleration} = F \sin \gamma$$

$$F = \text{acceleration produced by the low-thrust motor}$$
$$b = \text{semiminor axis} = a\sqrt{1 - P_1^2 - P_2^2}$$
$$h = \text{angular momentum} = n\,a\,b$$
$$n = \text{mean motion} = \sqrt{\frac{\mu}{a^3}}$$

$$\frac{a}{a+b} = \frac{1}{1 + \sqrt{1 - P_1^2 - P_2^2}} \tag{13.4}$$

$$\frac{p}{r} = 1 + P_1 \sin L + P_2 \cos L$$

$$\frac{r}{h} = \frac{h}{\mu(1 + P_1 \sin L + P_2 \cos L)}$$

The control variables in the problem are the thrust pointing angles β and γ. Angle β is the in-plane thrust pointing angle; it is measured from the normal to the radius vector and is a positive angle if it yields a component of thrust pointing radially-outward. Angle γ is the out-of-plane thrust pointing angle. It is positive if it yields a component of thrust in the direction of the orbital angular momentum.

The true longitude L is obtained from the mean longitude ℓ by first solving Kepler's equation in equinoctial variables,

$$\ell = K + P_1 \cos K - P_2 \sin K \tag{13.5}$$

for the eccentric longitude K. Then the ratio r/a may be found as

$$r/a = 1 - P_1 \sin K - P_2 \cos K \tag{13.6}$$

The functions of eccentric longitude appearing in the variational equations are then determined as:

$$\sin L = \frac{a}{r}\left[\left(1 - \frac{a}{a+b}P_2^2\right)\sin K + \frac{a}{a+b}P_1 P_2 \cos K - P_1\right]$$
$$\cos L = \frac{a}{r}\left[\left(1 - \frac{a}{a+b}P_1^2\right)\cos K + \frac{a}{a+b}P_1 P_2 \sin K - P_2\right] \tag{13.7}$$

The thrust acceleration F varies as propellant is consumed; assuming a constant thrust motor,

$$\frac{dF}{dt} = \frac{F^2}{c} = \frac{F^2}{g\,I_{sp}} \tag{13.8}$$

where g is the acceleration of gravity at the Earth's surface, c is the motor exhaust velocity, and I_{sp} is the motor specific impulse (in seconds).

The problem is then to choose the time history of the thrust pointing angles β and γ in order to minimize the performance index, which is the time of flight, subject to satisfaction of the system variational Eqs. (13.3), the system initial condition constraints,

$$(a, P_1, P_2, Q_1, Q_2, \ell) \text{ must satisfy (at } t = 0): \tag{13.9}$$
$$\bar{r} = \bar{r}_E$$
$$\bar{v} = \bar{v}_E + \bar{v}_{\infty/E}$$

(where $\bar{r}_E$ and $\bar{v}_E$ represent the position and velocity of the Earth; $\bar{v}_{\infty/E}$ is the hyperbolic excess velocity of the spacecraft with respect to the Earth) and the terminal constraint (of interception):

$$(a, P_1, P_2, Q_1, Q_2, \ell) \text{ must satisfy (at } t = t_{\text{final}}): \tag{13.10}$$
$$\bar{r} = \bar{r}_A$$

where the subscripts E and A refer to the Earth and the asteroid respectively. The optimizer is free to choose two parameters which do not explicitly appear in the variational equations; two pointing angles, in-plane (β_0) and out-of-(ecliptic)-plane (γ_0) pointing angles which describe the direction of $\bar{v}_{\infty/E}$ following the impulsive Δv which allows the vehicle to escape from low-Earth orbit.

The initial condition constraints Eqs. (13.9) yield the following six scalar equations:

$$a_E \cos \theta_E - \frac{r[\cos L + (Q_2^2 - Q_1^2) \cos L + 2Q_1 Q_2 \sin L]}{1 + Q_2^2 + Q_1^2} = 0$$

$$a_E \sin \theta_E - \frac{r[\sin L - (Q_2^2 - Q_1^2) \sin L + 2Q_1 Q_2 \cos L]}{1 + Q_2^2 + Q_1^2} = 0$$

$$\frac{2r[Q_2 \sin L - Q_1 \cos L]}{1 + Q_2^2 + Q_1^2} = 0$$

$$\frac{\sqrt{\mu/p}[\sin L + (Q_2^2 - Q_1^2) \sin L - 2Q_1 Q_2 \cos L + P_1 - 2P_2 Q_1 Q_2 + (Q_2^2 - Q_1^2)P_1]}{1 + Q_2^2 + Q_1^2}$$
$$- v_E \sin \theta_E + v_{\infty/E} \cos \gamma_0 \sin(\beta_0 - \theta_E) = 0 \tag{13.11}$$

$$\frac{\sqrt{\mu/p}[-\cos L + (Q_2^2 - Q_1^2) \cos L + 2Q_1 Q_2 \cos L - P_2 + 2P_1 Q_1 Q_2 + (Q_2^2 - Q_1^2)P_2]}{1 + Q_2^2 + Q_1^2}$$
$$+ v_E \cos \theta_E + v_{\infty/E} \cos \gamma_0 \cos(\beta_0 - \theta_E) = 0$$

$$\frac{2\sqrt{\mu/p}[Q_2 \cos L + Q_1 \sin L - P_2 + P_1 Q_1 + Q_2 P_2]}{1 + Q_2^2 + Q_1^2} - v_{\infty/E} \sin \gamma_0 = 0$$

where all unsubscripted variables and elements refer to the orbit of the interceptor spacecraft, θ_E is the true longitude of the Earth, v_E is the circular velocity of the Earth, and all quantities are evaluated at $t = 0$.

The terminal constraint (13.10) yields the following three scalar equations:

$$r_A[\cos\theta_A \cos\Omega_A - \cos i_A \sin\Omega_A \sin\theta_A]$$
$$- \frac{r[\cos L + (Q_2^2 - Q_1^2)\cos L + 2Q_1 Q_2 \sin L]}{1 + Q_2^2 + Q_1^2} = 0$$
$$r_A[\cos\theta_A \cos\Omega_A - \cos i_A \sin\Omega_A \sin\theta_A]$$
$$- \frac{r[\sin L - (Q_2^2 - Q_1^2)\sin L + 2Q_1 Q_2 \cos L]}{1 + Q_2^2 + Q_1^2} = 0$$
$$r_A \sin i_A \sin\theta_A - \frac{2r[Q_2 \sin L - Q_1 \cos L]}{1 + Q_2^2 + Q_1^2} = 0 \tag{13.12}$$

where all unsubscripted variables refer to the orbit of the interceptor spacecraft and all quantities are evaluated at t_{final}.

2.1 Numerical optimization

The problem is solved using the method of direct collocation with non-linear programming (DCNLP) (Hargraves and Paris 1987; Herman 1995; Herman and Conway 1996). In this solution method the continuous problem is discretized by dividing the total time into "segments" whose boundaries are termed the system "nodes." Each state is known only at discrete points; at the nodes and, depending on how the problem is formulated, at zero, one, or more points interior to a segment. Perhaps the best-known collocation method is that of Hargraves and Paris (1987), employed in their trajectory optimization program OTIS. This version of the collocation method is both useful in itself and a good introduction to the process of discretizing a continuous problem and optimizing via non-linear programming. We assume that the time history of the problem has been divided into N segments, each of width Δt_i. We then assume that the state within a segment is approximated by a cubic polynomial (in time). The polynomial is uniquely determined by the endpoint information available; the parameter values of the states, x_i and x_{i+1}, and the system equation values, $f_i(x_i, u_i, t_i)$ and $f_{i+1}(x_{i+1}, u_{i+1}, t_{i+1})$, where the us are control variables. It can be shown, using this cubic interpolant, that

$$x_C = -\frac{1}{2}(x_i + x_{i+1}) + \frac{\Delta t_i}{8}(f_i - f_{i+1}) \tag{13.13}$$

is an approximate value for $x(t_C)$ where t_C is the center time $(t_{i+1} + t_i)/2$ and $\Delta t_i = t_{i+1} - t_i$. Similarly, it can be shown that the first-time derivative of the cubic interpolant, evaluated at t_C is

$$x'_C = -\frac{3}{(2\Delta t_i)}(x_i - x_{i+1}) - \frac{1}{4}(f_i + f_{i+1}). \tag{13.14}$$

In the Hargraves and Paris method x'_C is equated to the system equation evaluated at the center, i.e.,

$$x'_C = f_C(x_C, u_C, t_C) \tag{13.15}$$

x'_C can be interpreted as the slope of the cubic at its center while the numerical value of f_C can also be interpreted as a slope. If (13.15) is satisfied the two slopes will be the same; i.e., the cubic interpolant will satisfy the system differential equations at three points, the left and right nodes and the center.

Substituting for Eq. 13.14 in Eq. 13.15 and isolating x_{i+1} yields

$$x_{i+1} = x_i + \frac{\Delta t_i}{6}[f_i + 4f_C + f_{i+1}] \tag{13.16}$$

which is Simpson's quadrature rule, and which must be satisfied for each of the n scalar state variables in the state vector x for each of the N segments into which the problem has been subdivided. The optimization problem has thus been converted into a non-linear programming (NLP) problem; find the discrete states and controls $[x_1, u_1, x_2, u_2, \ldots, x_{N+1}, u_{N+1}]$ satisfying given initial and final state boundary conditions and $n \cdot N$ non-linear constraint equations of form (13.16). Of course there may be upper and lower bounds on all of the state and control variables.

The preceding paragraphs describe just one elementary method for converting a continuous optimization problem into a NLP problem. There are many different ways of constructing the discretization, but all require some kind of implicit integration scheme, analogous to that of Eq. 13.16, i.e., some means of insuring that successive values of the states closely approximate those which would result from a forward integration of the system equations for the given controls. Herman and Conway (1996) have shown how the implicit integration can be accomplished more accurately using constraints which are derived from integration rules of higher order than, and hence having smaller truncation error than, Simpson's rule. In this work a fifth-degree Gauss–Lobatto integration rule,

$$\int_{t_i}^{t_{i+1}} f(t)\,dt = \frac{\Delta t_i}{180}[9\,f(t_i) + 49\,f(t_C - \sqrt{3/7}\Delta t_i) + 64\,f(t_C)$$

$$+ 49\,f(t_C + \sqrt{3/7}\Delta t_i) + 9\,f(t_{i+1})] \tag{13.17}$$

with an error function depending on $(\Delta t_i)^9$, as opposed to Simpson's rule which has an error depending on $(\Delta t_i)^5$, is used implicitly. By analogy with the case for Simpson rule-derived constraints, the constaints which result from using (implicitly)

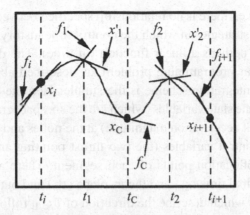

Figure 13.1 Cartoon showing the discretization of the state and control variables when using Gauss–Lobatto collocation.

the fifth-degree Gauss–Lobatto integration rule can be shown to be equivalent to approximating the state within the segment by a fifth-degree polynomial. Six quantities are needed to determine this polynomial; they are chosen to be $x_i, f_i, x_C, f_C,$ $x_{i+1}, f_{i+1},$ (where $x_i = x(t_i), f_i = f(t_i),$ etc.), i.e., the values of the state and system equation at the left node, center, and right node of the ith segment. Forcing the system differential equation to equal the slope of the polynomial at the two internal collocation points,

$$x_1 = x[t_1 = t_C - \sqrt{3/7}(\Delta t_i/2)], \quad x_2 = x[t_2 = t_C + \sqrt{3/7}(\Delta t_i/2)], \quad (13.18)$$

(rather than just the one, center point, in the method used by Hargraves and Paris) yields two system constraint equations per segment:

$$C_{5,1}(x_{i+1}, x_C, x_{i+1}) = 0 = \frac{1}{360}\{(32\sqrt{21} + 180)x_i - 64\sqrt{21}x_C$$
$$+ (32\sqrt{21} - 180)x_{i+1} + \Delta t_i[(9 + \sqrt{21})f_i$$
$$+ 98\ f_1 + 64\ f_C + (9 - \sqrt{21})f_{i+1}]\}$$
$$C_{5,2}(x_{i+1}, x_C, x_{i+1}) = 0 = \frac{1}{360}\{(-32\sqrt{21} + 180)x_i + 64\sqrt{21}x_C$$
$$+ (-32\sqrt{21} - 180)x_{i+1} + \Delta t_i[(9 - \sqrt{21})f_i$$
$$+ 98\ f_2 + 64\ f_C + (9 + \sqrt{21})f_{i+1}]\} \quad (13.19)$$

Note that the sum of these two constraint equations recovers the original integration rule (Eq. 13.17).

The optimizer is free to choose, for each segment, five control variables; $u_i, u_1,$ $u_C, u_2, u_{i+1},$ i.e., the controls at the nodes and at the collocation points (the points corresponding to the states $x_i, x_1, x_C, x_2, x_{i+1}$) as shown in Fig. 13.1. In this work the controls are "free"; i.e., there is no relationship specified between them as opposed to some collocation strategies in which the control time history can be represented by a polynomial or perhaps a linear function (in time). The discretized problem becomes a non-linear programming problem, as described above for the case in which the implicit integration scheme is the simpler Hermite–Simpson rule. The parameters are thus the state variables (which are the six spacecraft orbit equinoctial variables + the thrust acceleration magnitude) at the nodes and center points of the segments and the control variables (the two thrust pointing angles) at the nodes, center point, and collocation points of each segment. There are three additional NLP variables; the final time t_{final} and the in-plane (β_0) and out-of-(ecliptic)-plane (γ_0) pointing angles which describe the direction of $\bar{v}_{\infty/E}$ following the impulsive Δv which allows the vehicle to escape from low-Earth orbit. The system non-linear constraints are the "defect" equations Eqs. (13.19) which enforce satisfaction of the differential equations, the initial condition constraints (Eq. 13.11) and the conditions for interception (Eq. 13.12).

The NLP problem is solved using the program NZSOL, an improved version of the program NPSOL (Gill 1993). For the two example trajectories determined in the next section 20 segments were used to discretize the trajectory yielding an NLP problem with 452 variables. More segments could be used but the results for the state and control variable time histories, presented in the next section, show that all vary slowly and only over a small range, so that 20 segments capture quite well the system time history. This program requires an initial guess of the vector of parameters. The numerical optimization method used here has been quite robust when applied to a variety of orbit transfer problems (Scheel and Conway 1994; Tang and Conway 1995; Herman and Conway 1998). The method was found to be less robust here than it has been for these other applications. However, it has been the experience of this researcher that once an initial guess which allows the optimizer to converge to an optimal trajectory has been determined, even one which has quite different initial and/or terminal conditions from the desired case, new optimal trajectories can be obtained using the converged solution as the new initial "guess." Here the desired problem is solved by solving a succession of problems in which the orbit elements of the target asteroid of the existing solution are "stepped" towards those of the actual asteroid. As an example, for the case described in the next section, the interception of asteroid 1991RB, which has an inclination of 19.58°, the first converged solution was found for a fictitious target having zero inclination but an orbit of the same semimajor axis and eccentricity. This is a significantly easier problem to solve since two of the seven state variables, i.e., all of the Q_1's and Q_2's, are zero and can be set to zero in the optimizer. This solution was then used as an initial guess for a new problem with an inclination of 2°, which converged quickly, and the process continued until the desired problem was solved.

Table 13.1 *Variation of flight time with initial thrust acceleration, specific impulse, and hyperbolic excess velocity (of Earth escape)*

Case	Initial acceleration	I_{sp} (s)	$v_{\infty/E}$(AU/TU)	Flight time (days)	Final acceleration
K	0.140	2000	0.0988	136.57	0.280
M	0.140	4000	0.0988	140.29	0.188
H	0.160	2000	0.0988	129.41	0.349
J	0.170	2000	0.0988	123.55	0.377
N	0.140	4000	0.0330	145.89	0.191
P	0.070	4000	0.0330	166.95	0.083

3 Examples

Several optimal trajectories have been found for the interception of actual Earth-crossing asteroids. Since dimensionless units are used in this analysis (i.e., $\mu = 1$ and the length unit is 1 AU) 1 time unit (TU) is equal to 1 year/2π.

An optimal trajectory has been found for the interception of Earth-approaching asteroid 1991RB. Its orbit elements are, as of September 15, 1991,

$$a = 1.4524 \, AU, \ e = 0.4846, \ i = 19.580°$$
$$\Omega = 359.599°, \ \omega = 68.708°, \ M = 328.08°$$

Its orbit semimajor axis, eccentricity, and inclination are all very representative of the asteroids in the catalog of PHO's. This asteroid approached the Earth to within 0.04 AU, or 15 lunar distances, on September 19, 1998. It is assumed for the following example that launch from Earth takes place 6 months prior to the close approach, that is, on March 19, 1998. The true longitude of the Earth on that date is 177.21°. For this example (referred to in Table 13.1 as Case N) we assume that

$$F_{initial} = 0.14 \, AU/TU^2 = 8.46 \cdot 10^{-5}g, \ I_{sp} = 4000 \text{ s,}$$
$$v_{\infty/E} = .0333 \, AU/TU = 0.9807 \text{ km s}^{-1}$$

This initial thrust acceleration and specific impulse are representative of current electric propulsion technology. The hyperbolic excess velocity, provided by the chemical rocket motor at departure, is equal to approximately 1/3 of what is required for a Hohmann transfer to Mars, so that it also is representative of current technology.

The resulting optimal trajectory is shown in Fig. 13.2. The time of flight is 2.5096 TU = 145.9 days. The in-(orbit)-plane and out-of-plane thrust pointing angles are shown in Figs. 13.3 and 13.4. Note that the out-of-plane thrust pointing angle, γ, is always negative; this is because, as can be seen in Fig. 13.2, the spacecraft must go below the ecliptic plane to intercept the asteroid. In these figures the abscissa

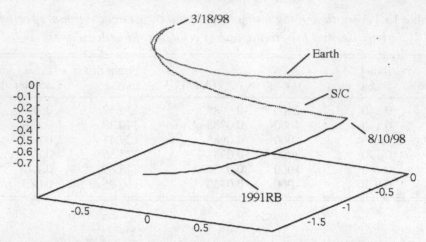

Figure 13.2 Optimal trajectory for the interception of asteroid 1991RB.

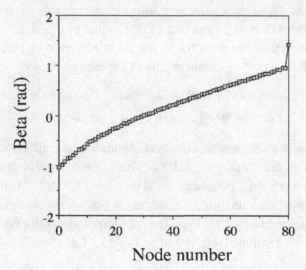

Figure 13.3 History of the in-plane thrust pointing angle for the interceptor orbit
shown in Fig. 13.2.

indicates node position, but this is linearly proportional to time, with the 80th node
corresponding to the final time of 2.5096 TU.

Optimal trajectories for interception of 1991RB have been determined for several different combinations of specific impulse, initial thrust acceleration, and
hyperbolic excess velocity. The results are summarized in Table 13.1. The initial acceleration (in units of AU/TU^2) is the initial value of F, whose time variation
is given by Eq. (13.8). The final acceleration provided in Table 13.1 allows the

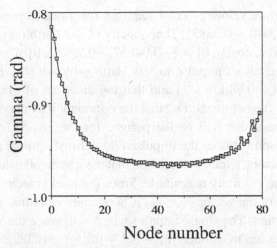

Figure 13.4 History of the out-of-plane thrust pointing angle for the interceptor orbit shown in Fig. 13.2.

amount of propellant used on each trajectory to be calculated, as shown in the next section. The $v_{\infty/E}$ or hyperbolic excess velocity of escape, is shown in each case, with $0.0988 = 2.94$ km s^{-1} just what is required for a Hohmann transfer to Mars, as mentioned previously. It results from an impulse applied in low Earth orbit, usually by the upper stage of the launch vehicle, and was chosen so that we may assume that the asteroid interceptor may be launched with an existing rocket. The smaller $v_{\infty/E}$ assumed for Cases N and P implies either launch with a smaller rocket or that the interceptor spacecraft is more massive than is assumed for the other cases.

4 Comparison with a one-impulse interception

The advantage of using continuous low-thrust propulsion is most apparent when the impulsive Δv required to duplicate the mission is determined. For the case described in the previous paragraph (Case N), the interception of 1991RB, the time of flight, the date of departure, (given in the previous section) and the positions of departure and arrival are known. Determining the conic that connects these two points, in the same time of flight as for the low-thrust trajectory, but using impulsive velocity changes, is a Lambert problem. Solving Lambert's equation yields the semimajor axes of the elliptic sections connecting these two points in the given flight time (Prussing and Conway 1993).

The trajectory with minimum required Δv has a semimajor axis $a = 1.3197$ AU. In terms of a Cartesian coordinate basis (in which the x–y plane is the ecliptic plane and the x-axis points toward the "first point in Aries") the heliocentric departure velocity vector required for a one-impulse interception is $\bar{v}_1 = [0.0931, -1.040, -0.3883]$. The velocity of the Earth on the date of departure is $\bar{v}_E = [-\sin\theta_E, \cos\theta_E, 0] = [-0.04857, -0.99882, 0]$. Case N assumes that the spacecraft is given an impulse in low Earth orbit which yields a $\bar{v}_{\infty/E}$ with magnitude $0.033 (= 0.98$ km s^{-1}) and that the direction of the v-infinity vector may be chosen by the optimizer; i.e., that the optimizer may choose the point over the Earth's surface which will be the perigee for the escape hyperbola. For an equitable comparison between the impulsive/low-thrust optimal trajectory and an all-impulsive trajectory, it must be assumed that the spacecraft using only impulsive thrust has the same v-infinity magnitude. Since the total velocity change required is in the direction of the vector $\bar{v}_1 - \bar{v}_E$, it is obviously optimal to have the small $\bar{v}_{\infty/E}$ in this direction. The single impulse which will place the spacecraft on the Lambert conic thus has magnitude $|\bar{v}_1 - \bar{v}_E| - |\bar{v}_{\infty/E}| = 0.382 = 11.37$ km s^{-1}.

This is a very substantial Δv, too large to be performed by a single-stage vehicle. As an example, if the impulse is provided by an optimal two-stage vehicle with engine specific impulses of 375 s and structural coefficient $\varepsilon = 0.12$ for each stage, the mass ratio at burnout of each stage becomes 4.689 and thus the first stage will have a mass of 79.59 times the payload mass and the second stage will have a mass of 8.43 times the payload mass (Prussing and Conway 1993). The rocket mass is thus 88.03 times the payload mass. For an optimal two-stage vehicle consisting of a first stage having the characteristics of a Centaur RL 10A-4 ($\varepsilon = 0.166$, $I_{sp} = 451$ s) and a second stage characteristic of a solid rocket motor ($\varepsilon = 0.10$, $I_{sp} = 290$ s) the first stage will have a mass of 68.98 times the payload mass and the second stage 5.12 times the payload mass. The rocket mass is thus 74.10 times the payload mass. These two examples are intended to show only that a two-stage vehicle performing the same mission as the low-thrust spacecraft will be very massive, on the order of 75–90 times the mass of the payload.

In comparison, for the low-thrust trajectory of Case N,

$$\frac{\text{final thrust accleration}}{\text{initial thrust accleration}} = \frac{m_0}{m_0 - m_{\text{fuel}}} = \frac{0.191}{0.140}$$

so that the mass of the fuel used by the low-thrust motor is approximately 27% of the total intial mass of the vehicle. The current technology for electric propulsion is exemplified by the Deep Space 1 spacecraft which produces 90 mN of thrust using 2.6 kW; it has a 71 kg power subsystem and 56 kg of ion propulsion hardware and structure, or approximately 1.4 kg of propulsion related hardware per mN of thrust. The initial thrust acceleration for Case N of Table 13.1 is 0.14 in dimensionless units or 8.46×10^{-5} g or 0.83 mN kg^{-1} of initial mass. Thus with the technology used in the Deep Space 1 spacecraft the asteroid interceptor would require

1.4 (0.83) = 1.17 kg of propulsion hardware plus 0.30 kg of fuel and tankage per kilogram of initial mass; i.e., the mission is infeasible.

A trajectory employing a smaller thrust, such as that of Case P of Table 13.1, is feasible today; it would require only 1.7 (0.415) = 0.71 kg of propulsion hardware plus 0.17 kg of fuel and tankage per kilogram of initial mass, leaving approximately 12% of the initial mass for payload. Since the flight time for Case P, which uses half of the intitial thrust acceleration of Case N, is only 11% greater than that for Case N, this is an attractive alternative. If we may assume that the technology will improve, reducing propulsion hardware to a mass fraction of perhaps 0.85 kg mN^{-1} from the 1.7 kg mN^{-1} needed for Deep Space 1, a mission such as that of Case N becomes just feasible and the Case P mission payload fraction increases from 12% to 48%. Thus, albeit for a simplified analysis, the low-thrust asteroid interception vehicle payload should be a minimum of 12% of initial (post-escape) mass with reasonable and even expected improvements in electric propulsion technology improving this to perhaps as much as 48%. The corresponding figure for chemical propulsion using a two-stage rocket is 1% to 2% with much less likelihood of significant improvement since the technology is mature.

5 Maximization of the deflection

Here the objective is to optimize the deflection of the dangerous asteroid, at the time of its close approach to Earth, by a given impulse applied at an earlier time (Conway 2001). We will again assume that a collision (or near-collision) is imminent, i.e., will occur before the asteroid has made another complete revolution about the sun. For most of the known Earth-approaching asteroids this means a period of a few years at most to take some type of preventative action.

If the time between application of the deflection impulse and close approach (Δt) is very brief, the deflection $\overline{\Delta}$ can be approximated assuming rectilinear motion, yielding $\overline{\Delta} = (\Delta \overline{v})(\Delta t)$. The opposite case, in which the asteroid will make more than one orbit of the Sun before its close approach to Earth, is an interesting one for which the relationship between the impulse and resulting deflection is much more complicated. Park and Ross (1999) show that for this case the best place to apply a deflecting impulse is at asteroid perihelion. Their analysis is one of very few to optimize explicitly the deflecting impulse, for arbitrary Δt, assuming Keplerian motion. However, their solution is two-dimensional, i.e., it assumes the asteroid orbits the Sun in the ecliptic plane. But more than 100 of the 461 known PHOs have inclination greater than 10°.

The intention of this work is to show that a near-optimal determination of the direction in which an impulse should be applied to the asteroid, as well as the resulting deflection, can be found without any explicit optimization. The method is easily applied to the true, three-dimensional geometry of the problem.

The objective is to determine the direction in which a given impulse should be applied to the asteroid at interception in order to maximize the subsequent close-approach distance of the asteroid (to the Earth). At the time of interception t_0 the system state transition matrix $\Phi(t, t_0)$ determines the perturbation in position ($\delta \bar{r}$) and velocity ($\delta \bar{v}$) which will result at time t due to a perturbation in position and velocity applied at t_0 (Battin 1987) i.e.,

$$\begin{bmatrix} \delta \bar{r} \\ \delta \bar{v} \end{bmatrix} = \Phi(t, t_0) \begin{bmatrix} \delta \bar{r_0} \\ \delta \bar{v_0} \end{bmatrix} = \begin{bmatrix} \tilde{R} & R \\ \tilde{V} & V \end{bmatrix} \begin{bmatrix} \delta \bar{r_0} \\ \delta \bar{v_0} \end{bmatrix} \qquad (13.20)$$

Therefore

$$\delta \bar{r}(t) = [R]\delta \bar{v_0}(t_0) \qquad (13.21)$$

with

$$[R] = \frac{r_0}{\mu}(1 - F)\left[(\bar{r} - \bar{r_0})\bar{v_0}^T - (\bar{v} - \bar{v_0})\bar{r_0}^T \right] + \frac{C}{\mu}\bar{v}\,\bar{v_0}^T + G[I],$$

$$F = 1 - \frac{r}{p}(1 - \cos\theta),\ \cos\theta = \frac{\bar{r} \cdot \bar{r_0}}{r r_0},$$

$$G = \frac{1}{\sqrt{\mu}}\left[\frac{r r_0}{\sqrt{p}}\sin\theta \right], \qquad (13.22)$$

p = orbit semilatus rectum,

$$C = \frac{1}{\sqrt{\mu}}[3U_5 - \chi U_4 - \sqrt{\mu}(t - t_0)U_2],$$

$$\chi = \sqrt{a}(E - E_0),$$

$$\alpha = 1/a,$$

where a is the orbit semimajor axis, E is the eccentric anomaly, μ is the gravitational parameter, and $U_1(\chi, \alpha)$, $U_2(\chi, \alpha)$, $U_3(\chi, \alpha)$, $U_4(\chi, \alpha)$, $U_5(\chi, \alpha)$, are the "universal functions" (Battin 1987).

We want to maximize $|\delta \bar{r}(t_c)| = \max([R]\delta \bar{v_0})$ where the time of interest, t_c, is the time of close approach to Earth. Equivalently we may maximize $\delta \bar{v_0}^T[R]^T[R]\delta \bar{v_0}$. This quadratic form is maximized, for given $|\delta \bar{v_0}|$, if $\delta \bar{v_0}$ is chosen parallel to the eigenvector of $[R]^T[R]$ that is conjugate to the largest eigenvalue of $[R]^T[R]$. This yields the optimal direction for the perturbing velocity impulse $\delta \bar{v_0}$, which will be expressed on the space-fixed basis since this is the basis on which $[R]$ is implicitly expressed. Then $\delta \bar{v_0}$ may be expressed in an asteroid fixed radial, transverse, normal

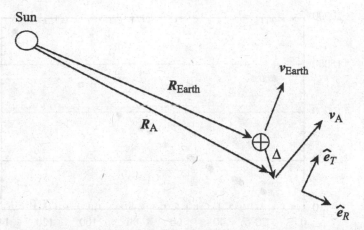

Figure 13.5 Position of the Earth and asteroid at the close approach.

basis as,

$$\delta \bar{v}_{0\ RTN} = \begin{bmatrix} c\theta c\Omega - c i s\Omega s\theta & c\theta s\Omega + c i c\Omega s\theta & s i s\theta \\ -s\theta c\Omega - c i s\Omega c\theta & -s\theta s\Omega + c i c\Omega c\theta & s i c\theta \\ s i s\Omega & -s i c\Omega & c i \end{bmatrix} \delta \bar{v}_{0\ XYZ} \qquad (13.23)$$

where $s\theta$ and $c\theta$ are abbreviations for $\sin \theta$ and $\cos \theta$ respectively.

It may be argued that, since in this approach the deflection $|\delta \bar{r}(t)|$ is maximized only at time t_c, which would be the time of collision absent the deflecting impulse, the subsequent motion of the asteroid and Earth may bring the two bodies closer together. That is, if we let

$$\bar{R} = \bar{R}_A - \bar{R}_{Earth} \qquad (13.24)$$

where subscript A stands for the asteroid and both position vectors are determined with respect to the Sun, as shown in Fig. 13.5, then the deflection $\bar{\Delta} = \delta \bar{r}(t_c) = \bar{R}(t_c)$. In the analysis of Park and Ross (1999) this is explicitly prevented by including necessary conditions for the NLP problem; $\dot{\bar{R}} = 0$ and $\ddot{\bar{R}} > 0$ at the time of close approach. The simplicity of the method described here for finding the near-optimal deflection precludes incorporating these necessary conditions; however, it is possible to guarantee that at time t_c the separation of the asteroid and the Earth is increasing.
Since

$$\frac{d\bar{R}}{dt} = \frac{d}{dt}(\bar{R}_A - \bar{R}_{Earth}) = \bar{v}_A - \bar{v}_{Earth} \qquad (13.25)$$

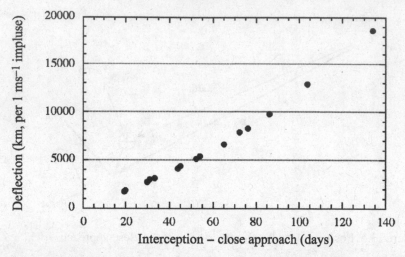

Figure 13.6 Specific deflection as a function of date of interception.

then

$$\dot{\overline{R}} = \frac{d\overline{R}}{dt} \cdot \frac{\overline{R}}{R} = (\overline{v}_A - \overline{v}_{\text{Earth}}) \cdot \frac{\overline{\Delta}}{\Delta} \qquad (13.26)$$

at time t_c, where $\overline{\Delta} = \delta\overline{r}(t_c) = \overline{R}(t_c)$, as illustrated in Fig. 13.5.

In order to have the separation of the bodies increasing, the scalar product in Eq. 13.7 must be positive. This can always be achieved; recall from Eq. 13.21 that $\overline{\Delta} = \delta\overline{r}(t) = R\delta\overline{v}_0(t_0)$ where $\delta\overline{v}_0$ is chosen parallel to the eigenvector of $[R]^T[R]$ that is conjugate to the largest eigenvalue of $[R]^T[R]$. Thus the magnitude of the deflection is unaffected by whether the deflection $\delta\overline{v}_0$ or $-\delta\overline{v}_0$ is applied to the asteroid at interception so long as this condition obtains; one of the two choices will cause the asteroid to miss the Earth by $\overline{\Delta}$ and separate further in their subsequent relative motion.

6 Examples

As an example, optimization of the deflection impulse has been applied to the case of the Earth-approaching asteroid 1991RB previously considered. Figure 13.6 shows the maximum amount of deflection that can be obtained, at what would otherwise be the time of close approach to the Earth, as a function of the interval between interception and close approach (on September 19, 1998). The impulse is assumed to be applied to the asteroid in the direction chosen, as described in the previous section, to maximize the deflection at the subsequent close approach. The figure shows that, if the asteroid is reached several months before the time of collision,

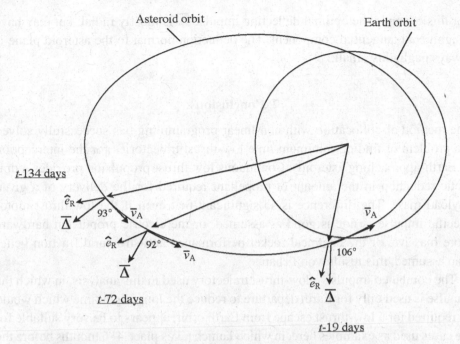

Figure 13.7 Optimal direction for application of deflection impulse at selected points on asteroid orbit. Time to crossing of Earth orbit is indicated.

each 1 m s^{-1} of velocity change imparted to the asteroid may yield a deflection distance comparable to the width of the Earth.

It has been consistently found, as expected, that at any given thrust acceleration level, the deflection is increased for an earlier launch (and corresponding earlier arrival). Similarly, for a given launch date the deflection increases with engine thrust acceleration level since higher thrust enables the spacecraft to intercept the asteroid more quickly. It is thus always the case, for this example where the close approach is imminent, i.e., will occur before the asteroid makes a complete revolution about the Sun, that an earlier interception, combined with a deflecting impulse directed "optimally," as defined in the previous section, always yields a larger deflection. This seems intuitively correct and agrees with the results of Park and Ross (1999: Figure 3) for the case of "impulse time" less than one period. The improvement in deflection distance obtained from increasing the engine thrust, and hence decreasing the time of flight, can be quite significant, e.g., increasing the thrust acceleration by approximately 60%, from 61 micro-*g* to 97 micro-*g*, increases the deflection obtained by almost 100%.

The direction in which the optimal deflection impulse is to be applied is shown in Fig. 13.7 at a few points on the orbit of the asteroid as it approaches its near collision with the Earth. The plane of the figure is the asteroid (1991RB) orbit plane.

For this example the optimal deflecting impulse $\overline{\Delta}$ is mostly radial, but may have a significant tangential component. The deflection normal to the asteroid plane is always negligibly small.

7 Conclusions

The method of collocation with non-linear programming has successfully solved the problem of finding minimum-time low-thrust trajectories for the interception of Earth-approaching asteroids. Continuous low-thrust propulsion provides a dramatic reduction in the amount of propellant required for the delivery of a given payload mass. The difference is so significant that even if the low-thrust motor specific impulse is not as good as assumed, or the electric propulsion hardware more massive, or the chemical rocket performance and structural fraction better than assumed, this result won't change.

The combined impulsive/low-thrust trajectory used in this analysis, in which the impulse is used only for Earth departure to reduce the long flight time which would be required for a low-thrust escape from Earth orbit, appears to be very suitable for the cases used as examples here, in which launch takes place 4–6 months before the predicted close approach of the asteroid. If the danger of collision is known much earlier, a trajectory that uses only low-thrust propulsion may be feasible and would likely have additional advantages in total propellant required.

Future research might include the variation with distance from the Sun of available power and hence specific impulse for a solar-powered spacecraft; the analysis could also be simply modified to require optimal rendezvous rather than interception.

Many of the strategies for amelioration of the danger of an asteroid's collision with the Earth involve, at the time of interception, applying a small impulsive velocity change to the asteroid. In this work we show how, using the asteroid orbit state transition matrix, this impulse should be applied in order to maximize the deflection of the asteroid at the time of close approach. The algorithm requires no explicit (numerical) optimization and is thus easily applied. We also find, for the case in which the asteroid will make less than one complete orbit of the Sun before its close approach or impact with the Earth, that the deflection resulting from a given magnitude of impulse increases the earlier the asteroid is intercepted. It is thus legitimate for this case to optimize the interception trajectory, minimizing the flight time, and then optimize the direction of the deflecting impulse separately, to maximize the deflection of the asteroid at what would otherwise be the impact time.

One important result, illustrated in Fig. 13.6 for a very representative Earth-crossing asteroid, is that, if the interceptor spacecraft is launched less than a year in advance of the asteroid's possible collision with Earth, the deflection obtained will be on the order of an Earth diameter for every 1 m s^{-1} velocity change applied to the asteroid. Considering how many millions of kilograms of mass some of the Earth-crossing asteroids possess it would be very difficult to change their velocity by even 1 m s^{-1}. Thus it may only be feasible to deflect asteroids of moderate size if they can be reached several years in advance of a potential collision.

Acknowledgments

Figures 13.1, 13.2, 13.3, and 13.4 and Table 13.1 were originally published in the paper "Optimal low-thrust interception of Earth-crossing Asteroids," *Journal of Guidance, Control, and Dynamics*, **20**, no. 5, pp. 995–1002. Copyright 1997 by the American Institute of Aeronautics and Astronautics, Inc. Figures 13.5 and 13.6 were originally published in the paper "Near-optimal deflection of Earth-approaching asteroids," *Journal of Guidance, Control, and Dynamics*, **24**, no. 5, pp. 1035–1037. Copyright 2001 by the American Institute of Aeronautics and Astronautics, Inc. Reprinted with permission.

References

Adushkin, V. V. and Nemchinov, I. V. 1994. Consequences of impacts of cosmic bodies on the surface of the Earth. In *Hazards due to Comets and Asteroids*, ed. T. Gehrels, pp. 721–778. Tucson, AZ: University of Arizona Press.

Ahrens, T. J. and Harris, A. W. 1994. Deflection and fragmentation of near-Earth asteroids. In *Hazards due to Comets and Asteroids*, ed. T. Gehrels, pp. 897–928. Tucson, AZ: University of Arizona Press.

Battin, R. H. 1987. *An Introduction to the Mathematics and Methods of Astrodynamics*. New York: American Institute of Aeronautics and Astronautics Publications.

Conway, B. A. 1997. Optimal low-thrust interception of Earth-crossing asteroids. *J. Guidance, Control, Dynam*. **20**, 995–1002.

2001. Near-optimal deflection of Earth-approaching asteroids, *J. Guidance, Control, Dynam*. **24**, 1035–1037.

Gill, P. E. 1993. *User's Guide for NZOPT 1.0: A Fortran Package For Nonlinear Programming*. Huntington Beach, CA: Boeing Aerospace Corp.

Hargraves, C. R. and Paris, S. W. 1987. Direct trajectory optimization using nonlinear programming and collocation. *J. Guidance, Control, Dynam*. **10**, 338–342.

Herman, A. L. 1995. Improved collocation methods used for direct trajectory optimization. Ph.D. thesis, University of Illinois at Urbana–Champaign.

Herman, A. L. and Conway, B. A. 1996. Direct optimization using collocation based on high-order Gauss–Lobatto quadrature rules. *J. Guidance, Control, Dynam*. **19**, 592–599.

1998. Optimal low-thrust, Earth–Moon orbit transfer. *J. Guidance, Control, Dynam*. **21**, 141–147.

Hills, J. G., Nemchinov, I. V., Popov, S., *et al.* 1994. Tsunami generated by small asteroid impacts. In *Hazards due to Comets and Asteroids*, ed. T. Gehrels, pp. 779–790. Tucson, AZ: University of Arizona Press.

Morrison, D., Chapman, C. R., Slovic, P., *et al.* 1994. The impact hazard. In *Hazards due to Comets and Asteroids*, ed. T. Gehrels, pp. 59–91. Tucson, AZ: University of Arizona Press.

Park, S. Y. and Ross, I. M. 1999. Two-body optimization for deflecting Earth-crossing asteroids. *J. Guidance, Control, Dynam.* **22**, 415–420.

Prussing, J. E. and Conway, B. A. 1993. *Orbital Mechanics*. New York: Oxford University Press.

Scheel, W. A. and Conway, B. A. 1994. Optimization of very-low-thrust, many revolution spacecraft trajectories. *J. Guidance, Control, Dynam.* **17**, 1185–1192.

Solem, J. C. and Snell, C. M. 1994. Terminal intercept for less than one period warning. In *Hazards due to Comets and Asteroids*, ed. T. Gehrels, pp. 1013–1034. Tucson, AZ: University of Arizona Press.

Tang, S. and Conway, B. A. 1995. Optimization of low-thrust interplanetary trajectories using collocation and nonlinear programming. *J. Guidance, Control, Dynam.* **18**, 599–604.

Yau, K. 1994. Meteorite falls in China and some related human casualty events. *Meteoritics* **29**, 864–871.

14

Close proximity operations at small bodies: orbiting, hovering, and hopping

Daniel J. Scheeres
University of Michigan

1 Introduction

Mitigation and detailed characterization of asteroids and comets require some period of close proximity operations about them. To support close proximity operations requires an understanding of dynamics of natural material on and about small bodies, and the dynamics, navigation, and control of artificial objects on and about small bodies. In this chapter we discuss some of the controlling issues that relate to close proximity operations, and draw connections between this issue and the design of spacecraft and mission concepts to carry out close proximity operations.

Since the field of astrodynamics and celestial mechanics is often considered to be a mature field, it is relevant to ask why the control of spacecraft about small solar system bodies is considered to be a difficult problem. There are a number of reasons for why this is the case, which we review here and explain in additional detail throughout this chapter. A clear rationale for why this is true can best be expressed through the following chain of facts.

(1) *Small bodies have large ranges in crucial physical parameters.* This result arises by the nature and number of small bodies. By themselves, small bodies fall into multiple classes, such as comets, rubble pile asteroids, monolithic asteroids, binary asteroids, and presumably other gross catagories that we have yet to discover. Within any of these classes there can be a diversity of different possible shapes, sizes, and densities, which immediately dictate the strength and distribution of the gravitational field of the body, which in turn has implications for the relevance of solar gravity and radiation pressure effects on trajectories near the body. Also, it is an established fact that small bodies can have a range of rotation states, from simple rotation about the maximum principal axis to complex rotation. Furthermore, the obliquity of their rotational angular momentum

Mitigation of Hazardous Comets and Asteroids, ed. M. J. S. Belton, T. H. Morgan, N. H. Samarasinha, and D. K. Yeomans. Published by Cambridge University Press. © Cambridge University Press 2004.

vectors are known to be distributed from $0°$ to $180°$, which can also have dynamical significance.

(2) *Each set of small body parameter values can have close proximity dynamics that are difficult in and of themselves.* As a matter of fact, most of the dynamical environments found at small bodies can exhibit strongly unstable and chaotic motion for trajectories of practical interest. Significantly, as the force parameters of small bodies are changed, the range of orbit elements that lead to unstable motion can change drastically. For example, for one class of bodies it is known that orbits with an inclination of $180°$ relative to their rotation pole can be very stable and safe. However, if the asteroid is relatively small these same orbits, due to solar radiation pressure effects, can be very unstable and lead to impact with the asteroid within a few orbits. Numerous other examples can also be found, indicating that small bodies, when taken together as a class, exhibit an extremely rich set of non-trivial dynamics.

(3) *Spacecraft designs and mission operation concepts can be driven in very different directions as a function of the close proximity dynamical environment.* As a case in point, the asteroid mission phase for the NEAR mission at Eros looks very different than the asteroid mission phase of the MUSES-C mission to 1998 SF36. For NEAR, there was no choice but to use an orbital approach, due to the large mass of Eros. However, due to its shape, rotation state, and rotation pole orientation, the orbital mission had to be designed carefully to avoid destabilizing interactions with the asteroid. For MUSES-C, due to the possible low mass of its target asteroid and the large mass to area ratio of the spacecraft, it could not be guaranteed that an orbital mission would even be possible. Thus, the entire mission consists of forcing the spacecraft to "hover" on the Sun-side of the asteroid (discussed later in this chapter), with an associated cost in terms of fuel and ability to measure the asteroid's gravity field. A crucial point to make is that the placement of instruments on the spacecraft bus are fundamentally different for each mission. The NEAR mission plan required instruments boresights to be placed orthogonal to the solar array normals, while the MUSES-C mission plan requires the boresights to be antiparallel to the solar array normals, a fundamentally different spacecraft design dictated by the type of dynamics about the small body, not dictated by an abstract "design philosophy."

(4) *Crucial small body parameters may not be known prior to rendezvous.* In light of the previous point, it is clear that this can lead to increased design costs and impacts during the construction phase of the spacecraft. If relatively nothing is known about the target body, it will be necessary to design a spacecraft that covers a wider range of possible orbit and close proximity strategies – which will invariably lead to higher design and fabrication costs and to increased spacecraft mass. This places a strong driver on discovering as much as possible about the physical characteristics of potential target bodies prior to spacecraft and mission design.

(5) *It is likely that vehicle designs and operations concepts that fit one small body may not fit another.* Again, it is instructive to compare the MUSES-C and NEAR spacecraft designs. The MUSES-C design is appropriate for very small bodies, and is designed to maximize scientific return from visiting such bodies. On the other hand, if the MUSES-C

spacecraft were sent to explore Eros, it is likely that the amount and quality of scientific measurements it could take would be significantly less than the NEAR spacecraft accomplished, and would require the spacecraft operations team to work much harder to accomplish them. The NEAR spacecraft design is appropriate for larger bodies such as Eros and was operated in that environment with a relatively small operations team. If, however, the NEAR spacecraft was used to explore a sub-kilometer asteroid it would have had profound difficulties in carrying out its mission and accumulating quality scientific measurements of that body, again at the expense of stressing the spacecraft operations team. Thus, spacecraft designs and mission scenarios that are optimized for one class of small body may not function well at all for a different class of small bodies. For missions that wish to visit multiple asteroids, this may force a difficult trade to be made between designing a spacecraft that can accomplish a mission to range of asteroid classes or designing a mission that only explores asteroids that fall within a restricted class of physical parameters.

Having sketched out our argument for why the issue of close proximity dynamics about small bodies is important, we will now explore some of these issues in more detail. First of all, it is important to note that a close coupling exists between the dynamics of natural material and artificial bodies about asteroids and comets. In recent years an appreciable body of literature has been built up in both of these areas, including a review chapter on the natural dynamics and modeling of natural material on and about asteroids (Scheeres *et al.* 2002). In the following we briefly review the main literature that has been published recently, and which is crucial for understanding the relevant details of close proximity dynamics at small solar system bodies.

One issue of importance is the modeling of the small body force environment. First consider the gravitational field of the body, the current state of the art is to describe these fields using spherical harmonic expansions (Miller *et al.* 2002). These are useful for estimation and orbital computation, but fail when within the circumscribing sphere about the body. A partial remedy is the use of ellipsoidal harmonic expansions (Garmier and Barriot 2001), which decrease the radius of convergence to the circumscribing ellipsoid. There may still be problems for use at the body surface, an approach remedied by using closed form expressions of polyhedral gravitational fields (Werner 1994; Werner and Scheeres 1997). A drawback with this approach is the requirement of constant density, although the evidence from Eros indicates that this is a good approximation for that body. The effect of solar radiation pressure and solar tides can be handled with standard approaches, applied to spacecraft in Scheeres and Marzari (2002). When considering the effect of cometary outgassing, models have not been developed to the same degree of fidelity. The basic model used for planetary science computations was first applied to spacecraft navigation in Miller *et al.* (1990), further refined in Weeks (1995), applied to spacecraft dynamical computations in Scheeres *et al.* (1998b), and used as the basis for a more accurate formulation in Scheeres *et al.* (2000a). A newly recognized area of concern is the electrostatic environment on the surface of an asteroid. This issue has been considered by Lee (1996), and has been suggested as a possible explanation for the dust ponds observed on the surface of Eros (Robinson *et al.* 2001).

The second issue of importance is natural and controlled dynamics of motion on and about small bodies, which includes studies of surface motion. The initial area of study concerned Phobos and Deimos (Dobrovolskis and Burns 1980; Thomas and Veverka 1980), and these asteroids are still a topic for research (Thomas 1998; Thomas *et al.* 2000). Studies of motion about more generic asteroids began with studies of the orbital environment about asteroids in support of the Galileo flybys of Gaspra and Ida (Hamilton and Burns 1991a, 1991b), closely followed by papers considering orbital dynamics close to asteroids (Chauvineau *et al.* 1993; Scheeres 1994) and studies of the orbital motion of Dactyl about Ida (Geissler *et al.* 1996; Petit *et al.* 1997). Using radar-derived and observationally based models of asteroids, a number of papers were published studying the basic physics of orbital motion about real asteroid shapes, separately considering the case of a uniformly rotating asteroid (Scheeres *et al.* 1996) and an asteroid in a complex rotation state (Scheeres *et al.* 1998a). A number of studies of orbital motion about 433 Eros were made, both before and after rendezvous (Scheeres 1995; Scheeres *et al.* 2000b; Thomas *et al.* 2001). More recently, Hu has focused on understanding the motion about asteroids over a wider range of parameter space (Hu 2002; Hu and Scheeres 2002). In terms of controlled motion about asteroids, recent work has broached the subject of directly controlled motion about asteroids (Scheeres 1999b; Sawai *et al.* 2002; Broschart and Scheeres 2003). There still exist new areas that demand further exploration, especially motion in binary systems. Based on a preliminary review of the basic properties of binary asteroids, the orbital environment about them appears to be quite complex as well (Scheeres and Augenstein 2003).

2 Defining the force environment

Given the large range of possible situations, and the diverse types of motion that can ensue, it is crucial that the specific force environment of a small body be defined as early as possible, as it is much more difficult to design a mission and spacecraft to a body with unknown characteristics. This means that ground-based characterization must play an important role in target selection and initial mission and spacecraft design. In the following we discuss what important physical characteristics of a small body can be estimated using ground-based measurements. Following that we discuss the measurements that should be taken following rendezvous in order to facilitate robust implementation of close proximity dynamics.

2.1 Pre-rendezvous characterization

There are a number of important small body parameters that are accessible using ground-based observations. Some of these measurements are routinely taken; however, a number of them require targeted observations that must be planned in advance and which require additional resources. It is fortunate to note that the majority of physical parameters needed to characterize the dynamical environment of a small body can at least be constrained by ground-based observations. The list of desired physical parameters needed to define the small body force environment is:

(1) Number of co-orbitals.
(2) Size.
(3) Density.
(4) Shape, gravity field, or density distribution.
(5) Surface and interior morphology.
(6) Spin rate and spin state.
(7) Orientation of rotational angular momentum relative to orbital plane.
(8) Heliocentric orbit.

The basic suite of ground-based measurements that is used to observe small bodies, and the physical parameters that they may be able to constrain, can be summarized as: astrometric measurements (items 1, 2, 8), intensity (lightcurve) measurements (items 1, 4–7), spectral measurements (items 2, 3), and range-Doppler radar imaging measurements (items 1–8). In the following we discuss each of these types of measurements and the physical parameters which they can determine, or at least constrain.

2.1.1 Astrometric measurements

These observations are generally associated with discovery, recovery, and ephemeris improvement campaigns. They serve as the basic data type on small bodies and directly contribute to determining the heliocentric orbit. With assumptions on asteroid albedo they can also constrain the body size. If extremely high-accuracy astrometry is used (e.g., Hubble), it is also possible to directly determine if co-orbitals exist and to measure size.

2.1.2 Intensity (lightcurve) measurements

Lightcurve measurements can directly determine the spin period of a body, detect a complex spin state, place direct constraints on the body shape, and can detect the presence of a co-orbital by observing an eclipse. With additional processing it is also possible to invert lightcurve data to estimate body shape and rotational angular momentum directions and constrain some aspects of surface morphology.

2.1.3 Spectral measurements

By measuring the spectra of a body it is possible to determine its "type" classification, which in turn places constraints on its composition, grain density, and surface albedo. These, in turn, can provide bounds on its density and can improve size estimates.

2.1.4 Radar measurements

Radar measurements can consist of detecting returns from the body, or if carried out at high enough signal strengths can provide resolution measurements in both Doppler and range. As has been demonstrated repeatedly, these data can be inverted to find the body shape and rotation state and can immediately detect the presence of co-orbitals. In addition to measuring the body shape, size, and spin state, radar measurements also provide the body's rotational angular momentum vector and significant improvements in its heliocentric orbit. In addition to these, characteristics of the return signal can be used to place constraints on the surface density of the body and on the roughness of the body at the scale of the radar wavelength.

While it is clear that radar observations of a body can provide the most comprehensive and detailed information, these are also the most restrictive measurements in terms of opportunities and are most effective when the body in question has been observed by all the other measurement types. It is also important to note that the ideal set of observations of a body would encompass all of the above types, as they each have unique measurement aspects and can often be combined together to complement each other in estimating physical properties.

2.2 Post-rendezvous characterization

Following rendezvous with the small body, it is necessary to develop precision models using navigation data, which generally consist of radiometric tracking data, optical observations, and altimetry. The specific physical parameters needed to support close proximity operations at the small body are its mass, gravity field, spin state, surface topography and roughness, and surface gravity field and density distribution.

The process of measuring these parameters using navigation data is rather involved, and ultimately relies heavily on combining the data to perform a joint solution for all of these parameters simultaneously. One of the main difficulties in performing these solutions is generating sufficiently accurate models to use as initial estimates. This is where the presence of existing models based on ground-based measurements can be crucial and can significantly cut the time required to estimate these models. Comprehensive discussions of these procedures for asteroids are documented in Miller *et al.* (1995, 2002) and for comets in Miller *et al.* (1990).

The only physical model that does not arise naturally out of the navigation measurement process is the surface gravity field and internal density distribution. While the estimated gravity field (usually parameterized as a spherical harmonic expansion) contains this information, the gravity field parameterization is generally not valid at the surface of the body and the density information cannot be uniquely extracted. One way to bypass the invalidity of spherical harmonic expansions at the surface of the body is to use ellipsoidal harmonics (Garmier and Barriot 2001), which are still formally divergent at the surface but which can be used much closer to the body surface. It is also possible to directly estimate ellipsoidal harmonic coefficients from existing spherical harmonic gravity field coefficients (Dechambre and Scheeres 2002). In contrast, the polyhedron gravity field potentials are valid up to and even within the body, but rely on constant density assumptions. It is possible to mimic density distributions by placing mass concentrations of different density within the body (Werner and Scheeres 1997), but this requires that the density distribution be estimated. While this is a non-unique process, a least-squares estimation technique that uses the measured shape model and gravity field coefficients as data may allow for a rational approach to estimating these distributions (Scheeres *et al.* 2001). It is important to note that the one small body for which we have an accurate gravity field and shape measurement, Eros, has only minimal density inhomogeneities (Yeomans *et al.* 2000). This provides the hope that constant-density surface gravity models may be sufficient for surface operations.

3 Close proximity dynamics and operations

In the introduction we gave a summary of past work on the question of close proximity dynamics of natural and artificial bodies, and we do not intend to review this material here. Rather, we will make a number of observations and discuss a number of scenarios where complications associated with close proximity dynamics can arise. In concert with these we will discuss various ways in which such complications can be mitigated, ultimately leading to the idea of controlled motions relative to a small body.

An important point to make is that the motion of particles in close proximity to a small body will usually deviate significantly from the familiar Keplerian motion due to perturbations from solar radiation pressure, solar gravity, small body shape/gravity, and small body rotation. Due to these perturbations it is very common to find trajectories that can escape, impact, or migrate substantially over only a few orbits about the body. Unlike most unstable and chaotic motion in the solar system, which has timescales on the order of thousands to millions of years, the timescales for these effects to act are very short, on the order of a few hours to days. Thus, these effects must be accounted for and understood in order to carry out close proximity spacecraft operations. It is important to stress that both orbital and surface motion must deal with these issues.

3.1 Complicating scenarios and possible resolutions

Based on past analyses of close proximity motion, there are a number of items that can be identified as being of specific concern to the implementation of close proximity operations on or about a small body. In this chapter they have been divided into three broad classes: dynamics of disturbed regoliths, orbit mechanics issues, and surface motion dynamics. Many of the complicating scenarios discussed below can be dealt with by the appropriate choice of orbiting strategy or spacecraft design. Thus, where appropriate we mention some known strategies for mitigating some of these adverse dynamical effects. These strategies are by no means exhaustive, but are representative of the types of approaches that can be used. Each of these strategies have their own drawbacks, making the design of a close proximity mission a challenging exercise in system optimization.

3.1.1 Natural dynamics of disturbed regoliths

An interesting idea, posed in Scheeres and Asphaug (1998) in the context of exploring small body surfaces, is that operations on the surface of a small body can excite the loose regolith, effectively creating a transient atmosphere. Since escape speeds on the surface of a small body are on the order of meters per second or less, it does not require much impulse to energize regolith. If the small body were a sphere, there would only be two outcomes for each particle, escape or re-impact. However, the small body will have an irregular shape, will be spinning, and is in the solar radiation pressure and gravitational field of the Sun; thus it is probable that the trajectory of non-escaping disturbed material can transition into a non-impacting orbit about the body, or at least into an orbit that will not re-impact for an extended period of time. In Scheeres and Marzari (2000) examples of this effect considering solar radiation pressure only are presented, while in Scheeres *et al.* (1998a) examples of this considering the gravity field and rotation state of the small body only are presented. In both cases, re-impact of disturbed ejecta may not occur for many months! When the particles do re-impact, they will in general have speeds up to local escape speed (which is computed taking body rotation into account; see Scheeres *et al.* (1996)).

While this would be an interesting effect to observe, it may not be a positive environment for a landed space vehicle or a vehicle in orbit about the small body. While the "density" of the transient atmosphere would likely be very low, there may be some increased risk associated with orbiting about a body surrounded by a dusty sphere. Similarly, a landed vehicle could be subject to re-impacting ejecta traveling at local escape speeds, with re-impact occuring long after the initial event. More serious issues could also involve the electrostatic potential on the asteroid surface in combination with an energized regolith and the introduction of a vehicle into that environment.

3.1.2 Orbit mechanics

The peculiarities of orbital motion about asteroids is a subject that is more fully understood than surface motion, and for which much recent work has been performed. For these situations the "complicating scenarios" are known to be a strong function of the body's heliocentric orbit, shape, size, rotation, and density, as well as the spacecraft's mass and area. The main complications are due to gravity, rotation, and solar radiation pressure effects. Each of these complications can be analyzed in isolation, but for some bodies they can play an important role in combination. We also briefly consider binary asteroids, which adds a new dimension to this problem.

Gravity and rotational effects First consider the effect of gravity and rotation. The simplest unifying dynamical idea for a rotating asteroid is its synchronous orbit radius, specifically the size and stability of the circular orbit that has the same period as the body's rotation. For an asteroid of a given mass, represented by its mass parameter μ, and rotation period T, its ideal synchronous radius is computed as: $r_s = (\mu T^2/4\pi^2)^{1/3}$. Due to the distributed mass of the body, an asteroid will in general only have four specific locations close to r_s where truly synchronous motion exists (Scheeres 1994; Scheeres *et al.* 1996), analogous to the Earth's case (Kaula 1966), located along the longest and shortest body axis in the equatorial plane. For the majority of asteroids with known shapes, all four of these synchronous orbits are unstable, and trajectories started in their vicinity generally lead to impact or escape within a few orbits. This simple result lies behind most of the difficulties encountered in orbiting small bodies. In fact, for asteroids with all of their synchronous orbits unstable, motion within two to three synchronous radii of the asteroid mass center tends towards instability, with escaping and impacting orbits being the rule. There is a strong inclination dependence on this instability, it being the most pronounced when motion is in the plane of rotation and in the same sense (i.e., zero inclination). As motion is considered at higher inclinations, the minimum radius for stable motion tends to decrease. In Fig. 14.1 we present a plot of stable and unstable regions as a function of radius and inclination, computed for the asteroid Eros (Lara and Scheeres 2002). As has been clearly established, when orbital motion is in the equatorial plane and opposite to the sense of rotation (i.e., retrograde), orbital motion is actually quite stable.

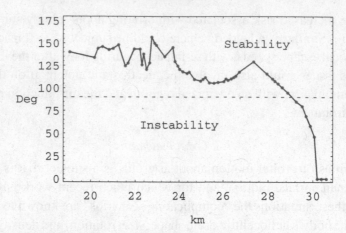

Figure 14.1 Stability regions for three-dimensional motion around the asteroid
Eros. From Lara and Scheeres 2002.

The basic complication from the gravity and rotation effect is that orbital motion
close to the body with low relative speeds is in general not an option. Unfortunately,
it is these orbits that are the most scientifically attractive for performing body-
relative measurements and deploying measuring devices. The remedy is to fly in a
retrograde orbit, which allows for very low altitudes relative to the long ends of the
body. This, however, results in relatively high speeds with respect to the asteroid
surface and places strict constraints on the geometry of close orbits.

Solar effects   While gravity and rotation work to destabilize motion close to the
body, the effect of solar radiation pressure (in tandem with solar gravitational tides)
is to destablize motion when relatively far from the body. The perturbative effect
of solar radiation pressure combined with solar gravitational attraction is a strong
function of the small body orbit, its overall mass, and of the spacecraft's mass to
projected area ratio. Clearly, a spacecraft that strays far enough from the attracting
asteroid will at some point fall under the influence of the Sun. An appropriate rule of
thumb for the distances at which these effects start to dominate can be approximated
by the sphere of influence about the body, or more dynamically appropriately by
the libration points of a particle between the asteroid and the Sun. Considering only
gravitational attractions, these will lie at a distance of $r_H \sim \pm(\mu/3\mu_S)^{1/3} d$ from
the asteroid along the asteroid–Sun line. Here, μ_S is the gravitational parameter
of the Sun, and d is the distance of the asteroid from the sun. For example, an
asteroid located at 1 AU from the sun would have its libration orbits located at $r_H \sim$
$\pm 133 \rho^{1/3} R$ km, where ρ is the body density in g cm^{-3}, and R is the mean radius of
the asteroid in km. Previous research has shown that circular orbits started with radii
less than one-half of this distance are generally stable against escape from the body

(Hamilton and Burns 1991a, 1991b). Furthermore, when relatively far from the body the orbital period can be very long, leading to relatively "slow" unstable dynamics that can be easily controlled. Thus, when viewed as a gravitational perturbation alone, we see that the Sun should not stress operations or spacecraft design except for the very smallest and least dense bodies.

The inclusion of solar radiation pressure can drastically alter the situation, however. Many spacecraft are moving towards the use of solar-electric propulsion technologies, which use the Sun's light to generate propulsion power. These designs naturally lead to spacecraft with relatively large solar arrays, and hence large surface areas. In addition, spacecraft are generally designed to minimize their total mass. The combined effect leads to spacecraft with relatively small projected area densitites, defined as the spacecraft mass divided by its total Sun-ward projected area. For example, the Rosetta and the MUSES-C spacecraft have area densities on the order of 20–30 kg m^{-2} (Scheeres and Marzari 2002).

The effect of the solar radiation pressure on spacecraft motion is complex, and different aspects of it are considered in Scheeres (1999a) and Scheeres and Marzari (2002), drawing on foundational work (Mignard and Hénon 1984; Richter and Keller 1995). Again, it is simplest to think of the effect in terms of the radius of libration points. Solar radiation pressure introduces an asymmetry between the Sun-side and anti-Sun-side libration points, with the Sun-side point being pushed further from the asteroid and the anti-Sun side point being pushed closer to the asteroid (Fig. 14.2). In Scheeres and Marzari (2002) the relative effect of the solar radiation pressure is parameterized by a quantity $\tilde{\beta}$ which is a function of the solar radiation pressure strength, the spacecraft area density, the Sun's gravitational parameter, and the asteroid's gravitational parameter. Leaving out the details of the derivation we find $\tilde{\beta} = 3.84/(B\mu^{1/3})$, where B is the spacecraft area density in kg m^{-2} and μ is the asteroid's gravitational parameter. When $\tilde{\beta} \leq \mathcal{O}(1)$ the Sun's gravitational effects predominate, however when $\tilde{\beta} \gg 1$ the solar radiation pressure force dominates. As an example again, the Rosetta and MUSES-C spacecraft will have estimated values on the order of $\tilde{\beta} \sim 30$. In this regime, the anti-Sun-side equilibrium point is at a distance of $\sim (\mu/\mu_S)^{1/3}d/\sqrt{\tilde{\beta}}$. For a spacecraft and asteroid with $\tilde{\beta} \sim 30$ this radius shrinks to $\sim 35\rho^{1/3}R$ km, which is 25% of the gravitational only libration point distance. If $\tilde{\beta} \gg 1$ the escape dynamics of the problem also change and we find that the maximum semimajor axis that a spacecraft can have and remain captured at the asteroid is approximately one-fourth the radius of the anti-Sun-ward libration point. These effects place a much more restrictive dynamical constraint on this system.

Furthermore, the dynamics themselves become much more complex, with orbital eccentricity experiencing large, periodic variations on the order of unity, which can lead to spacecraft impact after a few orbits. To avoid exciting such orbit eccentricity

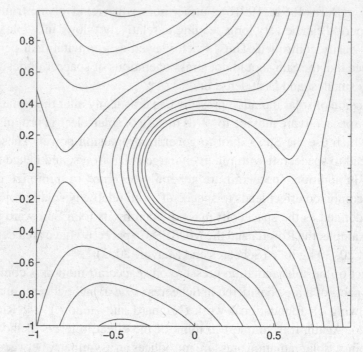

Figure 14.2 Generic zero-velocity curve for $\beta > 1$. From Scheeres and Marzari (2002).

oscillations it is necessary for the spacecraft orbit plane to be perpendicular to the Sun-line. Such a configuration, it turns out, yields a Sun-synchronous orbit in that the solar radiation pressure will force the orbit plane to always lie perpendicular to the Sun-line. When flying in this geometry, the spacecraft is in a terminator orbit and can have a constant eccentricity on average if its initial periapsis vector is chosen appropriately. Indeed, for small asteroids this may be the only orbital geometry that can yield feasible motion over long time-spans as orbits that deviate significantly from the terminator plane may impact with the asteroid in only a few orbits. These limits and constraints can all be predicted using analytical results (Scheeres 1999a; Scheeres and Marzari 2000).

A more significant issue arises when the asteroid has a non-spherical mass distribution and rotation. At the least, such mass distributions can cause precession of the orbit plane (analyzed in Morrow *et al.* (2002)) which in turn can excite the eccentricity oscillations that can lead to impact. If the maximum semimajor axis also happens to be within a few synchronous radii, the orbit may also be subject to destabilization from the gravity and rotation effects alone. The combination of all these effects will lead to difficult challenges for designing orbital missions to very small or underdense bodies.

Binary asteroids   All asteroids will be subject, to some extent, to the force perturbations discussed above. Further complicating the issue, current estimates state that
roughly 20% of the near-Earth asteroid (NEA) population may be binary asteroids.
While much is not known yet about the force environment of a binary asteroid,
current indications are that it will be relatively difficult to find safe and navigable
orbits close to either body. Approaching the problem from a conservative point of
view, we can state that safe orbits will generally lie outside the secondary, so long
as solar radiation pressure effects are not too severe, and will preferentially orbit
retrograde to the binary system's orbital plane. Orbits close to or within the orbit
of the secondary must contend with third-body forces in addition to the gravity and
rotation perturbations discussed above.

   One potentially attractive approach would be to place a spacecraft in one of the
triangular synchronous orbits of the binary system (note, these will be synchronous
with the secondary moving about the primary, but not synchronous with the rotation
of the primary in general). A recent survey of known binary asteroid systems in
the solar system (Scheeres and Augenstein 2003) shows that the majority of these
systems have mass ratios small enough for the classical stability to exist. For many
of these potentially stable environments, however, the gravitational disturbance of
the solar tide is sufficient to destabilize their motion, such as occurs in the Earth–
Moon system. If the mass ratio of the binary asteroid system is $\rho = M_2/(M_1 + M_2)$,
then the classical constraint for the triangular points to be stable is: $27\rho(1 - \rho) \leq 1$.
In Scheeres (1998) a simple criterion for when an external perturbation (such as the
Sun) can destabilize such a three-body system is found. Define the sidereal period of
the binary system to be T_b and the period of the binary system's orbit about the sun
to be T_o. Then the condition for the triangular points to be stable when subject to the
solar perturbation is: $T_b < 27/8T_o\rho(1 - \rho)$. The unstable motion associated with
the solar perturbation will place a spacecraft on a potentially impacting trajectory
within a few periods of the binary system. Thus, even though some NEA binaries
may have stable triangular libration points, which may be a suitable place to park
a spacecraft in, there will be many that have no stable solutions. Due to this, use
of the triangular libration points cannot be relied upon as a generic strategy for
exploring binary asteroids.

3.1.3 Surface motion

Finally, we can consider the motion of a vehicle (i.e., a rover) over the surface of a
small body. Due to the weak gravitational attraction of the body and the uncertain
properties of asteroid surfaces, the most feasible method of locomotion appears to be
hopping. Several simple designs are possible, including the use of internal flywheels
(Yoshimitsu *et al.* 2001). When viewed as a means for controlled locomotion over
an asteroid, however, some serious issues arise. First, let us consider some of the

peculiarities of motion over the surface of an asteroid. The defining quantity for characterizing surface motion on an asteroid is the local escape speed. On the surface of a spherical, non-rotating asteroid with gravitational parameter μ and radius R this speed is just $v_{esc} = \sqrt{2\mu/R}$. Once we consider a rotating body, however, we see that the escape speed relative to the asteroid surface will now vary with latitude and with the direction of the surface relative velocity vector. At the equator, a particle at rest on the surface will have an inertial speed of ωR, where ω is the asteroid rotation rate. Surface motion in the direction of rotation at a speed of $\sqrt{2\mu/R} - \omega R$ will be escape speed, while motion in the opposite direction of rotation at a speed larger than $\sqrt{2\mu/R} + \omega R$ will lead to escape. A velocity in any other direction will have to be added (vectorially) to the velocity due to the asteroid rotation to find the total speed in inertial space, and to ascertain what this speed is relative to escape speed. Finally, if the asteroid has a non-spherical shape, we also see that the local gravitational attraction may be greater or lesser at different points on the surface of the asteroid as well as the surface speed due to rotation, leading to additional variations in escape speed. For definiteness, local escape speed is generally defined as the speed perpendicular to the surface necessary to achieve escape speed relative to the body (see Scheeres *et al.* (2002) for a more complete definition). Such local variations can be extreme; for the asteroid Eros the local escape speed varies between 5 and 15 m s^{-1} depending on surface location (Miller *et al.* 2002).

Local escape speed is a useful characterization as it provides an easily computable limit on surface speeds; however we must note that speeds less than local escape speed may also end up escaping the asteroid. In Scheeres *et al.* (2002) a classification of surface ejecta is given in terms of their final state. It is useful to use periapsis passage relative to the asteroid to delimit between different classes of motion. At launch a particle starts from an initial radius $r_0 \geq q_0$, since in general the initial periapsis (q_0) lies beneath the body's surface. In the absence of perturbations the next periapsis passage q_1 will either equal q_0, and thus will be an impact, or will never occur, indicating escape. When perturbations are incorporated, or even if non-spherical shapes are allowed, it becomes possible for multiple periapsis passages to occur. We denote these as a series q_i; $i = 0, 1, 2, \ldots$. Associated with each periapsis passage is the periapsis vector, $\mathbf{q}_i$, representing the periapsis location in the asteroid-fixed space. If we denote the set of points that constitute the asteroid body as $\mathcal{B}$, then if $\mathbf{q}_i \in \mathcal{B}$ the sequence stops and an impact has occured. Conversely, given a periapsis passage q_i, if q_{i+1} does not ensue, then the ejecta have escaped. Finally, if the sequence never terminates ($i \to \infty$), then the ejecta are in a stable orbit about the asteroid. Figure 14.3 gives a graphical representation of these classes.

Based on this understanding, the following classifications can be defined (Scheeres *et al.* 2002):

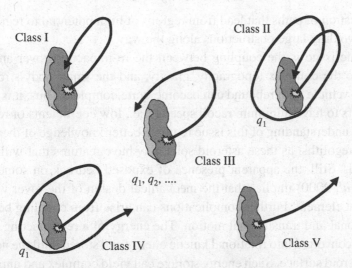

Figure 14.3 The five classes of ejecta fate. From Scheeres *et al.* (2002).

Class I Immediate reimpact: ejecta reimpact with the surface prior to first periapsis passage.

Class II Eventual reimpact: ejecta do not reimpact at the first periapsis passage, but eventually reimpact in the future.

Class III Stable motion: ejecta are placed into a long-term stable orbit about the asteroid.

Class IV Eventual escape: ejecta have at least one periapsis passage by the asteroid before escape.

Class V Immediate escape: ejecta escape from the asteroid prior to first periapsis passage.

Local escape speed really defines Class V ejecta. For motion relative to the surface, this is a sobering thought as an apparently "safe" trajectory that does not immediately escape may enter orbit and eventually escape. For controlled surface motion, the goal should be to place a rover trajectory firmly into Class I, which will immediately re-impact with the surface. Examples of all classes can be easily found; for example, in Scheeres *et al.* (1998a) a Class II trajectory that does not reimpact for almost a year is described. Such motion is clearly not suitable for controlled surface exploration using a rover.

These considerations force rover motion to be conservative, with only relatively small "hopping" speeds acceptable. This can lead to potential difficulties, however, depending on the topology of the asteroid surface. It is possible that a rover may be caught in a local potential well or gravitational basin, and that escape from this basin may require speeds large enough to place the vehicle into a Class II or higher trajectory. Being in such a situation would clearly be unacceptable, and implies that careful planning of surface motion trajectories are necessary prior to placing a rover on the surface of a small body. Specific strategies would probably involve the identification of paths that lead from regions of high potential to regions of low potential, avoiding large obstructions along the way.

A separate issue is the coupling between the re-impacting rover and the surface. Due to the complex topography, gravity, and the small body's rotation the motion of a vehicle with rebound can become quite complex. Thus, it is desired to design rovers to have minimum recoil speeds (i.e., low coefficients of restitution). A complete understanding of this issue requires better knowledge of the properties of asteroid regoliths, as these asteroid soils may have features that will minimize recoil speeds. Still, the apparent presence of exposed bedrock on some asteroids (Hudson *et al.* 2000) implies that the mechanical design of the rover will still be an important element. Further complications can arise from coupling between the rover rotational and translational motion. The energy of a re-impacting body may be partially converted into rotational kinetic energy and stored until the next impact with the asteroid surface. Such energy storage can yield complex and unpredictable motions of a body across the surface of an asteroid, and is an aspect of dynamics that is poorly understood and modeled (Sawai *et al.* 2001).

Finally, the coupling of surface motion with the disturbance and electric potential of surface regolith mentioned previously could potentially create a poor environment for the mechanical and electrical operation of a sophisticated rover vehicle. Initial designs of rover vehicles may be biased towards self-contained designs with a minimal number of sensor portals, such as the proposed Minerva rover for the MUSES-C mission (Yoshimitsu *et al.* 2001).

3.2 *Active control strategies for close proximity dynamics*

An exciting possibility for close proximity motion relative to small bodies is to take advantage of the relatively weak gravitational accelerations and actively control a spacecraft's motion relative to the body. Indeed, in the MUSES-C mission such strategies have already been developed at a basic level and will be used to circumvent the problems associated with trying to follow an orbital approach for that mission (Hashimoto *et al.* 2001; Kubota *et al.* 2001). It must be noted that active control techniques may not always be suitable for specific bodies and specific missions; still they can form a very important class of motions that will surely be applicable to the exploration of smaller asteroidal and cometary bodies.

There are two general approaches to controlled motion: near-inertial hovering and body-fixed hovering. In near-inertial hovering the spacecraft is stationed at a fixed location relative to the asteroid in the Sun–asteroid frame, the asteroid rotating beneath the spacecraft. Such a situation is useful for the initial characterization and measurement of the body. In body-fixed hovering the spacecraft is stationed at a fixed location relative to the rotating asteroid, implying that the spacecraft is rotating with the asteroid in inertial space. This situation is useful for sampling a small body surface. Both of these ideas, and their generalizations, are discussed in more detail below.

For the implementation of either approach, some minimal levels of sensing capability are needed on board the spacecraft. First is the ability to directly sense altitude, either using a laser altimeter or by the efficient processing of stereoscopic optical measurements. This measurement type forms the backbone of an automatic control system to maintain altitude and position relative to an asteroid. In addition to this, it is ideal for the vehicle to have the ability to sense its location relative to the asteroid surface. This can be implemented by optical sensors or scanning lasers; these technologies are in different stages of development. These are not the only types of measurements available or useful, but they are the most essential. The efficient measurement of altitude allows for the implementation of automatic control algorithms that stabilize the spacecraft hovering position, while measurements of body-relative location allows for an expanded capability for the control and motion of the spacecraft. For the latter case, the spacecraft must be able to correlate measured features with a global topography map in order to locate its current location. For some specific applications it may only be necessary to measure and detect lateral motion in addition to vertical motion; however, for the most general applications the ability to determine its global location on the asteroid is necessary. This implies that a global map of the asteroid surface has been created at some point, probably using the same instruments to be used for the relative navigation, although hopefully also using higher-resolution scientific instruments. The development and implementation of such sensor systems is a technology that is currently being developed, and should be available for use in the future.

In addition to the above sensing and estimation capability, the spacecraft will also require precise 6 degrees-of-freedom control capability. This implies a full set of thrusters for executing arbitrary control moves, perhaps augmented by momentum wheels for fine attitude control. It may be feasible to use more restrictive thruster configurations for the control of the spacecraft, although these would have to be carefully designed for specific implementation approaches. Finally, some, but not all, of these active control approaches imply that the vehicle may be out of sunlight for considerable periods of time. Thus, such power considerations should be factored into the development and design of space vehicles for these advanced approaches.

3.2.1 Near-inertial hovering

In this approach the spacecraft fixes its location relative to the body in the rotating body–Sun frame, creating an artificial libration point in this frame. A useful way to think about this approach is to first consider the Sun–asteroid libration point.

A spacecraft placed in this location will, ideally, remain fixed in its position. If, however, the spacecraft adds a constant thrust acceleration away from the asteroid, it would have to move its location closer to the asteroid in order for the forces to balance again. If a sufficiently large acceleration is added, it could conceivably hold its position relatively close to the asteroid. If it is close to the asteroid, it would have to supply an acceleration of $\sim \mu/r^2$ to "hover" at a radius of r from the attracting asteroid. Considering the more general case, it is possible to specify the necessary control thrust to maintain position at an arbitrary location in the asteroid–Sun rotating frame. Due to the relatively slow motion of the body about the Sun (on the order of degrees per day at fastest), this position can be considered to be nearly inertial over relatively short periods of time.

One complication with this approach to hovering is that these artificial equilibrium points are unstable, meaning that a small error in the open-loop thrust acceleration (or a small error in positioning the spacecraft) will cause the spacecraft to depart from its desired hovering point at an exponentially increasing rate. For example, assume a spacecraft is hovering above a spherical asteroid with a constant, open-loop thrust acceleration $\alpha = \mu/R^2$, but is positioned at a distance of $R \pm \delta R$ from the asteroid. If the spacecraft is at a slightly higher distance, $R + \delta R$, its thrust is greater than the gravitational attraction and it will start accelerating away from the asteroid. Conversely, if it is at a slightly lower distance, $R - \delta R$, its thrust is less than the gravitational attraction and it will start to accelerate towards the asteroid. Thus, practical implementation requires the addition of a closed-loop feedback control that senses the altitude or distance deviation of the spacecraft from its ideal hovering point. The necessary control loop to stabilize this motion is actually quite simple, and can be implemented in an automatic way using minimal spacecraft resources (Broschart and Scheeres 2003). There are limits to this approach, however. A spacecraft cannot inertially hover within the maximum radius of the asteroid at its hovering latitude, due to obvious physical constraints. Additionally, as the radius of hovering becomes closer to the body, the automatic control approach described here can become unstable, potentially leading to difficulties in implementation.

It is not necessary, however, to force the spacecraft to be fixed precisely at one location. A generalization of this idea places the spacecraft in an elliptic or hyperbolic orbit relative to the asteroid, but has its velocity vector "reflected" whenever it gets within a certain distance to the asteroid (for an elliptic orbit) or gets a certain distance away from the asteroid (for a hyperbolic orbit), forcing the spacecraft to travel back on, or close to, its original path but in the opposite direction. This approach can be thought of as hovering with a relatively large dead-band control about the nominal hovering point, and requires essentially the same control and sensing capability on board the spacecraft. This is essentially the approach to be used by the MUSES-C spacecraft when it arrives at its target asteroid (Kubota *et al.* 2001). In this approach the time between control maneuvers can be made arbitrarily long by increasing the size of the dead-band box.

Inertial hovering has several attractive attributes which may make it a mainstay approach of future exploration. There are also a number of drawbacks and limitations, however. On the positive side, this approach can be applied to any small body, and the cost of inertial hovering can theoretically always be driven to zero (not accounting for the statistical control to stabilize the hovering point). However, the position where hovering is feasible may be far from the body, and may not afford the optimal viewing geometry. For example, if the NEAR spacecraft had taken a hovering approach to its mission to Eros and implemented inertial hovering at a distance of 50 km from the asteroid (which was the nominal orbit radius for most of the mission), it would have required over 15 m s^{-1} per day to maintain this position, or for its prime 9-month mission would have required a total ΔV on the order of 4 km s^{-1}. Contrasted with the actual fuel usage (on the order of a few tens of meters per second), hovering was clearly not a reasonable approach for that body. Thus, to gain high-resolution scientific measurements this approach is largely limited to smaller bodies with their associated smaller hovering cost. A related drawback pertains to the ability of the spacecraft to measure accurately the mass and gravity field of the asteroid, and hence to compile an accurate global topographic map. When using such a controlled hovering mode, errors in the spacecraft thrusters and solar radiation pressure parameters will dominate over the signature of the asteroid gravity field acting on the spacecraft. While it might be possible to extract some averaged results on the total mass of the asteroid, these results would be corrupted by many different uncertainties that will not be uniquely separated from the gravity signature. Additionally, higher-order gravity fields will be nearly impossible to extract. This is a serious limitation, as it deprives the mission of essential scientific data and may make it difficult to subsequently transition to a body-fixed hovering exploration of the asteroid. It is possible to enact a few ballistic flybys of the asteroid to gain an improved estimate of its mass, but this estimate will not be as accurate as a mass estimate obtained by tracking the spacecraft while in orbit about the body.

3.2.2 *Body-fixed hovering*

In this approach the spacecraft now fixes its position relative to the rotating body. A natural way to visualize this approach is to imagine using a "jet-pack" to levitate off the surface of a rotating body, such as the Earth or an asteroid. Since the gravitational attraction is relatively weak at asteroids, it is possible to implement such hovering trajectories for extended periods of time (hours) with total costs on the order of tens of meters per second. This approach to controlling motion in close proximity to small bodies has been analyzed in detail (Scheeres 1999b; Sawai *et al.* 2002) and a detailed simulation of this approach has been developed for analysis of hovering over arbitrary models of asteroids (Broschart and Scheeres 2003). Resulting from this work, body-fixed hovering has been shown to be feasible from a dynamics and control perspective. The approach is essentially identical to inertial hovering, except everything is now done relative to the asteroid-fixed frame, which generally has a rotation period on the order of hours to days at most. Thus, the spacecraft acceleration must accommodate both the gravitational and centrifugal accelerations, although there are locations where the hovering cost is zero (at the synchronous orbits). This body-fixed hovering approach also suffers from the same basic instability noted for the inertial hovering case, although there are regions where this approach yields completely stabilized motion (Scheeres 1999b). A similar control strategy, using altimetry to maintain a fixed altitude, can stabilize a hovering point so long as it is located within the synchronous radius of the body. This result holds approximately true over the entire body and places an altitude "ceiling" on hovering for a simple control law to be able to stabilize its location. Some important points relative to this approach can be noted. First is that this technology can be seen as a precursor to a controlled landing on an asteroid surface, and is probably necessary for sample return missions. Second, the MUSES-C spacecraft is, in essence, using this approach when it performs its surface sampling runs (Hashimoto *et al.* 2001).

The implementation of body-fixed hovering can also be accomplished using a simple dead-band controller acting on the altitude of the spacecraft, in conjunction with a single thrust direction properly aligned relative to the asteroid gravity field. Thus, the technology to implement this is clearly available now. If, in addition to maintaining a single altitude at a specific location, it is desired for the spacecraft to translate in the asteroid-fixed frame, moving from one site to another, and to descend and ascend from the surface, additional sensing and control technology will have to be used. First, the spacecraft must maintain its attitude in the body-fixed frame – we should note that it will not do so naturally, as its attitude will remain fixed in inertial space and will want to spin in the asteroid-fixed space. Second, for it to perform translational motions will require that the spacecraft have the capability to locate itself relative to the surface and perform some higher-level control to null out transverse oscillations about the hovering point.

The development of this surface relative motion capability is perhaps the most advanced concept tendered here. This idea also solves the problem of rover loco-motion over an asteroid surface, as instead of relying on natural trajectories induced by mechanical "jumpers," we have controlled motion from one location to another. There are a number of interesting peculiarities associated with such surface relative motion, such as the fact that there is a preferred direction of motion about an asteroid in this mode (Sawai *et al.* 2002). Motion in the same direction as asteroid rotation can actually destabilize the dynamics of the spacecraft control, while translational motion in the opposite direction will tend to stabilize the control system. Other than this observation, which can be easily proven, there is little known about the stability and control of surface relative motion at small bodies, making it an essential topic for future research.

There are a number of drawbacks related to body-fixed hovering as well. First, it is essential that a fairly accurate model of the asteroid spin, topography, and gravity field be available. The gravity field must be defined down to the surface of the body as well, something which is not always easy to do (as described previously). Thus, body-fixed hovering should be preceeded by a period of characterization at a relatively high level of accuracy. In the future, it may be possible to dispense with this requirement, but that would only be after the basic technology and approach has been proven. Next, there is an altitude limit at which the conventional (and simple) stabilizing control becomes unstable, which could conceivably limit its application, although this does not seem to be so strong a drawback. Most importantly, however, is that a body-fixed hovering vehicle will undoubtably experience periods of solar occultation, making the presence of batteries or non-solar power generation essential for long-term operations at the surface of a body. Additional operations issues also exist, such as communications, attitude determination, and mechanical interfaces with the surface.

4 Conclusions

Ultimately, close proximity missions are driven by the small body environment that they will encounter. This is an important point that leads to a strong emphasis on pre-launch and pre-design observation campaigns for target bodies. Once the small body environment has been constrained, there are a number of approaches that can lead to a successful mission. None of these approaches is without drawbacks, but the wealth of options ensures that there exists a rational and reasonable strategy for exploring and characterizing any small solar system body. This being said, it must also be realized that there is a real diversity among these approaches, and a spacecraft designed to optimize one specific mission may not be suitable for a different mission. This is a serious issue and should be understood by spacecraft and mission designers alike. This fact leads to two different approaches to the problem:

(1) The target bodies are sufficiently characterized prior to mission and spacecraft design, allowing for a clearly delineated set of mission operations approaches.
(2) The spacecraft and mission is designed to handle a range of possible situations, out of which most asteroids of the class being visited would fall.

The latter approach is clearly more risky as it relies on unknown properties falling within a certain range, and in general will lead to a more expensive mission as this fundamental parameter and mission uncertainty must be accounted for at every step in the design. On the plus side, it opens up the number of target bodies for a single mission.

Within the realm of possible mission concepts, some are better understood than others. The issues that currently demand further study are the basics of dynamics about binary asteroids, the dynamics of rovers on the surfaces of asteroids, and the dynamics of transient atmospheres and re-impacting ejecta. In terms of controlled motions, there is still much work to be done on all aspects of hovering, with a special emphasis on surface relative translations, as this is the most demanding application of this approach.

Finally, as has been pointed out repeatedly, missions to asteroids with known physical properties are simpler, and allow the mission and spacecraft design to be optimized for the more precisely known parameters. While there is a large range of possible physical parameters which could be measured from the ground, there is a smaller set that yield the most important physical characteristics needed to decide which close proximity approach should be used. These are: number of co-orbitals, body size and type, body spin rate and spin state, and body shape.

Acknowledgments

The research reported here was supported in part by the Inter-planetary Network Technology Program by a grant from the Jet Propulsion Laboratory, California Institute of Technology which is under contract with the National Aeronautics and Space Administration, and by the Planetary Geology and Geophysics Program by a grant from the National Aeronautics and Space Administration.

References

Broschart, S. and Scheeres, D. J. 2003. Numerical solutions to the small-body hovering problem. *13th AAS/AIAA Space Flight Mechanics Meeting*, Ponce, Puerto Rico, February 2003, paper no. AAS 03-157.

Chauvineau, B., Farinella, P., and Mignard, F. 1993. Planar orbits about a triaxial body: application to asteroidal satellites. *Icarus* **105**, 370–384.

Dechambre, D. and Scheeres, D. J. 2002. Transformation of spherical harmonic coefficients to ellipsoidal harmonic coefficients. *Astron. Astrophys.* **387**, 1114–1122.

Dobrovolskis, A. R. and Burns J. A. 1980. Life near the Roche limit: behavior of ejecta from satellites close to planets. *Icarus* **42**, 422–441.

Garmier and R. and Barriot, J.-P. 2001. Ellipsoidal harmonic expansion of the gravitational potential: theory and application. *Celest. Mechan. Dynam. Astron.* **79**, 235–275.

Geissler, P., Petit, J.-M., Durda, D. D., *et al.* 1996. Erosion and ejecta reaccretion on 243 Ida and its moon. *Icarus* **120**, 140–157.

Hamilton, D. P. and Burns, J. A. 1991a. Orbital stability zones about asteroids. *Icarus* **92**, 118–131.

1991b. Orbital stability zones about asteroids. II. The destabilizing effects of eccentric orbits and of solar radiation. *Icarus* **96**, 43–64.

Hashimoto, T., Kubota, T., Kawaguchi, J., *et al.* 2001. Autonomous descent and touch-down via optical sensors. *Adv. Astron. Sci.* **108**, 469–480.

Hu, W. 2002. Orbital motion in uniformly rotating second degree and order gravity fields. Ph.D. thesis, The University of Michigan.

Hu, W. and Scheeres, D. J. 2002. Spacecraft motion about slowly rotating asteroids. *J. Guidance, Control, Dynami.* **25**, 765–775.

Hudson, R. S., Ostro, S. J., Jurgens, R. F., *et al.* 2000. Radar observations and physical modeling of asteroid 6489 Golevka. *Icarus* **148**, 37–51.

Kaula, W. M. 1966. *Theory of Satellite Geodesy*. Waltham, MA: Blaisdell.

Kubota, T., Hashimoto, T., Uo, M., *et al.* 2001. Maneuver strategy for station keeping and global mapping around an asteroid. *Adv. Astron. Sci.* **108**, 769–780.

Lara, M. and Scheeres, D. J. 2002. Stability bounds for three-dimensional motion close to asteroids. *J. Astron. Sci.* **50**, 389–409.

Lee, P. 1996. Dust levitation on asteroids. *Icarus* **124**, 181–194.

Mignard, F. and Hénon, M. 1984. About an unsuspected integrable problem. *Celest. Mechan.* **33**, 239–250.

Miller, J. K., Weeks, C., and Wood, L. J. 1990. *J. Guidance, Control, Dynam.* **13**, 775–784.

Miller, J. K., Bollman, W. E, Davis, R. P., *et al.* 1995. Navigation analysis for Eros rendezvous and orbital phases. *J. Astron. Sci.* **43**, 453–476.

Miller, J. K., Konopliv, A. S., Antreasian, P. G., *et al.* 2002. Determination of shape, gravity, and rotational state of asteroid 433 Eros. *Icarus* **155**, 3–17.

Morrow, E., Scheeres, D. J., and Lubin, D. 2002. Solar sail orbital operations at asteroids. II. Exploring the coupled effect of an imperfectly reflecting sail and a non-spherical asteroid. *Astrodynamics Specialist Meeting*, Monterey, California, August 2002, paper no. AIAA 2002–4991.

Petit, J.-M., Durda, D. D., Greenberg, R., *et al.* 1997. The long-term dynamics of Dactyl's orbit. *Icarus* **130**, 177–197.

Richter, K. and Keller, H. U. 1995. On the stability of dust particle orbits around cometary nuclei. *Icarus* **114**, 355–371.

Robinson, M. S., Thomas, P. C., Veverka, J., *et al.* 2001. The nature of ponded deposits on Eros. *Nature* **413**, 396–400.

Sawai, S., Kawaguchi, J., Scheeres, D. J., *et al.* 2001. Development of a target marker for landing on asteroids. *J. Spacecraft Rockets* **38**, 601–608.

Sawai, S., Scheeres, D. J., and Broschart, S. 2002. Control of hovering spacecraft using altimetry. *J. Guidance, Control, Dynam.* **25**, 786–795.

Scheeres, D. J. 1994. Dynamics about uniformly rotating triaxial ellipsoids: applications to asteroids. *Icarus* **110**, 225–238.

1995. Analysis of orbital motion around 433 Eros. *J. Astron. Sci.* **43**, 427–452.

1998. The restricted Hill four-body problem with applications to the Earth–Moon–Sun system. *Celest. Mechan. Dynam. Astron.* **70**, 75–98.

1999a. Satellite dynamics about small bodies: averaged solar radiation pressure effects. *J. Astron. Sci.* **47**, 25–46.

1999b. Stability of hovering orbits around small bodies. *AAS/AIAA Spaceflight Mechanics Meeting*, Breckenridge, CO, February 1999, AAS Paper no. 99–159.

Scheeres, D. J. and Asphaug, E. I. 1998. Debris and sample transport about asteroids. In *Space 1998*, eds. R. G. Galloway and S. Lokaj, pp. 340–346. Reston, VA: American Society of Civil Engineering.

Scheeres, D. J. and Augenstein, S. 2003. Spacecraft motion about binary asteroids. Paper presented at the *Astrodynamics Specialist Conferences*, Big Sky, Montana, August 2003.

Scheeres, D. J. and Marzari, F. 2000. Temporary orbital capture of ejecta from comets and asteroids. *Astron. Astrophys.* **356**, 747–756.

2002. Spacecraft dynamics far from a comet. *J. Astron. Sci.* **50**, 35–52.

Scheeres, D. J., Ostro, S. J., Hudson, R. S., *et al.* 1996. Orbits close to asteroid 4769 Castalia. *Icarus* **121**, 67–87.

1998a. Dynamics of orbits close to asteroid 4179 Toutatis. *Icarus* **132**, 53–79.

Scheeres, D. J., Marzari, F., Tomasella, L., *et al.* 1998b. ROSETTA mission: satellite orbits around a cometary nucleus. *Planet. Space Sci.* **46**, 649–671.

Scheeres, D. J., Bhargava, S., and Enzian, A. 2000a. A navigation model of the continuous outgassing field around a comet. Telecommunications and Data Acquisition Progress Report 42–142. Available online at http://tda.jpl.nasa.gov/progress_report/index.html

Scheeres, D. J., Williams, B. G., and Miller, J. K. 2000b. Evaluation of the dynamic environment of an asteroid: applications to 433 Eros. *J. Guidance, Control, Dynam.* **23**, 466–475.

Scheeres, D. J., Khushalani, B., and Werner, R. A. 2001. Estimating asteroid density distributions from shape and gravity information. *Planet. Space Sci.* **48**, 965–971.

Scheeres, D. J., Durda, D. D., and Geissler, P. E. 2002. The fate of asteroid ejecta. In *Asteroids III*, eds. W. M. Bottke Jr., A. Cellino, P. Paolicchi, and R. P. Binzel, pp. 527–544. Tucson, AZ: University of Arizona Press.

Thomas, P. C. 1998. Ejecta emplacement on the Martian satellites. *Icarus* **131**, 78–106.

Thomas, P. C. and Veverka, J. 1980. Downslope movement of material on Deimos. *Icarus* **42**, 234–250.

Thomas, P. C., Veverka, J., Sullivan, R., *et al.* 2000. Phobos: regolith and ejecta blocks investigated with Mars Orbiter Camera images. *J. Geophys. Res.* **105**, 15091–15106.

Thomas, P. C., Veverka, J., Robinson, M. S., *et al.* 2001. Shoemaker crater as the source of most ejecta blocks on the asteroid 433 Eros. *Nature* **413**, 394–396.

Weeks, C. J. 1995. The effect of comet outgassing and dust emission on the navigation of an orbiting spacecraft. *J. Astron. Sci.* **43**, 327–343.

Werner, R. A. 1994. The gravitational potential of a homogeneous polyhedron. *Celest. Mechan. Dynam. Astron.* **59**, 253–278.

Werner, R. A. and Scheeres, D. J. 1997. Exterior gravitation of a polyhedron derived and compared with harmonic and mascon gravitation representations of asteroid 4769 Castalia. *Celest. Mechan. Dynam. Astron.* **65**, 313–344.

Yeomans, D. K., Antreasian, P. G., Barriot, J.-P., *et al.* 2000. Radio science results during the NEAR-Shoemaker spacecraft rendezvous with Eros. *Science* **289**, 2085–2088.

Yoshimitsu, T., Kubota, T., Nakatani, I., *et al.* 2001. Robotic lander Minerva, its mobility and surface exploration. *Adv. Astron. Sci.* **108**, 491–502.

15

Mission operations in low-gravity regolith and dust

Derek Sears, Melissa Franzen, Shauntae Moore, Shawn Nichols,
Mikhail Kareev, and Paul Benoit

University of Arkansas

1 Introduction

The method to be used for mitigating the impact of an asteroid on Earth depends on the nature of the asteroid. A compact rock would react very differently to almost any violent mechanical event than would an object that consisted of unconsolidated dust and fragments. A water-rich, comet-like object would react very differently to laser heating than a completely hydrated object. Thus, impact mitigation begins with scientific investigation.

We have been investigating physical processes likely to be occurring on asteroids in connection with our efforts to understand the origin and history of meteorites and their relationship to asteroids. In this connection, we have been developing proposals for a near-Earth asteroid sample return mission called Hera (Sears *et al.* 2002c) (Fig. 15.1). Hera will visit three near-Earth asteroids, spend 3 months to 1 year in reconnaissance, and then nudge itself gently down to the surface to collect three samples from each asteroid at geologically significant sites (Britt *et al.* 2001). By returning weakly consolidated surface samples, the Hera mission will clarify many issues relating to the asteroid–meteorite connection and the origin and evolution of the solar system (Sears *et al.* 2001). In addition, interstellar grains in the samples will shed light on the relationship between our Sun and other stars.

The major challenge of the Hera mission is the design of the collector and this depends on a knowledge of the nature of the surface. We now have many images of asteroid surfaces from the Galileo and NEAR missions, some of them at high resolution (Fig. 15.2), and it seems clear that a surface of unconsolidated material is to be expected on all asteroids greater than about 200 m in size (Whiteley *et al.* 2000). It is also clear that there are a variety of surface features to be expected and collecting samples from different geological contexts will help in the interpretation

Mitigation of Hazardous Comets and Asteroids, ed. M. J. S. Belton, T. H. Morgan, N. H. Samarasinha, and D. K. Yeomans. Published by Cambridge University Press. © Cambridge University Press 2004.

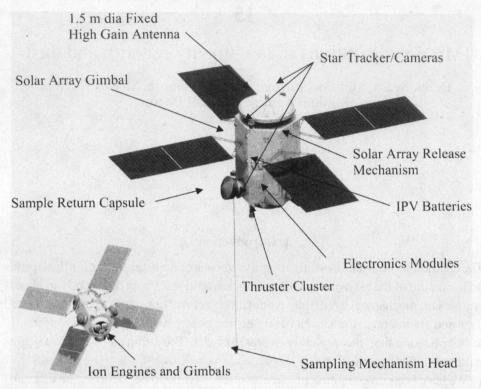

Figure 15.1 The Hera spacecraft according to concept drawings by SpaceWorks Inc. (courtesy of Jeffrey Prebble) and mission design by Leon Gefert (Glenn Research Center). The spacecraft has three sample return capsules, each with its own sample collection device, which is lowered to the surface without the spacecraft landing.

of the returned samples and meteorites already in our collections (Sears *et al.* 2002b). A great many collectors have flown on missions (Sears and Clark 2000), but all involve landers and are, therefore, impractical for our first low-cost sampling missions to asteroids. Three "touch-and-go" collectors have been described in the literature: the Lockheed-Martin auger attached to a large momentum wheel (Nygren 2000), the Honeybee counter-rotating cutters that are part of the present study (Sears *et al.* 2002a), and the MUSES-C percussion method (Yano *et al.* 2000) (Fig. 15.3).

We have performed experiments using simulated regoliths of sand, iron filings, gravel, and concrete (Akridge and Sears 1999; Bogdon *et al.* 2000). We have performed experiments in the laboratory under Earth's gravity and we have performed three sets of experiments (which we will refer to as campaigns I, II, and III) on NASA's KC-135 aircraft. The aircraft flies about 40 1-min "parabolas," rising steeply to 9750 m, tipping over and descending to 7300 m to create about 25 s of microgravity comparable to that on the surface of an asteroid. Each parabola

Figure 15.2 The surface of Eros as imaged by the *NEAR–Shoemaker* spacecraft. Top left and right, the regolith from orbit and on close approach during landing. Bottom left, a boulder-rich region on Eros viewed. Bottom right, four craters containing fine-grained deposits sometimes referred to as "ponds" (see Robinson *et al.* 2002, for details).

can be considered as having four phases, positive gravity (during climb), microgravity (as plane turns over the top of the parabola), transitioning from microgravity to negative gravity, and recovery (as the plane comes out of descent). We flew twice during each campaign, for a total of ~240 parabolas. It is particularly helpful to compare the test results in microgravity with the results in the laboratory. Our experiments have enabled us to gain some insights into the behavior of surface materials on asteroids and into mission operations in the presence of microgravity regolith and dust.

2 Ground-based experiments

For several years we have been investigating the behavior of mineral mixtures in fluidized beds (bed mobilized by the upward flow of gas) on the centimeter scale. This produced a realization that experiments relating to the behavior of materials

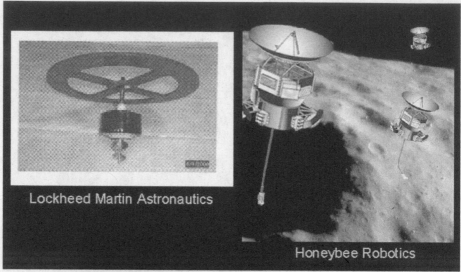

Lockheed Martin Astronautics

Honeybee Robotics

Figure 15.3 Sample collectors that have been proposed for use on asteroids that do not require the spacecraft to land. Lockheed-Martin has proposed a large auger attached to a momentum wheel, Honeybee Robotics has proposed a device with two counter-rotating cutters, while MUSES-C fires projectiles into the surface and collects the ejecta. SpaceWorks have also proposed sticky footpads on the end of articulated arms.

on the surface of asteroids needed to be performed analogous to the large volume of work performed simulating the surface conditions of comets (Sears *et al.* 1999). We have, therefore, constructed the Andromeda planetary environmental chamber. The chamber has been described elsewhere (Sears *et al.* 2002b) but essentially consists of a stainless steel cylinder 70 cm in diameter and 2 m long which is wrapped in heating elements and three sets of cooling coils, which may contain liquid nitrogen, methanol-CO_2 slurry, or chilled water. A stainless steel container

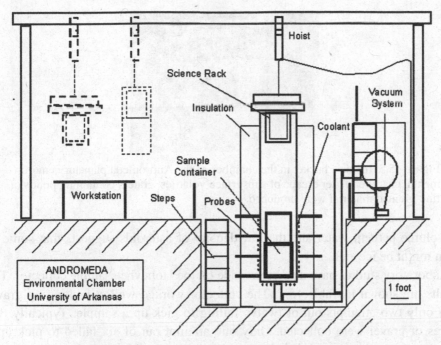

Figure 15.4 The Andromeda planetary environmental chamber at the University of Arkansas. This apparatus is being used to simulate conditions on the surface of asteroids except for gravity. Three sets of coolant coils are wrapped around the chamber, one for liquid nitrogen, one for methanol-CO_2 slurry, and one for chilled water. In addition, the chamber is wrapped with heater cable.

that can hold 1 m^3 of soil simulant can be lowered into the chamber and a lid holding a number of environmental simulators (solar lamps, power laser) and measuring instruments seals the chamber. A visible spectrometer, mass spectrometers, gas chromatograph, and closed circuit digital television cameras are also installed in the chamber (Fig. 15.4).

As an example of the sort of observations that can be made in the chamber Fig. 15.5 shows a small crater containing a "pool" of fine-grained material. The whole structure resembles the ponds on Eros. These features were made when frozen cultures were buried in the chamber as part of an experiment to investigate the survivability of anaerobic microorganisms on Mars. Upon exposure to vacuum, the water evaporated and the flow of gases upward produced the crater and size-sorted the grains to produce the fine-grained pools. Previously it had been suggested that the Eros ponds are the result of electrostatic levitation of grains and flow into small craters, or that they are the result of seismic shaking of regolith on the asteroid. Our results suggest an additional possibility which is that the ponds are due to the mobilization of dust in the surface by escaping fluids. Some meteorites are up to

Figure 15.5 Inside a bucket in the chamber of the Andromeda planetary environmental chamber. After escape of subsurface volatiles, craters containing ponds of
fine-grained material were produced.

20 volume percent water and the low density of asteroids suggests that some of
them might be ice-rich.

Laboratory studies have also been made on the Honeybee sample collector. The
results are shown in Table 15.1. The collector worked well at picking up gravel
with only two attempts out of twelve failing to pick up a sample. Typically two
pieces of gravel were collected. Only one attempt out of six failed to pick up a
sample of gravel and sand, but the reproducibility in picking up sand and gravel
in the original proportions was poor. The efficiency of picking up sand and iron
filings was also very high and the reproducibility of picking up iron and sand in the
original proportions was excellent.

3 Campaign I

During the first campaign, 310 Plexiglas tubes 2.5 cm in diameter and 10 cm long
were filled with various sand and iron mixtures and flown (Fig. 15.6). The sand
grains had mesh sizes of 149–250, 250–300, 300–425, and 425–600 and the iron
filings had mesh sizes of 74–105, 105–149, and 149–250. The sand to iron volume
ratio was 19 : 1. These values of grain size and sand–metal proportions are guided
by the values observed in chondritic meteorites. Air was passed through the beds
at a predetermined rate during the period of microgravity to simulate the release
of volatiles or mechanical disturbance and a plunger was depressed to freeze the
columns before the plane pulled out of free fall. About 25% of the tubes that acted
as controls were released and sealed without any air flowing through the tubes.
Separation of iron and sand was determined from image analysis of photographs
of the tubes after flight and from the measurements of removed samples.

An example of the results from campaign I is shown in Fig. 15.7. Many of the
tubes showed a tendency for the sand to iron ratio to be smaller in the lower
half of the 10-cm bed than in the upper part of the bed. Experiments in the

Table 15.1 *Collection details for sand and gravel in the laboratory tests of the sampler*

Sample	Number of chips	Gravel g	Sand g	Iron g	w/w
Gravel	1	0.219	–	–	–
Gravel	0	–	–	–	–
Gravel	4	2.978	–	–	–
Gravel	0	–	–	–	–
Gravel	5	5.05	–	–	–
Gravel	1	1.313	–	–	–
Gravel	2	2.912	–	–	–
Gravel	5	7.756	–	–	–
Gravel	2	2.812	–	–	–
Gravel	2	4.047	–	–	–
Gravel	2	2.633	–	–	–
Gravel	3	2.975	–	–	–
Sánd+Gravel	2	1.203	1.387	–	0.87
Sand+Gravel	1	0.638	0.519	–	1.23
Sand+Gravel	0	–	0.12	–	–
Sand+Gravel	2	1.397	0.058	–	24.1
Sand+Gravel	3	2.886	2.092	–	1.38
Sand+Gravel	1	0.114	0.536	–	0.21
Sand+Gravel[a]	–	–	–	–	0.72
Sand	–	–	8.319	–	–
Sand	–	–	9.719	–	–
Sand	–	–	7.382	–	–
Sand	–	–	9.354	–	–
Sand	–	–	7.782	–	–
Sand	–	–	5.118	–	–
Sand+iron	–	–	5.836	2.929	1.99
Sand+iron	–	–	3.249	1.433	2.27
Sand+iron	–	–	0.689	0.382	1.80
Sand+iron	–	–	1.792	0.806	2.22
Sand+iron	–	–	2.468	1.174	2.10
Sand+iron	–	–	2.211	1.055	2.10
Sand+iron[a]	–	–	16.479	8.335	1.98

[a] Control sample.

laboratory also showed that under certain circumstances metal rises to the surface due to its smaller grain size and greater tendency to be lifted in the gas flow (Akridge and Sears 1999). Ground tests also showed that this tendency for metal to rise increased with increasing size of the larger grain (in this case sand). We tested this by plotting the sand-to-metal ratio at 6–10 cm divided by the sand-to-metal ratio at 1–5 cm against sand particle size. Figure 15.7 shows that the tendency for metal and sand to segregate is related to the relative size of the grains.

Figure 15.6 Undergraduate students Ryan Godsey and Katrina Bogdon with the apparatus for campaign I. The apparatus consists of 310 2.5-cm Plexiglas tubes containing sand and iron files of chondritic grain sizes and in various proportions. The white "pistons" were depressed while the beds were under microgravity conditions in order to "freeze-in" any segregation that may have occurred during flight.

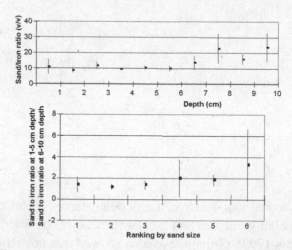

Figure 15.7 Examples of results from campaign I. In the upper graph, the sand-to-iron volume ratio was plotted against depth. The lower graph shows how the increase in sand-to-iron ratio with depth depends on the sand grain size. Tubes in which the sand-to-iron ratio is largest show the greatest tendency for segregation to occur.

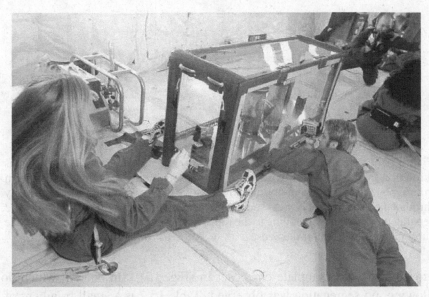

Figure 15.8 Undergraduate students Amber Holley and Mike Meyer operating the apparatus for campaign II. The apparatus consists of two 15-cm Plexiglas cylinders containing sand and iron filings with similar grain sizes and in similar proportions to those of chondritic meteorites. The experimenters recorded the behavior of the beds using digital cameras as the KC-135 went through nearly 80 parabolas of positive gravity, microgravity, and negative gravity and then positive gravity again.

Further details of these experiments can be found in Franzen *et al.* (2003).

4 Campaign II

For the second campaign, the alternative approach of observing a limited number of tubes very closely was taken. Two 15-cm diameter Plexiglas cylinders containing sand–iron mixtures in approximately chondrite grain sizes and proportions (300–425 mesh sand and 74–105 mesh iron) were mixed in 20 : 1 proportions (by volume) and observed with digital cameras (Fig. 15.8). The behavior of the mixture, in particular the separation of iron and sand was noted and any surface structures were recorded digitally. An experimenter then viewed the data filling out a questionnaire.

Some sample results appear in Table 15.2 and Fig. 15.9. The behavior of the beds as the plane went through the gravity cycles was very reproducible. A few anomalies were associated with unusual accelerometer data and therefore with unusual parabolas. Nothing would happen during positive gravity, but as gravity lessened air flowing through the bed would cause the bed to rise and metal to migrate fairly quickly to the surface. Then negative gravity would cause the surface of the bed to rise in a swirl but the metal–sand stratigraphy would prove very resilient

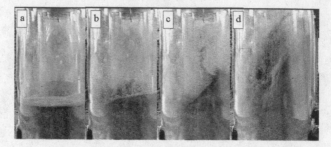

Figure 15.9 One of the experimental beds in campaign II as the KC-135 goes through a cycle of (a) positive gravity, (b) microgravity, (c) transitioning from microgravity to negative gravity, and (d) negative gravity while air flows vertically through the beds from below. As soon as the bed became mobilized, metal and sand separated with the metal moving to the top. This segregation survived considerable subsequent movement of the beds.

only to be lost when complete chaos set in (Fig. 15.9). In beds in which no air was flowing, no segregation was observed. Table 15.2 is a small fragment of our observations, but it illustrates the behavior of the beds and the contrast between cases in which air flowed and did not.

Further details of these experiments can be found in Moore *et al.* (2002).

5 Campaign III

The third campaign was essentially a test of the Honeybee Robotics touch-and-go surface sampler being considered as a possible sample collector for the Hera mission (Rafeek and Gorovan 2000). This device consists of two counter-rotating cutters that eject material into a cylindrical container with front doors, to allow collection, and a trap door below to allow ejection into the spacecraft container (Fig. 15.10). The collector was mounted on a vertical rail inside a double-walled enclosure and attempts were made to sample four surface stimulants: sand, sand and iron mixtures, sand and gravel mixtures, and concrete.

The samples consisted of 500 μm sand, 500 μm sand and 200 μm iron filings mixture (sometimes with 10% by volume 1–2-cm sized gravel pieces representing "clasts"), gravel, and concrete. The tests were designed to address several specific questions. Did the collector work under microgravity conditions? How much mass can be transferred by a 1–2-s touchdown? Did the collector change the metal–silicate ratio? Did the collector change the size distribution of the surface materials? What is the largest particle the collector could collect?

The results are shown in Table 15.3. Attempts to collect gravel and concrete were very successful, and although the concrete mass recovered was small it would still be enough for several laboratory measurements. The difficulties were with sand and iron filings, where a large number of attempts failed to collect material and where

Table 15.2 *Sample of data[a] obtained from campaign* II

Parabola	Gas flow	Description of bed at reduced gravity	MSF[b]	Flotsam
1	Yes	Fluid, turbulent (wave-like), spouting	Yes	Metal
2	Yes	Fluid, turbulent (wave-like), spouting (went to the side)	Yes	Metal
3	Yes	Was well mixed but data was inconclusive	Yes	Metal
4	Yes	Fluid, turbulent (wave-like), spouting in middle, right twist	Yes	Metal
5	Yes	Fluid, turbulent (wave-like), spouting in middle, right twist	Yes	Metal
6	Yes	Fluid, turbulent (wave-like), spouting in middle, right twist	Yes	Metal
7	Yes	Fluid, turbulent (wave-like), spouting in middle, right twist	Yes	Metal
8	Yes	Fluid, turbulent (wave-like), spouting in middle, right twist	Yes	Mixture
9	Yes	Fluid, turbulent (wave-like), spouting in middle, right twist	Yes	Metal
10	Yes	Fluid, turbulent (wave-like), spouting in middle, right twist	Yes	Metal
17	No	Hard to see trends, top layer rises	No	Mixture
18	No	Hard to see trends, top layer rises	No	Metal
19	No	All fell to the left of the cylinder	No	Unclear
20	No	Hard to see trends, top layer rises, metal spouts in middle	No	Unclear
21	No	Hard to see trends, top layer rises, metal spouts in middle	No	Unclear
22	No	Inconclusive	No	Metal
23	No	Inconclusive	?	?

[a] The full data set consists of about 80 parabolas (see Moore *et al.* 2002).
[b] Metal silicate fractionation, i.e., did the metal separate from the silicates?

the collector repeatedly jammed. After the second day of the flight tests, when the collector was stripped down for shipping, an iron-filing/sand mixture was found in the collector space with exactly the proportions of the original sample.

6 Summary of major findings as they concern mission operations

The major results of the three campaigns, in terms of implications for mission operations on the surfaces of asteroids and comets were several.

Figure 15.10 A sample collector developed by Honeybee Robotics, New York, in test apparatus designed by SpaceWorks aboard the KC-135 during campaign III. Regolith simulants consisting of sand, gravel, sand and gravel, sand and iron filings, and concrete were used for the tests.

Particle size sorting of the surface material occurs readily

While there are many details yet to be understood, it seems clear that segregation of minerals with different physical and chemical properties will occur readily on asteroid surfaces when the surface is disturbed. Aerodynamic sorting is an important process when the surface is degassing as a result of the disturbance, but mechanical sorting alone will also segregate grains with different physical properties like size and density. Furthermore, these segregations, once produced, are quite resistant to mixing by subsequent activities.

Segregations that occurred early in the process are retained during considerable amounts of subsequent activity

We were able to observe repeatedly that the stratigraphy introduced by the initial stages of microgravity, with a dark layer of metal at the surface of the beds, persisted as the bed was lifted by the negative gravity pulse and "faulted" as part of the process of swirling to the top of the chamber. While the surface grains are being subjected to considerable and fairly complex forces they are all suffering the same processes and tend to move together.

Table 15.3 *Collection details for gravel and concrete for the KC-135 tests of the sampler*

Sample	Number of chips	Mass (g)
Gravel	9	9.034
Gravel	2	3.356
Gravel	1	0.879
Gravel	4	~4
Gravel	1	~1
Gravel	1	0.422
Gravel	1	0.483
Concrete	–	0.041
Concrete	–	0.021
Concrete	–	0.053
Concrete	–	0.128
Gravel	1	1.595
Concrete	–	0.335
Concrete	–	0.026
Gravel[a]	19	25.4

[a] Control.

> *It was difficult to "see-through" the periods of negative g, which are an artifact of the KC-135 tests and would not be present during sample collection on an asteroid*

However, perhaps the major lessons learned from the experiments are the problems posed by negative gravity on the present KC-135 tests. This negative gravity pulse would not occur during surface operations on an asteroid.

Laboratory versus flight tests

A most significant result is that though the collector worked well on the ground, it worked far less well under microgravity conditions. Under microgravity conditions there was movement of the disturbed surface in all directions but mostly away from the collector and this was a big problem. In addition, once material had been thrown into the collector by the component engaging the surface, it would tend to move out again. An important message here is that intuition and experience applies to the Earth-gravity environment and we take much for granted. Surfaces do not move away from us on Earth and material thrown into a container stays there. For this reason and others, clogging of moving parts in such a dusty environment was also a problem, especially since the dust was mobile in a fashion different from that in the terrestrial environment.

7 Implications

Implications for science

The tendency of the material to segregate easily has implications for both the science of performing surface operations on asteroids, and for the engineering. First, it means that the science value of the returned materials will be compromised if the sample collection procedure alters their surface proportions. For example, the ratio of silicates to metal is an important characteristic of the major chondrite classes and the cause of the metal–silicate fractionation is the subject of considerable discussion by cosmochemists. Similarly, an important and highly enigmatic component in primitive meteorites are the chondrules, $\sim$200-μm crystallized glass beads whose origin has perplexed scientists for 200 years. Chondrules will readily segregate from other smaller components under microgravity conditions, especially since one of the smaller components is metal which also has a higher density that could enhance its tendency to segregate. Finally, but related to the previous two points, is that cosmochemists have developed a number of ideas and theories relating to even small (10–20%) deviations in bulk composition. All of these fundamental studies in cosmochemistry would be jeopardized if the material returned had suffered segregations during collection. The irony is that the predicted segregations concern the very components that are at the heart of understanding meteorite genesis.

Implications for engineering

For engineering there are two implications. First, chondritic metal is very ductile, and will readily clog moving parts. If segregation causes localized enrichments of metal on the surface of the asteroid then these would be a hazard to equipment that collects the samples with an instrument dependent on moving parts. More significantly, science calls for a collector design that does not cause segregations of the surface materials. Several options are possible. A collector might be designed that identifies an area and then picks up everything in that area regardles of physical properties. Alternatively, a collector might be designed that picks up coherent rocks, or soil clods, and not unconsolidated material. This also jeopardizes the value of the samples somewhat, since significant science related to space weathering and the interpretation of astronomical spectra, and the whole field of solar interactions, requires the fine-grained surface materials.

Implications for future work

It is clear that surface operations on asteroids require an apparatus that will disturb the surface as little as possible, and should collect rocks and clods as well as unconsolidated material. We do not yet have such a collector, but at least we have defined its chief characteristics. It also seems clear that we should be cautious in relying on experience and intuition, because they are not only of minimal relevance but may actually be misleading. Instead, we must rigorously test equipment in a reasonable microgravity environment. There are serious difficulties in using the KC-135 as a test bed for microgravity operations. Although we have not yet explored options for handling these problems, some improvements might include a test rig that "floats" rather than being anchored to the floor of the aircraft, or an apparatus that covers the simulated surface materials during negative gravity and reveals them only during microgravity. A more promising approach might be to use drop towers, the Shuttle, or the International Space Station. Simple collectors with a minimum of moving parts and with as much dust protection as possible are required. Collectors that cover or retain the surface materials as they are collected stand the best chance of success for recovering the most scientifically valuable samples.

Acknowledgments

We are grateful to the organizers of the Workshop on Scientific Requirements for Mitigation of Hazardous Comets and Asteroids for inviting us to participate in the workshop and write this book chapter; to Jeffrey Preble, John DiPalma, and Space-Works for organizing campaign III and constructing the test rig; to Steve Gorovan, Paul Bartlett, and Honeybee Robotics for supplying a prototype of their collector; to the NASA Reduced Gravity Students' Opportunity Program for funding campaigns I and II; to the National Science Foundation and the University of Arkansas for funding the Andromeda laboratory and campaign III; and to the National Science Foundation for funding nine undergraduate students over 3 years through their Research Experience for Undergraduates Programs. We also thank the University of Arkansas for funding our work on the Hera mission.

References

Akridge, D. G. and Sears, D. W. G. 1999. The gravitational and aerodynamic sorting of meteoritic chondrules and metal: experimental results with implications for chondritic meteorites. *J. Geophys. Res.* **104**, 11853–11864.

Bogdon, K., White, C., Godsey, *et al.* 2000. The origin of chondrites: metal–silicate separation experiments under microgravity conditions. *Meteorit. Planet. Sci.* **35** (supp.), A30.

Britt, D. T., Sears, D. W. G., and Cheng, A. F. 2001. Asteroid sample return: 433 Eros as an example of sample site selection. *Meteorit. Planet. Sci.* **36**, A30–31.

Franzen, M., Nichols, S., Bogdon, K., *et al.* 2003. The origin of chondrites: metal–silicate separation experiments under microgravity conditions – I. *Geophys. Res. Lett.* **30**, issue 14, SSC7–1.

Moore, S. R., Franzen, M., Benoit, P. H., *et al.* 2002. The origin of chondrites: Metal–silicate separation experiments under microgravity conditions – II. *Geophys. Res. Lett.* **30**, issue 10, 29–1.

Nygren, W. D. 2000. Rapid remote sample collection mechanism for near-Earth asteroid sample return. In *Near-Earth Asteroid Sample Return Workshop (abstracts)*, Lunar and Planetary Institute, Houston, TX, LPI Contribution no. 1073.

Rafeek, S. and Gorovan, S. 2000. NEA touch and go surface sampler. In *Near-Earth Asteroid Sample Return Workshop (abstracts)*, Lunar and Planetary Institute, Houston, TX, LPI Contribution no. 1073.

Robinson, M. S., Thomas, P. C., Veverka, J., *et al.* 2002. The geology of 433 Eros. *Meteor. and Planet. Sci.* **33**, 1651–1648.

Sears, D. W. G. and Clark, B. C. (2000) Sample collection devices for near-Earth asteroid sample return. In *Near-Earth Asteroid Sample Return Workshop (abstracts)*, Lunar and Planetary Institute, Houston, TX, LPI Contribution no. 1073.

Sears, D. W. G., Kochan, H. W., and Huebner, W. F. 1999. Laboratory simulation of the physical processes occurring on and near the surface of comet nuclei. *Meteor. and Planet. Sci.* **34**, 497–525.

Sears, D. W. G., Allen, C., Britt, *et al.* 2001. Near-Earth asteroid sample return missions. In *Space 2001 Conference and Exposition*, Albuquerque, NM, Paper no. 2001-4728.

2002a. Near-Earth asteroid sample return. In *The Future of Solar System Exploration, 2003-2013: Community Contributions to the NRC Solar System Exploration Decadal Survey, 2003-2013*, pp. 111–139.

Sears, D. W. G., Benoit, P. H., McKeever, S. W. S., *et al.* 2002b. Investigation of biological, chemical and physical processes on and in planetary surfaces by laboratory simulation. *Planet. Space Sci.* **50**, 821–828.

Sears, D. W. G., Franzen, M., Bartlett, P. W., *et al.* 2002c. The Hera mission: laboratory and microgravity tests of the Honeybee Robotics touch-and-go sampler. *Lunar Planet. Sci. Conf.* **33**, Abstract no. 1583.

Whiteley, R. J., Tholen, D. J., Bell, J. F., *et al.* 2000. Sample return from small asteroids: mission impossible? In *Near-Earth Asteroid Sample Return Workshop (abstracts)*, Lunar and Planetary Institute, Houston, TX, LPI Contribution no. 1073.

Yano, H., Fujiwara, A., Hasegawa, S., *et al.* 2000. MUSES-C's impact sampling device for small asteroid surfaces. In *Near-Earth Asteroid Sample Return Workshop (abstracts)*, Lunar and Planetary Institute, Houston, TX, LPI Contribution no. 1073.

16

Impacts and the public: communicating the nature of the impact hazard

David Morrison
NASA Ames Research Center

Clark R. Chapman
Southwest Research Institute

Duncan Steel
University of Salford

Richard P. Binzel
Massachusetts Institute of Technology

1 Introduction

In the twenty-first century, we must consider the asteroid and comet impact hazard in a context in which citizens of many nations are apprehensive about hazards associated with foods, disease, accidents, natural disasters, terrorism, and war. The ways we respond psychologically to such threats to our lives and well-being, and the degrees to which we expect our societal institutions (both governmental and private) to respond, are not directly proportional to actuarial estimates of the causes of human mortality, nor to forecasts of likely economic consequences. Our concerns about particular hazards are often heavily influenced by other factors, and they vary from year to year. Citizens of different nations demonstrate different degrees of concern about risks in the modern world (for example, reactions to eating genetically modified food or living near a nuclear power plant). Yet one would hope that public officials would base decisions at least in part on the best information available about the risks and costs, and scientists have a responsibility to assist them to reach defensible conclusions.

Objective estimates of the potential damage due to asteroid impacts (consequences multiplied by risk) are within the range of other risks that governments often take very seriously (Morrison *et al.* 1994). Moreover, public interest is high, fueled by increasing discovery rates and the continuing interests of the international

Mitigation of Hazardous Comets and Asteroids, ed. M. J. S. Belton, T. H. Morgan, N. H. Samarasinha, and D. K. Yeomans. Published by Cambridge University Press. © Cambridge University Press 2004.

news media. In this chapter we consider the past, present, and future of interactions by scientists with the public and media on the subject of the impact hazard.

Today it is commonplace to realize that rocks from space bombard the Earth, and that a strike by a big one could end civilization. This was not true when the *NASA Spaceguard Survey Report* (Morrison 1992) was released, providing the first quantitative estimate of the impact hazard and concluding that we are as much at risk from impacts as from other better-known hazards such as earthquakes and severe storms. Some in the media treated this original report with derision. At a minimum, there was a "giggle factor" associated with claims that "the sky is falling."

A decade later, the impact hazard is fairly well understood within the science community and is increasingly being accepted by the public at large. Yet it is still a difficult concept, because the danger from impacts occurs in ways that are different from anything in our experience. People will rarely be killed individually, or by the hundreds or even the thousands, by impacts. Rather, the primary risk is from a global environmental catastrophe that might happen only once in a million years, and yet would kill a substantial fraction of the Earth's population. Impacts are an extreme example of a hazard of very low probability but very great consequences. Nobody has ever been killed in such a major impact event in recorded history, yet we recognize this possibility as a serious challenge for individuals and governments.

The best news a decade after the *Spaceguard Survey Report* is not that the hazard is better understood, but that we are actually doing something about it. Unlike any other natural hazard, impacts can potentially be predicted with high precision and prevented (at least in principle) by the application of space technology to deflect or destroy a potential impactor.

The first step in any effort to mitigate the impact danger is to find out whether the Earth is – or is not – the target for such a collision within our lifetimes, or perhaps the lifetimes of our grandchildren. Sometimes we talk as if this were a statistical question. In fact, discussions of probabilities can be a way of covering our ignorance. We know enough about probabilities to say that any impact in our lifetimes, whether it is from an auditorium-sized "Tunguska-class" asteroid or a much more massive "extinction level" asteroid or comet, is highly unlikely. Nevertheless, we want to know whether, against the odds, our generation will need to prepare to defend the planet from this threat. Scientists have therefore shifted from an emphasis on understanding the probabilities to a straightforward program to find the potentially threatening asteroids and compute their orbits, one at a time, to see if any will hit us. That is what people and governments need to know: not "What are the odds?" but "Will we be hit soon?" Today we can change a probabilistic question into a deterministic answer for the larger asteroids, and eventually it should be possible to do the same for sub-kilometer impacts as well.

In the United States the program to search for potentially hazardous asteroids is funded by NASA and the US Air Force. The present Spaceguard goal is to find

90% of the near-Earth asteroids (NEAs) larger than 1 km in diameter (thought to cover the lower limit for a global catastrophe) by 2008, which is ten years from the adoption of the goal. Due to the phenomenal productivity of the handful of active search programs in the USA, the current count of known large NEAs is already more than halfway to the goal, with nearly 700 found out of an estimated total of 1100 ± 100 such asteroids (Stuart 2001; Morrison *et al.* 2002; Stokes 2003). The good news is that while many of these will hit the Earth eventually, none seems to be on a collision course in the short term (meaning the next century). By finding and eliminating these asteroids as possible short-term impactors, the risk is being lowered that we will be hit unawares by a global impact catastrophe. And when, eventually, all the NEAs bigger than 1 km have been found, the most likely outcome will be a demonstration that none is due to hit us soon, thereby removing for our generation concern about this particular threat to the survival of civilization. Of course, if we do find one on a collision course, we will have a very different challenge: to mount a mitigation campaign that will divert the asteroid. If we have several decades of warning, we may be confident that such an effort would succeed, given the high motivation that an impending impact would provide.

While NEAs dominate the impact hazard, comets also pose a danger. Comets have a different size distribution from asteroids; it may be that the largest potential impactors are comets (e.g. Comet Hale-Bopp, which is 40 km in diameter and passed perihelion within the terrestrial orbit in 1997), whereas apparently there is a near-absence of comets 1 km and smaller. We currently do not have the technology either to predict an impact of a comet long in advance, or to do much about it if we did. At the small end of the size distribution, comets are not a significant hazard, comprising less than 1% of near-Earth objects (NEOs) <1 km in diameter (D. K. Yeomans, personal communication). For these reasons, the present discussion is focused on the hazard of asteroid impacts.

A large cosmic impact would represent the ultimate environmental catastrophe, and the chance of such a calamitous event occurring soon is far from negligible, although it is small. If the frequency of major impacts were not low, we would not be here to discuss them; but if the occurrence rate were zero, then we would not be here either, because the dinosaurs would likely still rule the Earth. How do we communicate with the media, public, and governments alike the nature and severity of a very infrequent but ultimately inevitable disaster that is outside our personal or historical experience?

2 Risk perception and the unique nature of the impact hazard

We all face potential hazards to life and limb each day, and these moderate our activities in a variety of ways. Individual responses to specific risks vary markedly, and not all hazards can be entirely avoided. Apart from human-made hazards such as automobile crashes or firearm accidents, there are many natural hazards that individuals may either ignore or take very seriously, depending on personal outlook. Some people may specifically avoid living near earthquake fault lines, while others eschew coastlines subject to hurricanes, or inland plains liable to be swept by tornadoes. While some people fear sharks to such an extent that they would never dip a toe into the sea, others avoid areas where snakes may lurk – yet the risks from both sharks and snakes are negligible.

All the above are examples of hazards that may be obviated to some degree, either by behavior changes or else avoidance of certain terrains or locations. There are, however, hazards that are unavoidable no matter where one lives. Global climate change, for example, must affect everyone, although by differing degrees, and with both winners and losers. This is also the case for cosmic impacts: no matter where one lives, every human would be affected by a large impact (roughly, with an energy greater than 1 million megatons: Toon *et al.* 1997), no matter where the collision occurred. This means that the frequent public question of "Where might it hit?" is somewhat misguided: it makes little difference, for a large impactor, although it is certainly true that smaller projectiles would have more localized effects.

Public perceptions of risk and benefit have been studied extensively by social scientists (e.g., Slovic 1987). Among the factors that enhance public concern about a hazard are its potential for catastrophe, dread, lack of control, and threat to future generations – all of which seem to be associated with the impact hazard. On these grounds, we might expect public concern to be high – higher than is suggested by the actuarial calculations of the hazard. Generally, however, this is not the case; if anything, the public seems to underestimate the impact threat. Presumably we are not presenting credible case, or else the extreme rarity of the event causes it to be dismissed as too improbable to be taken seriously.

NEO impacts are qualitatively different from other hazards in terms of the risk they pose to individual humans. They are also quantitatively different, in that the numbers of people killed by such a horrible event would be far larger than in any natural disaster that has occurred during historical times, and may approach the whole population of the planet. Mass extinction events due to hugely energetic impacts have occurred in the past, and will occur in the future unless we intervene. The risk to individuals, however, is dominated by smaller, but still very substantial, impacts that might cause the death of a quarter to a half of all humankind. Everyone on the planet is at risk from impacts across the entire size range from Tunguska-class impacts up to extinction level events. The one exception is the risk from impact-induced tsunamis, which is limited to the small fraction of the world's population that lives close to unprotected coasts.

Although some aspects of the impact hazard (e.g., its predictability) are unusual or unique, many of the associated destructive effects resemble those of tsunamis, earthquakes, atomic bomb and volcanic explosions, sudden climate change, wild-fires, and so on. To some extent the effects of impacts may be explicable in terms of phenomena with which the public has familiarity, even if only through the mass media.

The only other catastrophe that might provide a similar outcome to an impact by an NEO with energy greater than 1 million megatons would be a global nuclear war. The chance of such a calamity is taken very seriously by many people, and governments. The question might therefore be asked: why is the impact hazard not taken more seriously? There are several reasons that contribute, many of them discussed in this chapter. One, of course, is a lack of awareness of the nature and level of the hazard, but another is the fact that blame might be attached to various people or countries in the case of a nuclear war, whereas an impact might be regarded as an Act of God. Until specific agencies within governments are made responsible for this matter, it may well be that progress in tackling this unique hazard will continue to be slow.

3 Hazard scales as a communication mechanism

The NEO community has taken several actions to facilitate communications with the media and the public. First, the nature of the hazard itself has been explained in a variety of public forums (for example, hearings in the United States Congress and documentaries produced for television broadcast). Second, the Internet has been widely used to explain the hazard and to provide up-to-date information on asteroid discoveries and orbits. The two National Aeronautics and Space Administration (NASA) websites, for example (the Impact Hazard website at http://impact.arc.nasa.gov and the NEO Program Office site at http://neo.jpl.nasa.gov) each receive roughly a million hits per month, and both have been recognized as among the most popular NASA websites. Third, the International Astronomical Union (IAU) has attempted to provide authoritative information on NEOs and possible future impacts with mixed results (as further described below).

Whether testifying before Congress, providing soundbites on television, or writing messages to post on the Internet, it is vital for scientists involved in disaster prediction to communicate their calculations in simple ways that can be understood by the news media, public officials, and disaster relief agencies, so that news of a potential disaster evokes consistent and appropriate responses. Most people are familiar with such scales as the Richter (or magnitude) scale for earthquakes, the Saffir–Simpson scale for hurricanes, and the terrorism scale adopted by the US Department of Homeland Security. Numerous other scales have been developed to communicate technical evaluations of computer virus threats, nuclear reactor security, space weather hazards (e.g., due to solar flares), generalized wind damage, etc. There are more prosaic scales developed along the same lines concerning UV dangers from sunlight, forest fire danger, etc. Many of these scales involve predicted future events such as weather phenomena, but others like the Richter scale define the severity of an event after the fact.

In nearly all such hazard scales, multi-dimensional parameter space is compressed into a single linear, often color-coded, scale from 1 to 5 or 1 to 10 (not usually including decimal points, despite the prominent exception of the earthquake magnitude scale). Development of these scales has often involved much behind-the-scenes controversy among specialists. The disputes have not generally entered the public sphere, which is fortunate because "battling scales" is a sure way to defeat the prime objective, which is simplification and consistent use by all concerned. With such motivations, and following some bad cases of sensationalistic news stories, some NEO researchers began to think about developing a hazard scale that science journalists and others could use to help the general public evaluate the seriousness of predicted NEO impacts.

Beginning with the premise that "collisions of asteroids and comets with the Earth is a topic so provocative and so prone to sensationalism that great care must be taken to assess and publicly communicate the realistic hazard (or non-hazard) posed by close approaches of newly discovered objects," Binzel (1997) proposed to the NEO community a 0 to 5 hazard index as a public communication tool. The timing of this proposal (at the 1995 United Nations International Conference on Near-Earth Objects) was deemed important because "with the advent of expanded near-Earth object (NEO) surveys, the number of known close approaches to Earth will increase dramatically." Multiple factors led to this proposed system not being adopted for use, however. These factors included the preliminary nature of the system and the general lack of appreciation by astronomers of the necessity for (and of the difficulty of) clear and effective hazard communication.

The discovery of asteroid 1997 XF11, the first object for which an orbit calculation gave a non-negligible probability for impact of globally catastrophic consequences, provided a baptism by fire for the NEO community. From the missteps in early announcements to the wide realization of the difficulty of making and communicating impact probability calculations, a maturation of ideas began for effectively conveying information on predicted close Earth approaches. This progress was apparent at the "International Monitoring Programs for Asteroid and Comet Threat" (IMPACT) workshop held in Torino, Italy, in June 1999. Key to this progress were advances in the ability to perform the complex orbital and probability calculations, as exemplified by Chodas and Yeomans (1999a, 1999b) and Milani *et al.* (2000). (These advances continue; see Chesley *et al.* 2002.) At the Torino workshop a color-coded 0 to 10 point hazard scale was brought forward by a communications subcommittee and through discussion and debate by all workshop attendees. With the subsequent endorsement by the workshop sponsors as well as top officials of the IAU and NASA, the Torino Impact Hazard Scale was announced simultaneously by the IAU and by NASA in July 1999. The name recognizes the workshop's endorsement and the historical contributions to asteroid science made by the Torino Astronomical Observatory. Binzel (2000a) describes the details of the scale itself while Beatty (1999) and Binzel (2000b) relate more of the history of its development.

The Torino Scale values range from 0 (no threat) to 10 (certain threat of bad global consequences), with inseparably reported essential information such as the name of the object and the date(s) of its close approach. It provides a simple vehicle for allowing the public to appreciate whether the object merits their concern. Certainly education and public familiarity are necessary to understand the scale, but even soundbite news reporting "The December 2037 encounter by object XYZ ranks only a 1 on the 10-point Torino hazard scale" correctly conveys a very low level of concern on a distant date without knowing anything else about the scale or going into details of probability calculations, error ellipses, orbital nodes, etc.

Concerns among professional astronomers are that the 0 to 10 point system is too simple and compresses too many important details into a single number. Herein lies the central perceptual divide between scientists and the lay public concerning hazard scales: a multi-dimensional problem is reduced to a single dimension. Important details are sacrificed for the sake of simple communication of the central message. What's more, there is no mathematically unique (or quantifiably best) solution for making this dimensional transformation; it depends on values and judgments. Thus any one-dimensional scale is demonstrably flawed. Overcoming these flaws requires viewing the hazard scale simply as a communication tool that is only one piece of the overall effort to inform the public about predicted NEO close encounters. As the scale value increases, there must be a corresponding increase in the amount of additional information conveyed to policy makers and the public.

For its construction, the Torino Scale considers a "hazard space" consisting of the kinetic energy of the encountering body plotted versus its probability of impact, as shown in Fig. 16.1. The annual probability for an object of any given size striking the Earth, such as presented by Chapman and Morrison (1994), defines a line in this space. The region defining category 1 is bounded on the upper side by a factor of 25 greater than this annual probability and on the lower side by a factor of 4 below, giving it a width of two orders of magnitude in probability. At the upper end of the scale, the region encompassing a >99% collision probability (for an object with greater than 10^5 megatons of kinetic energy) is denoted by 10. Such an event would cause devastation on a global scale. Categories 9 and 8 similarly correspond to certain (>99% chance) collisions, at energies greater than 100 megatons

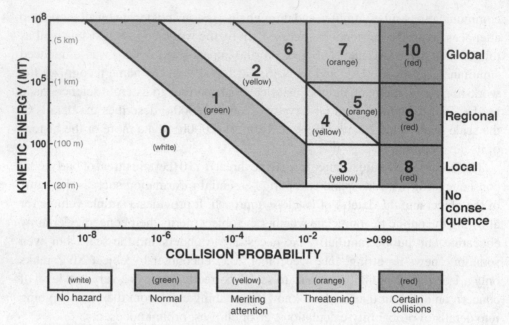

Figure 16.1 Hazard assessment space used in defining values on the Torino Scale. The space is comprised by plotting the kinetic energy of a potential impactor versus its collision probability for an encounter on a specific date. The category boundaries remain unchanged from Binzel (2000a).

(regional devastation) and 1 megaton (local damage), respectively. These two fundamental design decisions using the slope of the annualized probability to define the 0 to 1 transition, and defining the highest values of the scale (8, 9, 10) to correspond to encounters that are certain to be Earth impacts, drive the construction of the remaining categories within the hazard space. Using the annualized probability and setting an upper limit for certain impacts are essential to the scale for it to have a logical basis for the transition from 0 to 1 and for it to have a fixed upper limit (10) that corresponds to the worst possible outcome over geologic timescales. The remaining categories (2 through 7) have their regions delimited by being two orders of magnitude wide in probability and shaped by lines parallel to the annual probability curve. Their numbering follows an assessment of their increasing hazard whereby: hazard increases as the probability increases; hazard increases as the kinetic energy increases. Color coding also denotes the assessment of hazard, ranging from no color (category 0), green (1), yellow (2, 3, 4), orange (5, 6, 7), to red (8, 9, 10).

While Fig. 16.1 depicts how an encounter has its Torino Scale value determined, this hazard space figure is not the public release document intended to describe the scale. The public description of the scale, shown in Fig. 16.2, is a text summary

THE TORINO SCALE
Assessing Asteroid/Comet Impact Predictions

No Hazard	**0** (white)	The likelihood of collision is zero, or is so low as to be effectively zero. Also applies to small objects such as meteors and bolides that burn up in the atmosphere as well as infrequent meteorite falls that rarely cause damage.
Normal	**1** (green)	A pass near the Earth is predicted that poses no unusual level of danger. Current calculations show the chance of collision is extremely unlikely with no cause for public attention or public concern. New telescopic observations very likely will lead to re-assignment to Level 0.
Meriting Attention	**2** (yellow)	A somewhat close but not highly unusual pass near the Earth meriting attention by astronomers. An actual collision is very unlikely, with no cause for public attention or public concern. New telescopic observations very likely will lead to re-assignment to Level 0.
	3 (yellow)	A close encounter, meriting attention by astronomers. Current calculations give a 1% or greater chance of collision capable of localized destruction. Most likely, new telescopic observations will lead to re-assignment to Level 0. Attention by the public and by public officials is merited if the encounter is less than a decade away.
	4 (yellow)	A close encounter, meriting attention by astronomers. Current calculations give a 1% or greater chance of collision capable of regional devastation. Most likely, new telescopic observations will lead to re-assignment to Level 0. Attention by the public and by public officials is merited if the encounter is less than a decade away.
Threatening	**5** (orange)	A close encounter posing a serious, but still uncertain threat of regional devastation. Critical attention by astronomers is needed to determine conclusively whether or not a collision will occur. If the encounter is less than a decade away, governmental contingency planning may be warranted.
	6 (orange)	A close encounter by a large object posing a serious, but still uncertain threat of a global catastrophe. Critical attention by astronomers is needed to determine conclusively whether or not a collision will occur. If the encounter is less than three decades away, governmental contingency planning may be warranted.
	7 (orange)	A very close encounter by a large object, which if occurring this century, poses an unprecedented but still uncertain threat of a global catastrophe. For such a threat in this century, international contingency planning is warranted, especially to determine urgently and conclusively whether or not a collision will occur.
Certain Collisions	**8** (red)	A collision is certain, capable of causing localized destruction for an impact over land or possibly a tsunami if close offshore. Such events occur on average between once per 50 years and once per several 1000 years.
	9 (red)	A collision is certain, capable of causing unprecedented regional devastation for a land impact or the threat of a major tsunami for an ocean impact. Such events occur on average between once per 10 000 years and once per 100 000 years.
	10 (red)	A collision is certain, capable of causing a global climatic catastrophe that may threaten the future of civilization as we know it, whether impacting land or ocean. Such events occur on average once per 100 000 years, or less often.

Figure 16.2 Graphic and wording for public release describing the Torino Scale categories. These descriptions have been updated and made more explanatory since the original release (Binzel 2000a).

of each of the categories. The descriptions we give here for each category are expanded from the original release (Binzel 2000a) in order to more directly convey to the public (and to observatory directors and policy-makers) what actions are being taken or should be considered. Naturally, actual policy decisions require much more information and analysis than provided by a Torino Scale value or its description.

For professional astronomers making and assessing probability calculations for their own analysis and discussion, the Torino Scale is not satisfactory owing to its discrete increments and non-uniform transitions. (For example, in Fig. 16.1, a very slight change in probability or calculated energy can cause the Torino Scale value to jump from category 2 to a value of either 4, 5, 6, or 7.) Binzel (2000a) recognized this problem of public versus professional use of the Torino Scale and proposed for professional use a normalized probability parameter (M):

$$M = P_i/(P_B \Delta T)$$

where P_i is the probability of the impact, P_B is the annual (background) probability for an object of the same kinetic energy, and ΔT is time (in years, $\Delta T > 0$) until the encounter. With extensive, detailed consideration of how to calculate the parameters that enter into this equation, Chesley et al. (2002) define $P = \log_{10}(M)$ which they dub the Palermo Technical Scale (PTS). A simple but approximate transformation between the Torino Scale and PTS exists for close encounters occurring within a few decades and for which the impact probability P_i is comparable to P_B: Torino Scale value = PTS value + 1. The PTS is an excellent tool for professional use while the Torino Scale serves its role in public communication. The challenge and responsibility for astronomers who seek clear public communication is consistently to use the Torino Scale when making announcements or providing media information on future close encounter events.

4 Role of the IAU Working Group on NEOs

In the absence of national or intergovernmental agencies to deal with the NEO impact issues, the IAU has assumed some of the responsibility by default. The IAU formed a Working Group on NEOs in the early 1990s to advise on coordination of NEO activities worldwide, on reporting of NEO hazards, and on research relevant to NEOs. In the wake of the media interest and widespread public confusion associated with asteroid 1997 XF11 (to be discussed in more detail below), the IAU also assumed a limited responsibility for providing authoritative information to the media and public on possible NEO impacts.

When someone predicts a close approach to Earth by an asteroid, a committee of the IAU Working Group can be convened to advise the IAU on the reliability of the prediction. This IAU Technical Review Committee of international specialists offers prompt, expert review of the scientific data, computations, and results on NEOs that might present a significant danger of an impact on Earth in the foreseeable future. The use of this review process is voluntary, and researchers worldwide remain free to publish whatever results they wish in whichever way they wish, at their own responsibility. In several cases the IAU has also seen fit to post a statement on its own website (http://web.mit.edu/rpb/wgneo/Public.html) discussing the reliability of impact predictions.

The initial purpose of the IAU Technical Review was to encourage scientists to check each other's data and calculations before making public statements about possible future impacts. Such a review has been invoked half-a-dozen times. However, with the advent of automated systems to calculate orbits and generate impact probabilities, the need for such human intervention has largely evaporated. In practice, if the NEODys and JPL-Sentry systems agree on a prediction, it is considered confirmed, and the IAU has no direct role to play.

Communication with the international scientific community and with the interested public represents an important part of the Working Group efforts. One tool for public communication is the Torino Scale described above, which was endorsed by the IAU. The IAU has also encouraged the use of several websites for improved communications. These include the NASA NEO Program Office (http://neo.jpl.nasa.gov), the NASA impact hazard website (http://impact.arc.nasa.gov), the UK NEO Information Centre (http://www.nearearthobjects.co.uk), the Spaceguard Foundation and its online magazine *Tumbling Stone* (http://spaceguard.ias.rm.cnr.it/SGF/), and two sites that post continuously updated orbital predictions: NEODys (http://newton.dm.unipi.it/cgi-bin/neodys/neoibo) and Sentry (http://neo.jpl.nasa.gov/risk/).

The role of the IAU is limited: it deals only with the discovery of NEOs, not with mitigation, and it has limited ability to respond rapidly to new discoveries. From the IAU perspective, it remains the responsibility of the individual science teams to decide whether to release information to the public and the press on NEO discoveries or orbital calculations.

5 Historical perspective on impacts

The history of alarms (false or otherwise) concerning possible cosmic impacts on the Earth extends back a surprisingly long time. Although the public often imagines the concept to have a very recent genesis, beginning perhaps with the seminal paper by Alvarez *et al.* in 1980, in fact the recognition that our planet must be dealt catastrophic blows from time to time stretches back several centuries.

The possibility of collisions was realized as soon as the nature of the heliocentric orbits of planets and comets was unveiled. In 1694 Edmund Halley described how his orbit calculations indicated that some comets (like that bearing his name) have paths crossing that of the Earth, making cataclysmic collisions feasible; he even suggested that the Caspian Sea might be a remnant scar from an ancient impact. Two years later William Whiston published his *New History of the Earth*, suggesting that a near-passage of our planet by a huge comet had cracked the crust and raised huge tides, explaining the biblical flood.

During the 1700s several other authors discussed the dangers posed by comet collisions. The fact that the possibility of cometary impacts on the Earth was widely known may be seen from the writings of Jonathan Swift. In 1726 Swift published *Gulliver's Travels*, which contains the following passage:

These people are under continual disquietudes, never enjoying a minute's peace of mind; and their disturbances proceed from causes which very little affect the rest of mortals. Their apprehensions arise from several changes they dread in celestial bodies. For instance . . . That the Earth very narrowly escaped a brush from the tail of the last comet, which would have infallibly reduced it to ashes and that the next, which they have calculated for one and thirty years hence, will probably destroy us.

The comet due to return 31 years later was Halley's, which did come back as Edmund Halley had anticipated, a few decades after his death.

Several comet scares, based on the words of scientists rather than fantasy writers like Swift, occurred during the following decades. For example, a public lecture in Paris by Lalande, in 1773, caused considerable alarm when people imagined that Comet Lexell, which had made the closest ever observed cometary approach to the Earth in 1770, was returning to strike the planet. This was a misinterpretation of the scientist's words, a recurrent phenomenon even today.

English poet Lord Byron was exposed to early ideas suggesting a long history of life on Earth, together with possible large extinctions. In *The Deformed Transformed* (1822), Byron mused upon the possible consequences of a future impact:

> *When I grow weary of it, I have business*
> *Amongst the stars, which these poor creatures deem*
> *Were made for them to look at. 'Twere a jest now*
> *To bring one down amongst them, and set fire*
> *Unto their ant hill: how the pismires then*
> *Would scamper o'er the scalding soil, and, ceasing*
> *From tearing down each others nests, pipe forth*
> *One universal orison! Ha! Ha!*

About the same time Byron suggested the notion that advancing technology might enable an impact by a threatening comet to be avoided:

Who knows whether, when a comet shall approach this globe to destroy it, as it often has been and will be destroyed, men will not tear rocks from their foundations by means of steam, and hurl mountains, as the giants are said to have done, against the flaming mass? – and then we shall have traditions of Titans again, and of wars with Heaven.

(Medwin's *Conversations of Lord Byron*, 1824)

In view of these comments we might regard Byron as being the father of planetary defense!

The above notions were based on knowledge of the existence of comets in Earth-crossing orbits. Asteroids, or minor planets, were unknown until the nineteenth century, and the first NEA not until 433 Eros was found in 1898. Eros does not cross our path at present, but over long timescales (of order a million years) its orbit may change so as to make a collision possible (Michel *et al.* 1996).

In the 1930s discovery of the first Earth-crossing asteroids (1866 Apollo, 2101 Adonis, and 1937 UB Hermes) alerted astronomers to the risk of NEA impacts. Finding a few of these objects, which are much more difficult to spot than bright comets, implied that there must be thousands of them, so that asteroid collisions with the Earth must happen much more often than the very infrequent comet impacts that had been perceived earlier. Fletcher Watson's book *Between the Planets* (1941) described the implications:

When these cosmic bullets swing past at a mere million kilometers we start worrying about the likelihood of collision . . . Close approaches by these flying mountains are rare and the earth probably goes at least a hundred thousand years between collisions with them. Yet there may be myriads of smaller bodies traveling in similar orbits. As we shall see in later chapters, sizable bodies do strike the earth every thousand years and millions of small particles dash into our atmosphere each day.

In 1942 Harvey Ninninger wrote a short piece in *Popular Astronomy* magazine in which he suggested that all mass extinctions and geological boundary events were the result of major asteroid collisions. The idea that such impacts might still be possible today was espoused by Ralph Baldwin in his book *The Face of the Moon* (1949), where he spelled things out: "The explosion that caused the crater Tycho [on the Moon] would, anywhere on Earth, be a horrifying thing, almost inconceivable in its monstrosity." In 1956 the paleontologist M. W. De Laubenfels proposed that the dinosaur extinction was due to asteroids, although his concept was for myriad small impacts, like the 1908 Tunguska event, rather than a single calamitous event.

The most detailed astronomical hypothesis providing a link with extinctions was that of Ernst Öpik, who published a paper on the subject in the *Irish Astronomical Journal* in 1958. Öpik made the first semi-quantitative estimates of impact rates. Better known was a brief letter to *Nature* by Harold Urey (1973), in which he drew attention to the apparent coincidence between tektite ages and faunal mass extinction events, suggesting that major impacts might be the cause of the latter. However, the key work was that of Alvarez *et al.* (1980), who identified the physical evidence to link the K/T extinction event to a massive asteroid impact (a result quickly confirmed by Smit and Hertogen (1980)) and boldly suggested that other mass extinctions as well could be the result of impacts. The headline stories that followed, and the ongoing debates over the next decade (well described by James Powell in his delightfully titled book *Night Comes to the Cretaceous*), firmly entrenched in the public mind the idea of NEO impacts grossly affecting past life on Earth, although they did not always make the obvious connection to possible future catastrophes. In fact the collision rate upon the Earth and Moon has been sensibly constant over the past several billion years; in essence, the impact danger is the same now as it has ever been.

Notwithstanding the release of the movie *Meteor* in the 1979, there was little overt public interest in the present impact hazard, and a handful of astronomers continued to search for NEOs in relative peace, spurred in the early 1980s by the Snowmass Workshop chaired by Gene Shoemaker. This age of innocence began to change in 1989, however, due to the well-publicized discovery of asteroids Toutatis and Asclepius (originally 1989 AC and 1989 FC). Each of these in turn was predicted to make the closest (to that point) flyby of Earth. The same year saw the publication of the book *Cosmic Catastrophes* (Chapman and Morrison 1989), which brought the matter again to the attention of the reading public including a few key staffers in the US Congress.

The fact that Asclepius was discovered 10 days after its passage by the Earth at a closest distance only about 50% further away than the Moon aroused some public dismay and prompted the American Institute of Aeronautics and Astronautics (AIAA) to draft a statement and begin to lobby the US Congress for a government response. In 1990 the Science Committee of the House of Representatives directed NASA to form two international committees to examine, first, how potentially hazardous NEOs might be detected (Morrison 1992) and, second, how an identified threatening NEO might be intercepted and deflected (Canavan *et al.* 1993). Reports from these studies were presented to Congress in 1993, and shortly thereafter the world was riveted by the prospect of watching the fragments of Comet Shoemaker-Levy 9 crash into Jupiter. The Great Comet Crash was well publicized on television, and that in turn led to a series of one-hour TV documentaries that appeared in 1997, and to the Hollywood feature films *Deep Impact* and *Armageddon* in 1998. The stage was clearly set for the current public and media interest in impacts.

6 Learning from experience: five NEAs that made the evening news

Much of the press coverage of the NEO impact hazard has been excellent. Some journalists have worked hard to master the intricacies of this low-probability risk and to translate the information into a form that is accessible to a wider public. However, reporting of individual cases has revealed problems in the communications between astronomers and the non-specialist media.

In this section we briefly review five cases in the past decade where newly discovered NEAs were widely reported (or misreported) in the press as posing a risk of impact with the Earth (see Chapman (2000) for more detailed reporting of the first two cases). Each case involves communications failures of a different kind. These examples illustrate the frustrating breadth of ways that things can "go wrong."

What constitutes an asteroid being newsworthy, and hence capable of spawning a scare, has changed gradually over the past decade. The selection criteria are such things as the probability and proximity in time of the potential impact. As new records have been set for each parameter, the media apply new standards to rumors and reports of potentially hazardous NEAs. Perhaps astronomers as well learn how to communicate better. Despite this, with the anticipated increasing sky coverage and deeper surveys, it seems inevitable that impact alarms will continue to make frequent appearances in the media.

(1) 1997 XF11

The first modern impact scare was associated with asteroid 1997 XF11 (now asteroid no. 35396), an approximately 1-km diameter NEA discovered on December 6, 1997. By early in 1998, preliminary orbital calculations showed a possible close pass by Earth in October 2028. Brian Marsden, Director of the Minor Planet Center at the Harvard-Smithsonian Center for Astrophysics, was following the orbit in early March when the most recent observations suggested that XF11 might pass well inside the orbit of the Moon. On March 11 he made this information available and quickly distributed a Press Information Sheet in which he stated that "the chance of an actual collision is small, but it is not entirely out of the question." From the calculated miss distance and its estimated error, Marsden and other astronomers suggested that the probability of an impact could be as high as 1 in 1000, and news media all over the world put the story on the front page.

Unlike Marsden, Don Yeomans and Paul Chodas at the Jet Propulsion Laboratory (JPL) had the software to estimate the actual odds of hitting, and within 2 hours of Marsden's announcement they had calculated that an impact was impossible – although the position of the asteroid in its orbit had a substantial uncertainty, there was zero possibility of a collision. (The error ellipse was effectively a line that passed well off the Earth, rather than a fat ellipse.) Marsden did not withdraw his comments, however, until pre-discovery observations from 1990 were found a day later that improved the orbit and demonstrated than the asteroid would not only miss, it would not even come close to the Earth in 2028. Only then did the media run retraction stories, mostly along the lines that new data had corrected the original prediction.

The XF11 episode demonstrated the need for rapid calculation of impact odds as well as nominal asteroid orbits and miss distances. In addition to the JPL team, Andrea Milani at Pisa and Karri Muinonen in Helsinki quickly demonstrated this capability. The second lesson was that astronomers should check their results with colleagues before "going public" if they wished to avoid the embarrassment of a scientific debate in the public eye. Third was a more general concern that astronomers looked rather foolish when a prediction of a possible impact was corrected or withdrawn within less than a day. It was suggested that perhaps the IAU should vet such predictions, and in this case the IAU issued a statement that "contrary to preliminary reports, there is no danger of its colliding with the Earth in 2028. Like all Earth-crossing asteroids, XF11 may someday hit our planet, but this seems to be an event for the distant future, and at present we are more at risk from some unknown asteroid colliding with the Earth than from XF11 or any other object already discovered."

In summary, the problem with XF11 was premature announcement without calculating a formal impact probability or consulting with colleagues. The solution seemed to be better software and more consultation before making announcements.

(2) 1999 AN10

On January 13, 1999, the LINEAR search program discovered a mile-wide NEA designated 1999 AN10. Milani and his colleagues at Pisa showed that there was a less than 1 in 1 000 000 chance of AN10 passing through a keyhole (see below) during a 2027 close approach so that it would impact the Earth several decades later. In the year since XF11, great strides had been made in orbit prediction, including the concept of "keyholes" defining a very restricted set of orbits at one pass that could lead to a resonant return at a subsequent pass. Although the result was unremarkable from the perspective of the impact probability, it was of interest to many dynamicists, and Milani prepared a manuscript and asked colleagues for an informal review within 2 weeks. Receiving no complaints, he posted his paper (Milani *et al.* 1999) with no fanfare on his website.

Benny Peiser of John Moores University, the moderator of the CCNet Digest Internet forum, discovered the unheralded paper on Milani's website and charged cover-up. The widely reported media story was not about the impact risk, but about the idea that astronomers would conspire to withhold from the public information on a possible (even though exceedingly unlikely) impact for weeks. As a result, Milani challenged his colleagues and the IAU to come up with a better way to review and (if appropriate) to publicize such cases. Shortly thereafter the IAU Working Group on NEOs set up its Technical Review Committee, committed to carrying out a review within 72 h. The Torino Scale was also adopted in June 1999 to aid in public understanding as described above (Binzel 2000b).

In summary, the problem with AN10 was that the media could read scientific websites and extract information on low-probability future impacts. As a result, the scientists lost control of the discussion and were subject to potential criticism for being either too open or for suppressing information, according to the taste of the critic. The solution seemed to be to formalize the review and let the IAU, through its Technical Review Committee, provide an international, professional context for any released information.

(3) 2000 SG344

The IAU review process was soon exercised for the very small object 2000 SG344, which was interesting primarily because of the possibility that it may be a spent rocket upper stage in an orbit very similar to that of the Earth. For a few days, however, it appeared that SG344 might have an impact probability as high as 1 in 500, triggering the IAU 72-h technical review. The review was completed on a Friday afternoon, confirming the relatively high impact probability, and accordingly JPL issued a press statement and the IAU posted the results on their webpage. Media interest was immediate, as this was the highest impact probability ever predicted and confirmed.

Within a few hours, new observations were available that showed no impact was possible. Unfortunately, on the weekend no one was correcting the statements of alarm posted by JPL and the IAU. Two mistakes thus compounded the problem: a statement was issued that was technically correct but did not anticipate the availability of new data, and no one was ready to issue a formal correction over the weekend. Once again the astronomers looked foolish.

In summary, the problem with SG344 was a literal application of the IAU guidelines, which called for release of information as soon as the technical review was complete, without recognizing that the data upon which the orbital calculations had been based were improving daily. The solution was, as a minimum, to relax the strict IAU guidelines. It was also suggested by some that no information should be released until all possible data were collected, even if this meant a delay of weeks or possibly months, since in nearly all cases the refinement of the orbit would eliminate the possibility of an impact.

(4) 2002 MN

Asteroid 2002 MN was a small asteroid that flew past the Earth at a distance of 0.0008 AU (one-third of the lunar distance) 3 days prior to its discovery. The media judged this to be a failure on the part of astronomers because there was no prior warning, whereas actually it was a success: the asteroid was found and catalogued.

Orbit calculation showed that it did not pose any danger of future impact (for the 50–100 year window that was used to search for future close passages).

In summary, the problem with 2002 MN was that it was discovered, as often happens, after closest approach rather than before. The community had difficulty explaining that the Spaceguard Survey was not designed to detect NEAs on their final plunge to Earth, and that finding an object after closest approach is just as useful as finding it before, and is in fact nearly as likely, since many asteroids move into the night sky from the sunward side of the Earth.

(5) 2002 NT7

Asteroid 2002 NT7, with the relatively large diameter of 2 km, was discovered in July 2002. By this time the calculation of impact probabilities was fully automated, and on July 18, NT7 was posted on the "risk page" of both the Pisa NEODys and JPL Sentry systems, showing a possible but very unlikely (of order 1 in 100 000) impact just 17 years in the future. Because new data were coming in and NT7 remained a zero on the Torino Scale (although very close to Torino Scale value of 1), it was decided not to call for a formal IAU technical review or to make any public statements, pending improvements in the orbit.

On July 24 this remote chance of impact became an international media story when the BBC picked the information up from the Internet and reported that the asteroid was "on a collision course with Earth." As expected, however, additional observations quickly eliminated the possibility of an impact. The "all clear" for any impact in 2019 was released on July 26, and by August 1 continuing orbital improvements also eliminated a lower-probability impact in 2060 – a progression of events that reflects the normal working of the Spaceguard system. So why all the media fuss about NT7, especially in the UK press? This is unclear. Especially provocative was the BBC story that called NT7 "the most threatening object yet detected in space." As the asteroid itself receded from interest, the media story focused on the sensationalist reporting.

Science journalist Robert Britt concluded in a story in Space.com that "The whole affair, over an asteroid that is almost certainly harmless, illustrates the stylistic ocean that separates American and British media and scientists' tactics in dealing with them." The following quotes are from his report. Duncan Steel suggested that asteroid stories have become so common that in his country they either make headlines or they're not used at all. Unless a reporter "makes it sensational, the editor will nix it. Ditto (especially) for the printed media." Don Yeomans said that he was unprepared when "the media blitz struck." "Most of the six interviews I did with BBC reporters Tuesday night began with their assumption that there would be a collision," Yeomans said. "One is then forced to back up and try to explain the real situation and the fact that there is not really a story here. They didn't wish to hear that." Yeomans later concluded that journalists and scientists both need to strengthen efforts to help the public understand how asteroid risks are determined. "There is plenty of blame to go around," he said.

In summary, NT7 became a media story in spite of the astronomers. The press obtained information directly from technical webpages and gave it a sensationalist spin. Astronomers then spent several days trying to correct the misimpressions. While there is no clear solution, most of us concluded that scientists should have responsible statements ready to provide to the press or post on websites as an antidote to this sort of outbreak. But no one suggested that we should try to keep impact predictions secret. That is a clear lesson from all of the encounters with the press from XF11 to NT7.

7 Sustaining public interest in this hazard: specific examples

Media coverage of asteroid and comet impact "scares," as discussed in Section 6 above, has been intense and not always accurate. Nevertheless the apparent fact that the media has yet to tire of the subject, with coverage accelerating rather than waning, surely reflects continued public interest in the impact hazard.

In order to make more concrete the nature of the impact hazard, what damage might be done, and what precautionary or after-the-fact measures might be taken to mitigate losses, here we present six different impact scenarios in detail. These scenarios were presented by author Chapman at the January 2003 Organization for Economic Cooperation and Development (OECD) workshop in Frascati, Italy. The examples differ greatly in their likelihood of occurrence, the magnitude of the destruction, the degree of predictability by scientists, and the kinds of prevention or mitigation that might be undertaken. The individual cases discussed here would affect various nations differently (depending, for example, on whether the country is coastal or landlocked). There are many other possible cases that could be drawn from within the range of possible impactor sizes and types; differences in the environmental effects and probable societal responses would also depend on where (or what country) the projectile struck, and also on other variables.

Each case scenario involves a body of approximately the size stated; for example, the two cases involving a "~200-m" NEA roughly characterize circumstances for impacts by bodies ~150 to ~300 m in size, and the quoted chance of happening refers to impacts of all bodies >200 m, dominated by those in the range 200–250 m.

Case A: Tsunami-generator: ~200-m asteroid impacts in ocean

Nature of the devastation

Imagine a flying mountain, larger than the world's biggest domed stadium (the New Orleans Superdome), crashing into the ocean at a speed a hundred times faster than a jet airliner. The resulting explosion, with an energy around 600 MT (MT = million tons of TNT equivalent), would be 10 times the yield of the largest thermonuclear bomb ever tested, and would loft about 10 km^3 of water high into the sky. By far the greatest danger would result from tsunamis, which would convey much of the impact energy toward far-distant coastlines. Typically, the resulting few-meter high tsunami in the open ocean would be amplified to a wave as much as 10 m high before it breaks and runs up on a coastline 1000 or more kilometers from the impact site (Ward and Asphaug 2000). Researchers are very uncertain about the importance of impact-generated tsunamis, and the effects would vary widely from place to place along the coast of the impacted ocean. Run-ups on coastal plains could range inland as far as kilometers; some low-lying plains (e.g., Bangladesh) would be affected in the same way as flooding by the greatest hurricanes and typhoons. In the worst case, millions might die. Consequences for nations without coastlines or on opposite sides of the planet would be restricted to indirect economic and political repercussions.

Probability of happening

Threatened countries are those with inhabited and/or developed ocean coastlines. A >200 m diameter asteroid impact has a 1 in 100 chance of happening, world-wide, during this century, but the probability of any given point on the shoreline experiencing a significant impact tsunami is in the once-in-a-million-year range. Unlike the other impact hazards discused here, which are generally distributed over the population, the tsunami hazard effects less than 1% of the Earth's population, namely those who live on an unprotected coast within a wave run-up range of about 10 m and a run-in range of a kilometer or two (Chesley and Ward, unpublished data).

Warning time

It is unlikely that astronomers will discover such a small impactor in advance, until better telescopic searches are under way. (If they do, there would likely be years or decades of warning.) Most likely, the tsunami warning would come from tsunami-warning infrastructures currently in place. Unfortunately, the Pacific Tsunami Warning Center's warnings are triggered by sensors for earthquakes that might not immediately recognize the seismic signature of an oceanic asteroid impact, and tsunami warning systems are less developed for coasts of other oceans. Many hours of advance warning might be possible, and coastal warning sirens and other protocols could assure evacuation and protection for many.

Post-warning mitigation possibilities

In the unlikely event that the approaching NEA were discovered in advance by astronomers, impact could be prevented by having spacefaring nations deflect (or destroy) the asteroid, a very costly but probably feasible option. It is much more likely that, if people have warning at all, it would come only hours before the tsunami arrives; the chief response would be to save lives by evacuation to higher ground. Given a functional warning system, the primary effects woud be destruction of economic infrastructure, not loss of lives. (In the words of Alan Harris, people might be wet and angry, but not dead.)

After-event disaster management

The immediate aftermath of such an impact-generated tsunami would resemble other major, localized civil emergencies, but distributed in many countries around the ocean rim. Many nearby locales, just kilometers inland from coastlines, would be unaffected, so could serve as centers for organizing relief. There would be minimal lingering after-effects that would hamper rescue and recovery.

Advance preparation

The most effective preparations would be the same ones that would protect coast-lines from normal tsunamis. However, many nations are insufficiently prepared for such tsunamis (Bryant 1991), quite apart from the additional dangers of asteroid-induced tsunamis, which might be larger and have different traits from familiar or historical tsunamis. Mitigation efforts might include hardening vital coastline infrastructures, limiting and/or hardening developments within several hundred meters or kilometers of the coast, and developing civil defense procedures that would be effective in evacuating endangered people (perhaps to distances or ele-vations not normally contemplated), especially along coastlines not frequently affected by natural tsunamis, like the Atlantic. Officials and personnel responsi-ble for national and international tsunami-warning systems should be appraised of differences between impact- and earthquake-generated tsunamis and should be linked into astronomical/military/maritime organizations that might report imme-diate information about an impact. National geologists and ocean scientists could research local circumstances, like bathymetric modelling of the effects on tsunami run-up of underwater, near-coastal topography.

Case B: ~200-m asteroid striking land

Nature of the devastation

As in Case A, imagine an enormous rock larger than any of the world's largest buildings crashing through the Earth's atmosphere in a few seconds, but striking land instead of an ocean (Toon *et al.* 1997). An enormous crater would form within seconds, ~4 km across, deeper than the Grand Canyon. Anything within several kilometers of the crater rim would be smashed and totally buried by flying material; all things/people would be destroyed/killed immediately within this city-sized zone. Serious devastation and death due to the blast would extend at least 50 km in all directions: trees would be toppled by the shock wave; wooden and unreinforced structures might implode, their debris blown about by a brief spell of super-hurricane-force winds. Fires and a damaging, local earthquake would add to the calamity. Even hundreds of kilometers from ground zero, rocks would fall from the skies, choking smoke and dust would blow downwind from the crater, and houses would be shaken and damaged. In short, such an impact would substantially destroy a region the size of a small nation or a modest-sized American state. The death toll would range from thousands to hundreds of thousands, depending on population density (possibly much higher if a city were struck or lower if an isolated desert were struck).

Probability of happening

The chance of a >200 m NEA striking land during the twenty-first century is about 1 in 1000. Cosmic impacts are not selective, so the world's largest countries (Russia, Canada, China, the United States, Australia, and Brazil) would be the likely target. A very small country would not likely be directly struck, but would be affected by an impact closer than several hundred kilometers away. (Countries may have greater or lesser chances of being struck due to differing land areas, but of course individuals all have the same probability of being near ground zero.)

Warning time

As in Case A, there is <20% chance (given current telescopic survey efforts) that such an NEA would be discovered before impact, but the situation is likely to change rapidly in the next two decades if new search telescopes are built. If such surveys exist, warning times would likely be decades or longer. If it struck unseen, the devastation would be immediate, unlike the hours of possible tsunami warning in Case A. Individuals who witness the terrifying, fiery plunge through the skies might have seconds to tens of seconds to take cover (e.g., hiding behind a strong wall or in an underground shelter) and lessen somewhat their personal risk from the blast; however, if they are closer than 10 km from ground zero, they have no time to react and minimal chance of survival.

Post-warning mitigation possibilities

Today there would be most likely no warning at all. If, thanks to future surveys, the object is discovered long before impact, then it would be possible to divert it so that it would miss the Earth (see Case A).

After-event disaster management

The destruction would be total within and near the enormous crater, and the severity would diminish with distance, out to several hundred kilometers. The causes of death, injury, and destruction would mainly be the same as those of some other natural disasters – earthquake, volcanic explosion, typhoon, firestorm – except that the effects of all four would be compounded. The disaster's extent would resemble that of the greatest localized natural disasters, like the 1882 explosion of Krakatoa. The disaster zone would cover few nations, unlike the tsunami case, but the zone's interior would be more difficult to reach and service because of its breadth. Unlike natural earthquakes or storms, there would be no lingering threats of more NEA impacts; fires and environmental toxicity might be the longest-lasting after-effects. The kinds of emergency management issues facing society in the aftermath of such an impact (public health issues, panic, etc.) are described by Garshnek *et al.* (2000).

Advance preparation

Unlike Case A, for which vulnerable zones (coasts) can be identified, there is no spot on Earth more likely to suffer direct blast damage from an NEA impact than any other. Normal mitigation and emergency management procedures designed to protect lives and infrastructure from extreme windstorms, fires, and earthquakes would also serve in case of an impact. Unless an incoming NEA is discovered before it hits, there is little justification for mounting asteroid-specific mitigation measures (except at the margins) in the face of the 1-in-1000 chance of this scenario playing out on some continent during this century.

Case C: Mini-Tunguska: once-in-a-century atmospheric explosion

Nature of the devastation

In this case, a massive rock the size of an office building (say 30–40 m across) streaks down through the atmosphere and is torn apart, exploding with the force of several megatons, perhaps 15 km up (near the bottom of the stratosphere). Case C represents roughly the minimum sized impact (except for rare iron projectiles) that can do significant damage on the ground (Chyba 1993). It would be dangerous to be within several tens of kilometers of such an event. Fires might well be ignited beneath the brilliant explosion, unless it were cloudy. Weak structures might be damaged or even destroyed within a 20 km radius by the shock wave and subsequent hurricane-force wind gusts. Exposed people and animals could be struck by flying objects. While the somewhat larger Tunguska blast in 1908 killed few people if any, the once-in-a-century class of asteroid airburst would be very frightening to witnesses and potentially deadly in a susceptible locality.

Probability of happening

Such an event happens about once a century. Since no location is favored or disfavored, the next one would probably occur over an ocean or desolate desert where its effects would be minor.

Warning time

It is quite unlikely that such a small asteroid would be discovered by astronomers or military surveillance prior to impact. The brilliant explosion would happen without warning, and its effects would be over within seconds to minutes, except for lingering fires and a stratospheric pall.

Mitigation issues

Nothing practical can be done about this modest hazard other than to clean up after the event. It would be very costly to build a telescopic search system that could find most 30 m bodies before one strikes. Once an atmospheric impact occurs, the usual disaster management protocols should be able to handle trauma and damage in the affected locality. It makes no sense to plan ahead for such a modest disaster, which could occur anywhere, other than educating the public about the possibility.

Case D: Annual multi-kiloton flash in the sky

Dangerous consequences

A rocky meteoroid the size of a bus explodes 20 km up in the stratosphere with the energy of a small atom bomb (2–10 kT), producing a brief, blinding flash much brighter than the Sun. While such an event could do no damage on the ground, there is concern that military commanders in a region of tension – unable immediately to verify the true cause of the explosion – might regard it as the hostile act of an enemy and retaliate dangerously. While the existence of meteoritic fireballs may be known by most military establishments of nuclear powers, the degree to which adequate command and control procedures are in place to handle such rare and frightening events is not known.

Probability of happening

Impacts of 3 m bodies happen annually, somewhere on Earth (Brown *et al.* 2002).

Warning time

Objects of these sizes strike without warning.

Mitigation issues

Currently, the US Departments of Energy and Defense regularly observe such events worldwide from geostationary surveillance satellites designed for other purposes. The signatures of such events are recognized as distinct from hostile military phenomena, so retaliation by the United States is unlikely. Normally, information about these events is released to the public days or weeks afterwards. Relaying information much more quickly to other nations would be required to prevent a mishap. Beyond that, raising public consciousness worldwide about rare, brilliant fireballs could only help.

Case E: Civilization destroyer: 2–3 km asteroid impact

Nature of the devastation

A 3 km diameter NEA would explode with the almost inconceivable yield of a million megatons of TNT, as though more than 1000 of the Case A or B impacts hit the same place simultaneously. The crater alone would engulf an area comparable to one of the world's largest cities. An impact into the deep ocean would penetrate into the seafloor, ejecting enormous quantities of oceanic crustal rocks plus perhaps 10 000 km^3 of ocean water; the resulting tsunamis would be of a scale unprecedented in human history. Localized devastation analogous to that of Case B would become regional in extent. New effects would add to such magnified, compound effects. Material thrown out of the Earth's atmosphere would rain back down, filling the sky with blazing fireballs and incinerating an area perhaps as large as India or Argentina. The Earth's ozone layer might be destroyed for several years, subjecting everyone to dangerous UV sunlight.

Such apocalyptic devastation pales compared with the worldwide death and economic calamity due to stratospheric contamination that would darken the Sun within weeks. Huge amounts of dust, water vapor, sulfate aerosols, and nitric oxide would transfigure stratospheric chemistry and block out most sunlight worldwide for months. A global "impact winter" would ensue, probably ruining one growing season worldwide. Without advance preparation, mass starvation might result in the deaths of perhaps a quarter of the world's population.

One can only speculate about secondary repercussions, such as disease, disruption of global economic interdependencies, perturbation of military equilibria, social disorganization, and so on. Depending on the robustness or fragility of modern civilization, the world might well be jolted into a new Dark Age (Chapman and Morrison 1994). Obviously, there are great uncertainties about whether it might take only a 1 km NEA or instead would require one >5 km to wreak such environmental disruption and kill so many; would it really destroy modern civilization? Still, such a calamity would surely be the most catastrophic in recorded human history. Ever since Chapman and Morrison (1994) identified these globally catastrophic impacts as the dominant impact hazard, they have received the greatest interest in the astronomical community, but not necessarily in the hazard mitigation community, where they exceed other more familiar catastrophes by a large factor.

Probability of happening

A >2 km NEA strikes Earth roughly every 2 million years. Most large NEAs have already been discovered by the Spaceguard Survey, none of which will hit soon. Therefore, the remaining threat of civilization-threatening impacts is mostly posed by long-period or "new" comets, whose numbers are poorly known and many of which are not discovered until a year or so before they enter the inner solar system. Very crudely, there may be a 1 in 50 000 to 1 in 100 000 chance that such an impact will happen this century.

Warning time

Such a large NEA would very likely be discovered long before it struck Earth, giving decades of warning. Comets, however, are found only months to a few years before entering the inner solar system. There is a small chance that such an impact would happen with little or no warning.

Post-warning mitigation possibilities

Such an NEA could perhaps be diverted using advanced space-based technologies. Unlike Case A, moving an object this large would be technically very challenging. But the motivation would be high so the challenge could probably be met, at Apollo Program costs or more, especially if design work had already begun to deflect smaller NEAs. If diversion cannot be done or if the warning time were only months or years, then mitigation would turn to (a) evacuation of the entire sector of the Earth where the impact's effects would be greatest, (b) optimal advance production and storage of food, and (c) "hardening" susceptible but vital elements of civilization's infrastructure (such as communications, transportation, and medical services). Surely all nations would help face this enormous challenge; if the warning time were just months, few effective efforts could be mobilized in time.

After-event disaster management

These issues have been briefly considered by Garshnek et al. (2000). In view of the wholly unprecedented nature of such a holocaust, one might gain as much insight from historical and even fictional accounts of past or imagined wars, disasters, and breakdowns of civilizations (e.g., Lucifer's Hammer, Niven and Pournelle (1977)).

Advance preparation

Disaster management agencies should consider this disaster case, if only to encourage "out-of-the-box" thinking. However, the chance of such an impact happening is very remote, and its probable consequences are too extraordinary to be substantially addressed by affordable, practical efforts (in the absence of a predicted, impending impact). Nevertheless, some level of strategic and systems planning should be done to understand the technological challenges of diverting large NEAs or comets, including taking first steps involved with moving much smaller NEAs. Also, some inexpensive measures, taken at the margins in the course of generic disaster and emergency planning, could cost-effectively prepare for Armageddon. For instance, as part of normal efforts at networking and coordination, national and international disaster management entities should develop communications channels with the astronomical and military projects that detect and track asteroids. All nations, in proportion to their capabilities, should consider their responsibilities for protecting their citizens in the face of a potential impact catastrophe.

Case F: Prediction (or media report) of near-term impact possibility

Nature of the problem

As asteroid detection programs improve and near misses are more frequently reported, the most likely aspect of the impact hazard that a public official will encounter is not the actual impact by a dangerous NEA but (a) the prediction of a possible impact or threatening near miss or (b) a serious mistake by scientists or, more probably, journalists. Human foibles are more likely than a rare impact, but they have real social and political consequences. Examples of likely possibilities are:

- The actual "near miss" by a bigger-than-Tunguska, >100 m NEA "just" 60 000 km from Earth, or closer, predicted with a few days' notice. This probably will happen during this century. The passing projectile would be visible to ordinary people with their naked eyes. Will people believe scientists or military officials who say it will miss?
- The prediction by a reputable, but mistaken, scientist – published in mainstream news media – that a devastating impact will occur, say, on 1 April, 2017, in a particular country. The prediction is not analyzed by other scientists and withdrawn for several days. Meanwhile, people in the affected country become very frightened, especially if rumors or sloppy journalism (see below) lead them to think that the disaster is imminent.
- The official prediction by astronomers, coordinated by the IAU, that a dangerous, multi-hundred meter NEA has an unusually large, one-in-several-hundred chance of impacting Earth on a specific date 10 or 20 years in the future. This would rate an extremely unusual

2 (in the yellow zone) of the Torino Scale (Binzel 2000a). It might take months for astronomers to rule out the chance of impact.

- An unusually grotesque example of media hype in which one of the above already worrisome examples is badly misreported (accompanied by even more exaggerated banner headlines) by mainstream wire services and cable TV news networks. At least three cases of prominent misreporting about NEAs by the worldwide news media happened during 2002 alone; more egregious cases could readily occur during the next decade.

Concerns by an agitated public about predicted impacts might be presented to national elected leaders, emergency management agencies, and military and space departments, but few governments have an informed official who could respond. In many localities, health agencies, school officials, and police might have to deal with panic by frightened people, especially children.

Probability of happening

Instances have already happened. All of these examples probably will happen during the next century, some (especially news media hype) many times.

Warning time

It is the nature of modern life, fueled by the Internet, that the cited examples could reach page-one status around the world within hours and catch officials totally by surprise.

Mitigation issues

An uninformed, apprehensive, risk-averse public combined with media hype are elements of the modern world, confounding many issues at the interface of science and society. The business goals and/or political agendas of the media often run counter to dispassionately educating and informing the public. Dialog among scientists, journalists, and public officials might change things for the better, but the problem seems to be getting worse. Better information exchange and coordination among relevant entities (astronomers, fledgling NEA information organizations (e.g., the Spaceguard Foundation, the British Near Earth Object Information Centre, and NASA's Near-Earth Object Program Office), national and international disaster management agencies, etc.) might reduce official mistakes and miscommunications. Use of the Torino Scale could help to ensure that impact predictions are interpreted by science journalists and the public within an increasingly familiar context. Improvements in education (chiefly involving science and rational thinking) can serve to minimize irrational and exaggerated responses to technology generally and to the impact hazard, in particular.

8 Discussion of examples

The divide between science and the public has never been greater, thanks to poor science education and news media hype. The impact hazard, according to studies of risk perception, is particularly vulnerable to misunderstanding and heightened public fear. There is much public misunderstanding that could be amplified in critical circumstances, preventing rational responses to real approaches or impacts. Different governments and societies have varied approaches to disseminating reliable information to citizens. It is not too early to consider ways to prepare citizens and emergency management organizations to respond appropriately to what might well be badly distorted information in the world news media about an impending impact. One element of such an approach is to develop a consistent protocol for placing predictions or warnings of potential impacts into context by utilizing the 10-point Torino Scale (Binzel 2000a).

Uncertainty is a fundamental attribute of the forecasting sciences. Moreover, it is notoriously difficult for technical experts to communicate uncertainty to public officials in ways that can be translated into practical measures (numerous examples are discussed by Sarewitz *et al.* (2000)). The impact hazard involves its own peculiar suite of uncertainties. While an NEA impact can be more reliably predicted than any other natural disaster, that is true only once the asteroid's orbit has been precisely determined, which may take months or even years after it is first discovered. In the interim, an arcane suite of uncertainties clouds the reliability of predictions, and the ongoing technical work is difficult for science journalists to understand or translate to the public. Simplified analogies, like throwing darts at a target, do not generally apply. Moreover, the impact hazard is especially prone to meta errors – perceptual mistakes, computer programming errors, inadequate modeling or extrapolations, miscommunications, and other confusions, amplified by the new and unfamiliar nature of the hazard. Unlike weather forecasters, there aren't legions of trained, practiced impact forecasters. Protocols for forecasting and communicating about impact events are rudimentary at best, increasing chances for error or miscommunication. Any actual impact, of course, will present itself as a unique case, with exceptional features never previously encountered, especially ripe for confusion. And predictions of the physical, environmental, social, and economic consequences of an unprecedented potential or actual impact will be made by supposed experts who are actually acting in unchartered waters. Thus decision-makers must expect a wider range of contingencies than would be true for more common scientific hazard predictions (e.g., for maximum river levels in a flood).

While the concept of stopping a natural disaster from happening is not unknown (e.g., avalanche control), most natural disasters are marginally or not at all preventable. The impact hazard is unique in this respect. Potentially available spacebased technologies could divert an NEA, causing it to miss the Earth, given years to decades of lead time. Spacefaring nations need to determine what priority should be given to budgeting mitigation-oriented activities before any NEA is known to be aimed for Earth. Consideration should be given to the "deflection dilemma" (Sagan and Ostro 1994) before fully developing an asteroid deflection technology in the absence of an impending impact. The need for full, open, international discussion of trustworthy deflection methodologies is illustrated by the argument of the B612 Foundation (Schweickart 2004) that the long-term (months to years), controlled, low-thrust pushing on an NEA will move ground zero across the Earth, perhaps alarming nations not originally targeted by nature, before the NEA is eventually assured of missing the Earth.

Especially in the United States, the terrorist attacks on September 11, 2001 have greatly affected the public's perception of personal safety and security in the face of unexpected disasters. Despite the malicious motivations of 9/11, there are obvious potential similarities between 9/11 and cosmic impacts. There was, or would be, little or no warning. The ~3000 9/11 deaths and direct physical damage to Lower Manhattan and the Pentagon were magnified enormously (in both social and economic terms) as the US government and citizens responded by minimizing travel, changing national budgetary priorities, attacking Afghanistan, etc., because of the unexpected, horrifying nature of the attack. Even a modest NEA impact might evoke similar reactions, according to research on risk perception. Connected viscerally to the event by TV news coverage, many people would fear that they could be the next random casualties. Victims might seek scapegoats, especially since an incoming impactor can, in principle, be diverted using existing space technology, and the reason it is not being implemented is an implicit political decision, questionable after the fact, about priorities.

As the OECD (2001) document *Identify Risk* notes, despite much individual, personal risk-taking behavior, "collective risks are barely tolerated, regardless of the anticipated degree of risk." Thus, it would be wise for "governments and other standard-setting organizations . . . to define a rational level of acceptable or tolerable risk" for the impact hazard and to do so, not by benign neglect, but rather by examining "scientific and socio-economic information in a public forum open to free communication and debate by all concerned parties." In this way, the development of appropriately prioritized approaches to responding to the impact hazard would have a rational legitimacy.

9 Responsibilities of the international scientific community

The present level of public and governmental recognition of the NEO impact hazard has come about from the individual and collective actions of various scientists, principally astronomers, over the past dozen years or so. This has resulted in enhanced research on aspects of NEOs, and especially in the implementation of the Spaceguard Survey to seek out unknown NEAs. The question remains whether the present level of activity is commensurate with the risk. Many scientists familiar with the nature and consequences of the impact hazard would argue that, regardless of their own involvement in NEO research, a much higher expenditure is justifiable, a conclusion also reached by a NASA-sponsored science study carried out since this chapter was written (Stokes 2003).

Although the NEO impact hazard is clearly a matter that affects all nations, to date it has been only the United States that has taken steps to tackle the problem directly through scientific research and astronomical observations. In Japan, a small observational program has recently been initiated. An NEA search and tracking project that operated in Australia between 1990 and 1996 was canceled by the government there. Italy has an especially active asteroid study community, although the level of expenditure is small.

In 2000 the government of the United Kingdom commissioned a report on the hazard, which was duly published later that year (Atkinson *et al.* 2000). A set of 14 recommendations, including some highlighting the need for international cooperation and action, were made. Although this report has yet to result in any specific research work on NEOs in the UK, it did bring about the formation of an NEO Working Group under the aegis of the OECD Global Science Forum. In January 2003, a workshop organized by the OECD approved a set of specific findings and conclusions, for presentation to the OECD member states for their consideration. In summary, these covered the following:

- The NEO impact hazard is real, and is at a level comparable to many other natural hazards that are taken very seriously (earthquakes, storms, etc.). There is a need for action at both national and international levels.
- Few governments have yet to make any specific office or administration responsible for assessing and dealing with the impact hazard, even though they may have a few scientists working on related scientific problems. Thus there is a need for each country to allocate responsibility to some specific branch of its government.
- There is a need for each country to assess its individual liability with regard to the impact hazard, because each faces a particular set of circumstances (e.g., nations with large coastal populations facing onto major oceans have an enhanced exposure to impact-induced tsunami inundation).
- More scientific research is needed in a number of specific areas connected with assessing the hazard (e.g., understanding the hazard posed by comets as opposed to asteroids), and international collaboration in this regard is especially desirable. There is a need for funding to be made available through avenues other than the usual highly competitive science research channels.

- Research is required on how impacts might be mitigated, ranging from possible technologies for the interception and deflection of any identified impactor through to procedures for emergency response to an unforeseen impact having occurred.
- There is a need for international cooperation over a wide range of matters connected with the NEO impact hazard, not just the present ad hoc arrangement whereby astronomers have discussed the subject for some time. A wide variety of scientists from other disciplines should also be involved, as should policy-makers. Matters needing discussion range from social responses to the recognition of the impact hazard through to understanding the way in which the biosphere might react to impact events at different energy levels.

In the absence of formal government recognition of the impact hazard, the responsibility for informing the public rests with the scientific community, although this is not a role most scientists are prepared for. On the one hand, many feel it necessary to alert people to the hazard, and explain its seriousness; on the other hand, it is recognized that discussion of the impact hazard in the mass media leads to the development of considerable disquiet and worry among many members of the public, especially those least able to understand such things as statements of probabilities.

This matter has been the subject of very considerable debate within the community of NEO researchers, with strongly held views being espoused in several directions. Almost every time an asteroid scare story appears in the media there are arguments over whether it deserved the coverage it obtained, whether the basis of the story was correct, whether it was good for science or for public relations that the story came out, and who was responsible/to blame for initiating or facilitating the media coverage. The public availability of the Pisa NEODys and JPL Sentry Internet sites, with robots providing continual assessments of specific asteroid's Earth-impact probabilities, means that to some extent there is no longer any individual responsibility on the part of scientists for the dissemination of the information used by journalists to prepare their stories, but the question still remains: what are the responsibilities of the scientific community in this regard? Should potential impacts that are identified soon after the discovery of an NEA be kept secret, on the basis that they will shortly (within a week or a month) be negated by additional position measurements, or should the information be maintained as being freely and immediately available? Is the concern expressed over worrying members of the public sufficient to warrant short-term restrictions on the circulation of information, given that such restrictions could lead to potential observers not being alerted, and thus vital data not being collected? The consensus among international NEO researchers is that, whatever the downsides, information should be made promptly and freely available to the scientific community and the public at large, and current procedures are in accord with this precept.

10 Conclusions: major communication challenges

In conclusion, we highlight some of the difficult concepts to communicate concerning the impact hazard. Most of these relate to people's unfamiliarity with this threat and the difficulty of understanding such an extreme example of a low-frequency but high-consequence hazard.

- The greatest hazard is from infrequent events. It is counterintuitive that the most likely impacts (Tunguska class events up to a few tens of megatons) are not the most dangerous. It is difficult to explain to an official, for example, that even though the Earth is 1000 times more likely to be hit by a Tunguska class event during his watch than by a civilization-threatening impact, it is nevertheless true that each of his constituents is 10 to 100 times more likely to die during his term in office from the rare civilization-threatening event than from a Tunguska.
- Disaster managers have valid reasons to worry about smaller impacts. Although as individuals we are more at risk from large impacts, a disaster manager or government official is more likely to be faced with a small impact within his jurisdiction, especially if it has a large area. In Australia, as an example, with nearly 2% of the Earth's area, there are likely to be about 20 Tunguska class impacts for every global catastrophe.
- We do not need more accurate estimates of impact probabilities. Many people, even government officials, emphasize that the Spaceguard Survey is improving our knowledge of the frequency of impacts, and hence of the statistical magnitude of the hazard. For purposes of public policy, however, we already know the frequency well enough (and besides our greatest uncertainties are not in the asteroid flux but in the consequences of an impact). The purpose of a survey like Spaceguard is to find individual objects. The public does not need refined estimates of impact rates; they want to know whether or not an impact will occur, and if so where and when.
- Asteroids pass close to the Earth many times before they hit. A survey picks them up decades (or more) before their final plunge toward impact, and they do not (except in Hollywood) change orbits capriciously. We have neither the desire nor the capability to find them a few hours, or even a few weeks, before impact.
- We are not especially concerned about survey "blind spots," whether they are due to the Sun (approaches from the day side), absence of a Southern-hemisphere search telescope (for objects in the far southern sky), or bad weather (current Spaceguard search telescopes are mostly wiped out by cloudy weather in the southwest USA during July and August). An asteroid missed one year will be picked up on a later pass when the geometry is better. Similarly, it is no failure of the system when we discover an asteroid after closest approach. It does not matter when we get it, as long as we do pick it up and acquire the data needed to determine its orbit prior to any impact.
- Impacts by small asteroids are not harder to predict than those by large ones. Even the best-informed people seem to assume that we will have less warning for a Tunguska class event than for a civilization-threatening impact. It is certainly true that we are more likely to have no warning at all for small ones, but once they are discovered in a survey,

their orbits can be projected forward just as well as those of large ones, yielding similar warning times. (But of course larger telescopes are required to obtain these orbital data, and as a practical matter small asteroids are more easily lost.)

• We are interested in protecting the present generation, not far-future generations. The surveys we carry out and the mitigation strategies we develop are directed toward a possible impact within the next century. If it falls on our generation to defend the Earth, we need to know this. But what we do now will not protect generations centuries in the future. They will need their own surveys and their own mitigation plans, which will likely be much more advanced than what we are doing today.

Communication is a two-way street. As scientists we have an obligation to explain the situation honestly as we understand it. We are also obliged to listen to others, and to recognize that they may have valid reasons for pursuing public policies that are not exactly congruent to our perceptions of the impact hazard.

Appendix I   Impacts and cancer

Duncan Steel

Broadly speaking, there are two sides to the communication task that need emphasis, in order to aid the public in developing an understanding of the NEO impact hazard. The first is connected with the nature and level of the hazard: how does one juxtapose the known catastrophic consequences of an NEO impact with the rarity of their occurrence, and most especially the lack of a truly major event during historical times? The fundamental public communication problem here stems largely from the apparent fact that the average person, even with a college education, tends to have a marked lack of understanding of two areas of scientific study beginning with a "P": physics and probability.

The other side to the communication task requiring careful attention is almost the contrary of the above. It is a matter of instilling in the public some confidence that the NEO impact danger is a problem that is solvable; indeed solvable at a cost that is far less than the expectation of loss, so that it makes no economic sense simply to ignore it.

As a suitable tool for addressing both these aspects – on the one hand the seriousness yet rarity of the hazard, on the other hand the fact that it is within our capabilities to fix it – the following metaphor has proved useful, based on a concern that faces all humans: the possibility of developing some form of cancer.

(1) Early identification is vital. Most cancers need to be picked up very early in their development if they are to be treatable. So it is with NEOs. We have no time to lose in identifying any potential Earth impactor.

(2) Cancer screening (and NEO surveillance) is cheap. The cost of screening is smaller than the cost of treatment, and much less than the cost of doing nothing.

(3) Everyone can be involved in some way. Self-inspection (e.g., for breast, skin, or testicular cancer) is simple, but detailed investigations (e.g., for brain tumors) are expensive. Similarly amateur astronomers can provide vital help in studying NEOs, although in the end the professionals will need to tackle most of the job.

(4) Identification of a real problem is unlikely. Individuals are unlikely to contract specific cancers for which screening is done, but we must aim to check everyone periodically. In the same way we need to seek out all NEOs and keep tabs on them.

(5) False alarms are common. Any indicator of a potential problem necessitates careful monitoring, and causes considerable worry. But one should be pleased when the tumor proves benign. Precisely the same applies to NEOs: asteroids and comets discovered and initially flagged to be potential impactors but later shown to be sure to miss our planet represent victories on our part.

(6) Tackling any confirmed cancer (NEO impact) is certain to be unpleasant. No one suggests that chemotherapy, radiotherapy, or surgical intervention is fun, but they are necessary, as would be the steps employed to divert an NEO, such as the nuclear option. Nor would they be cheap: but the cost would be of no consequence, as with a serious cancer.

(7) Just because we don't yet know the cure for cancer does not mean that we should give up looking and trying. Where there is life, there is hope. If we should find an NEO destined by the clockwork of the heavens to impact the Earth in the near future (within the next few decades), and using our advanced science and technology we manage to divert it and so save ourselves, this will rank as perhaps the greatest achievement of modern-day civilization.

(8) Just because there are more significant problems facing the world does not mean that we should ignore this one. Having a bad cold or influenza does not mean that you should neglect to have the lump in your breast checked out. If there is a substantial NEO due to strike our planetary home soon, then we face no greater problem: not terrestrial disasters, not terrorism, not wars, not disease, not global warming, not unemployment nor economic downturns. The most likely result of a proper study of the impact hazard is that it will go away. But the converse is also true: what we now see as a slim chance (low probability of a large impact) may turn into a virtual certainty, which would then supplant our other earthly concerns.

(9) Just because we don't yet know a cure for the common cold does not mean that we cannot find the solution for this disease. Some of the greatest dangers we face on a daily basis have quite simple solutions, like imposing speed limits to cut down road fatalities. Conceptually, planetary defense against NEO impact is a far simpler problem than, say, trying to stop major earthquakes or volcanic eruptions, or halting a hurricane in its path.

(10) While searching for the cure for cancer we may anticipate discovering many other useful things. It is the very nature of scientific inquiry that discoveries are made that could not have been imagined prior to beginning the project. In the case of NEOs, it is

already known that among their number are the most accessible objects in space, easier to get to than the surface of the Moon, and they contain the metals, the water/oxygen, and the other materials that we will need for our future exploitation of the high frontier.

Acknowledgments

We are grateful to very many colleagues with whom we have discussed communication issues over the past decade as we all struggled to explain the impact hazard to the public. We especially thank Kelley Beatty, Bob Britt, Andrea Carusi, Steve Chesley, David Chandler, Alan Harris, Andrea Milani, Oliver Morton, David Roepik, Paul Slovic, and Don Yeomans. Alan Harris also provided a helpful referee report. Clark Chapman thanks the JPL NEO Program Office and the Global Science Forum of the Organization for Economic Cooperation and Development.

References

Alvarez, L. W., Alvarez, W., Asaro, F., *et al.* 1980. Extraterrestrial cause for the Cretaceous–Tertiary extinction. *Science* **208**, 1095–1108.

Atkinson, H., Tickell, C., and Williams, D. 2000. *Report of the Task Force on Potentially Hazardous Near Earth Objects*. London, UK: British National Space Center. Available online at http://www.nearthearthobject.co.uk

Beatty, J. K. 1999. The Torino Scale: gauging the impact threat. *Sky and Telescope* (October issue), 32–33.

Binzel, R. P. 1997. A near-Earth object hazard index. *Ann. New York Acad. Sci.* **822**, 545–551.

2000a. The Torino Impact Hazard Scale. *Planet. Space Sci.* **48**, 297–303.

2000b. Assessing the hazard: the development of the Torino Scale. *Planet. Report* **19**, 6–10.

Brown, P., Spaulding, R. E., ReVelle, D. O., *et al.* 2002. The flux of small near-Earth objects colliding with the Earth. *Nature* **420**, 294–296.

Bryant, E. A. 1991. *Natural Hazards*. Cambridge, UK: Cambridge University Press.

Canavan, G. H., Solem, J. D., and Rather, J. D. G. (eds.) 1993. *Proceedings of the Near-Earth Object Interception Workshop*, Los Alamos National Laboratory, Los Alamos, NM, publication no. LA-12476-C.

Chapman, C. R. 2000. The asteroid/comet impact hazard: *Homo sapiens* as dinosaur? In *Prediction: Science, Decision-Making, and the Future of Nature*, eds. D. Sarewitz, R. A. Pielke, Jr., and R. Byerly, Jr., pp. 107–134. Washington, DC: Island Press.

Chapman, C. R. and Morrison, D. 1989. *Cosmic Catastrophes*. New York: Plenum Press.

1994. Impacts on the Earth by asteroids and comets: assessing the hazard. *Nature* **367**, 33–39.

Chesley, S. R., Chodas, P. W., Milani, A., *et al.* 2002. Quantifying the risk posed by potential Earth impacts. *Icarus* **159**, 423–432.

Chodas, P. W. and Yeomans, D. K. 1999a. Orbit determination and estimation of impact probability for near-Earth objects. 21st Annual American Astronomical Society Guidance and Control Conference, Breckenridge, CO, Paper AAS no. 99-002.

    1999b. Predicting close approaches and estimating impact probabilities for near-Earth objects. American Astronomical Society/American Institute of Aeronautics and Astronautics Astrodynamics Specialists Conference, Girdwood, AK, Paper AAS no. 99-462.

Chyba, C. F. 1993. Explosions of small Spacewatch objects in the Earth's atmosphere. *Nature* **363**, 701–703.

Garshnek, V., Morrison, D., and Burkle, F. M., Jr. 2000. The mitigation, management, and survivability of asteroid/comet impact with Earth. *Space Policy* **16**, 213–222.

Milani, A., Chesley, S. R., and Valsecchi, G. B. 1999. Close approaches of asteroid 1999 AN10: resonant and non-resonant returns. *Astron. Astrophys.* **346**, L65-L68.

Milani, A., Chesley, S. R., Boattini, A., *et al.* 2000. Virtual impactors: search and destroy. *Icarus* **145**, 12–24.

Michel, P., Farinella, P., and Froeschlé, C. 1996. The orbital evolution of the asteroid Eros and implications for collision with the earth. *Nature* **380**, 689–691.

Morrison, D. (ed.) 1992. *Report of the Near-Earth-Object Interception Workshop: The Spaceguard Survey*. Pasadena, CA: Jet Propulsion Laboratory.

Morrison, D., Chapman, C. R., and Slovic, P. 1994. The impact hazard. In *Hazards due to Comets and Asteroids*, ed. T. Gehrels, pp. 59–91. Tucson, AZ: University of Arizona Press.

Morrison, D., Harris, A. W., Sommer, G., *et al.* 2002. Dealing with the impact hazard. In *Asteroids III*, eds. W. F. Bottke, Jr., A. Cellino, P. Paolicchi, and R. P. Binzel, pp. 739–754. Tucson, AZ: University of Arizona Press.

Napier, W. M. and Clube, S. V. M. 1979. A theory of terrestrial catastrophism. *Nature* **282**, 455–459.

Niven, L. and Pournelle, J. 1977. *Lucifer's Hammer*. Del Rey.

Organization for Economic Cooperation and Development. 2001. *Identify Risk*. OECD Public Management Committee, PUMA/MPM(2001)2.

Sagan, C. and Ostro, S. J. 1994. Long-range consequences of interplanetary collisions. *Issues Sci. Technol.* (Summer 1994), 67–72.

Sarewitz, D., Pielke R. A., Jr., and Byerly, R., Jr. 2000. *Prediction: Science, Decision Making, and the Future of Nature*. Washington DC: Island Press.

Schweickart, R. 2004. A system for trustworthy mission design and execution. B612 Foundation Special Paper, in press.

Slovic, P. 1987. Perception of risk. *Science* **236**, 280–285.

Smit, J. and Hertogen, J. 1980. An extraterrestrial event at the Cretaceous–Tertiary boundary. *Nature* **225**, 198–200.

Stokes, G. (ed.) 2003. *Study to Determine the Feasibility of Extending the Search for Near-Earth Objects to Smaller Limiting Diameters*. NASA Report, Washington, DC.

Stuart, J. S. 2001. A near-Earth asteroid population estimate from the LINEAR survey. *Science* **294**, 1691–1693.

Toon, O. B., Zahnle, K., Morrison, D., *et al.* 1997. Environmental perturbations caused by the impacts of asteroids and comets. *Rev. Geophys.* **35**, 41–78.

Urey, H. 1973. Cometary collisions and geological periods. *Nature* **242**, 32–33.

Ward, S. N. and Asphaug, E. 2000. Asteroid impact tsunami: a probabilistic hazard assessment. *Icarus* **145**, 64–78.

Yeomans, D. K. 2003. Population estimates. In *Study to Determine the Feasibility of Extending the Search for Near-Earth Objects to Smaller Limiting Diameters*, ed. G. Stokes, pp. 7–19, Washington DC: NASA.

17

Towards a national program to remove the threat of hazardous NEOs

Michael J. S. Belton

Belton Space Exploration Initiatives, LLC

1 Introduction

It is a demonstrable fact that asteroids of all sizes and less frequently cometary nuclei suffer collisions with the Earth's surface. The impact hazard, which is defined in Morrison *et al.* (2002) as "... the probability for an individual of premature death as a consequence of impact," has undergone considerable analysis with the conclusion that the greatest risk is from the very rare collisions of relatively large asteroids that can create a global scale catastrophe in the biosphere (Chapman and Morrison 1994). In the last decade, the question of how to deal with the hazard has led to considerable activity and advocacy on the part of the interested scientific community, and activity at government level has been stimulated in the United States, Europe, and Japan (a detailed overview is given by Morrison *et al.* 2002): there are now survey programs to search for objects that could be potentially hazardous; there are high-level calls for increased observational efforts to characterize the physical and compositional nature of near-Earth objects (NEOs) (e.g., The UK NEO Task Force report: Atkinson *et al.* 2000); an impact hazard scale has been invented to provide the public with an assessment of the magnitude of the hazard from a particular object; there have been considerable advances in the accuracy of orbit determination and impact probability.

Nevertheless, it seems that the question of how governments should go about preparing to mitigate the hazard needs some further attention. It has been advocated, as reflected in the review of Morrison *et al.* (2002), that because of long warning times (decades to hundreds of years have been suggested) we should simply wait until an actual impactor is identified to develop a mitigation system for asteroidal collisions. In the meantime, or so it is presumed, surveys to reach

Mitigation of Hazardous Comets and Asteroids, ed. M. J. S. Belton, T. H. Morgan, N. H. Samarasinha, and D. K. Yeomans. Published by Cambridge University Press. © Cambridge University Press 2004.

ever-smaller objects, scientific research and exploration characterizing these objects, basic research, etc., would continue to be supported by government agencies much as they are today. Such presumptions are, in my opinion, dangerous and, unfortunately, a high priority for these activities relative to other future scientific endeavors cannot always be guaranteed. Productive programs that enjoy adequate support today may face dwindling support in the future simply because of changing national priorities and interests. In addition, waiting an indeterminate amount of time for an impactor to be found invites, at least in my opinion, neglect, particularly at the level of government.

To resolve these problems in the United States an affordable and justifiable national plan is needed, which incorporates the above scientific research and exploration and that is focused on the technical goal of mitigating the most probable kind of impact that can cause serious damage to the social infrastructure in the lifetime of the current population. Such an approach requires redefining the hazard in terms of cost rather than deaths together with a demonstration that the expected cost of the plan is commensurate with the losses that would most likely be incurred in the impact. This approach also builds into a mitigation program the notion of scientific requirements. An operational mitigation system or device can still wait until an impactor is identified, but meeting the scientific requirements for that system is something that ought and, I believe, should proceed now. There are other benefits to this approach:

(1) By defining this program as a technical imperative rather than a scientific one the element of direct competition with established science goals is removed – even while significant elements of the program remain scientifically productive.
(2) By focusing on the most probable impacts, i.e., smaller asteroids, a process of learning and gaining experience is implied that might, unless fate and statistics defeat us, allow us to deal more effectively with the larger and less probable objects further into the future.

2 Goals

The probability of impact appears to be random and the average impact rates of the dominant component – the near-Earth asteroids (NEAs) – are reasonably well known. In this chapter I will consistently use impact rates estimated by a power-law distribution in Morrison *et al.* (2002). In other recent, but unpublished, work it is pointed out that the observed rates for objects near 50 m in size may be less by a factor as large as 2 (Harris 2002). If these new rates are substantiated it should be a straightforward task to adjust the relevant numbers given in this paper with little change to the argument.

Asteroids larger than 50 m across, roughly the minimum size that could cause calamitous effects at the surface, collide with the Earth on average once every 600 years. This is equivalent to roughly a 0.3% chance that United States territory could be hit in the lifetime of its population ($\sim$100 years). With a typical relative velocity near 20 km s^{-1} (Morrison *et al.* 2002) the impact will almost instantaneously release an energy of 10^{16}–10^{17} J into the local environment, i.e., roughly the equivalent of a 10 megaton bomb or about half the energy that the US Geological Survey estimates was released in the Mt. St. Helens volcanic event. I have chosen to deal with objects of this size because they are the most likely impactors that present-day American public officials may have to deal with. Also the effects of such natural disasters are close to the realm of contemporary public experience, e.g., the effects of the 1908 Tunguska meteor explosion over the Siberian wilderness where the blast severely affected an area of 2000 km^2 of forestland are widely known (Vasilyev 1998). Impacts by much larger objects, i.e., larger than about 1 km that can cause global scale catastrophes, will, by definition, also affect United States territories whatever the location of the impact (Chapman 2001). But these less frequent collisions occur at a global rate of about 1 per 500 000 years, which translates into a 0.02% chance during the lifetime of the current population of the United States. While I include these kinds of impacts in the argument below, it does not depend upon them. At the present time no government agency in the United States has been given the responsibility to deal with these potentially hazardous collisions. The National Aeronautics and Space Administration (NASA) exercises a mandate from the US Congress to locate 90% of the objects greater than 1 km that exist in near-Earth space by 2008 but has no existing authority to act if an object on a collision trajectory is found (Weiler 2002). Given the above collection of facts, it would seem that the primary issues that confront society with respect to mitigation are:

- When is the best time to invest in the research and development that would make it practical to mitigate the effects of such hazardous collisions in the future?
- Who should be responsible?
- What is the best way to go about it?

One can anticipate that achieving resolution on such issues will be a controversial task and each of the above questions could stimulate wide discussion. In this chapter I will simply assume that, if the justifications outlined below hold up, most United States citizens will want their government representatives to support the development of a system that could prevent the impact of a dangerous asteroid (i.e., one greater than 50 m in size) found on a collision course with United States territory, or a $\sim$1 km asteroid found on a collision course with the planet at large, particularly if it were to occur during their lives. The prevention of such collisions I take to be the goal of the national mitigation program.

3 Justifications

There is a set of conditions that I expect would have to be satisfied in order to justify
the expenditure of United States national treasure on an asteroid mitigation system.
These conditions reflect the kinds of questions that I believe any reasonable citizen
might ask before agreeing to proceed, e.g., why are such low-probability events
worth worrying about? Is today's technology up to the job? Will the result of this
effort be useful to us even in the absence of a collision in our lifetimes? Will this
effort to protect our lives and property create collateral problems we don't need? I
have tried to capture the essence of these questions in the following statements:

(1) The public would need to view the prospect of an impact by a 50 m asteroid within
    the territorial boundaries of the United States, or 1 km object impacting anywhere on
    Earth, as a serious concern.
(2) Our technical ability to create a reliable mitigation system would need to be reasonably
    assured, and it should be possible to build it in time to give a fair chance that the next
    hazardous object to threaten the territories of the United States could be dealt with.
(3) The net cost of creating a reliable mitigation system should be no more than typical
    losses that might be incurred if an impact of a 50 m object were to happen within the
    territorial boundaries of the United States.
(4) The implementation of a mitigation system must not create more dangers than already
    exist.

It seems self-evident that the first step towards a national program would be a
high-level, government-sponsored, study of such issues. This would be followed,
if warranted, by the assignment of responsibility and the establishment of a funded
program perhaps along the lines of existing community recommendations (e.g.,
those in the report of Belton (2003)).

The first condition involves the perception and assessment of risk by the public.
This is apparently a topic with few experts (cf. Chapman 2001) and may be impossible to quantify. In my view, it is essentially a political issue and any assessment
is almost certainly made best by politicians currently in office, e.g., by relevant
congressional committees or in the administration itself. I have already noted that
the impact rate for 50-m and larger objects give about a 0.3% chance of an asteroid collision on United States territory during the lifetime of the population. The
chances that any particular location in the United States would be directly affected
are approximately 5000 times less. These chances have to be modified for coastal
cities (where much of the population resides) since they could be seriously inun-
dated by a tidal wave, say 5 m high or greater, caused by asteroids that impact
in the ocean. Ward and Asphaug (2000) have considered such impacts, but their
impact rates for the most efficient impactors for this process are about six times
too high relative to those in Morrison *et al.* (2002). Correcting for this I find the

respective chances of this happening are about 0.07%, 0.03%, and 0.1% for San Francisco, New York City, and Hilo in a 100-year period. To make it clear that these are small probabilities, I note that the chance that the population will not experience the effects from a collision in its lifetime is about 99.6%. Such small chances are, I believe, unlikely to raise much public concern even though the threat is real. It is only when palpable knowledge of the level of destruction that a random 10 megaton explosion could cause on a particular area, i.e., the combined energy released by more than 770 Hiroshima bombs, or roughly half the energy of the Mt. St. Helens disaster, or roughly ten times the energy radiated by the largest earthquake ever recorded in the United States, is pointed out to the public that notice might be taken. When knowledge of this level of destruction is combined with an awareness that a reliable defense could be built for a relatively modest cost, and that some significant fraction of the costs could themselves be mitigated through productive applications to science and space exploration, then I believe there is a chance that the need for a mitigation effort now could become justified in the public mind.

It is interesting to speculate on how typical individuals in the population might view these risks and trades. I would imagine that such persons would quickly conclude that an impact would be very unlikely to have any direct affect on them, their family, or their livelihood. I would expect that they would quickly lose interest and presume that if something should be done about such rare and terrible events then "someone" in government would be taking care of it. They might be surprised to learn that the "someone" in government they assumed to be taking care of things doesn't exist and that, in fact, no one in government presently has any responsibility to do anything about it. Certainly, in the aftermath of a random 10 megaton explosion somewhere in the United States, or a 5 m tsunami wave inundating a coastal city, they would be both pleased at the performance of disaster relief and tsunami warning organizations but sorely perplexed by the lack of preparedness in government organizations that might have prevented the disaster.

The second condition addresses whether the construction of a reliable mitigation system can be assured and whether it would be timely. There appear to be four essential elements in such a system. First, there must be an assured ability to locate and determine the orbit of the impactor with sufficient accuracy and warning time; second, it must be possible to reliably deduce the general physical properties of the impactor so that planning for a mitigation system can achieve a reliable result; third, we must have the ability to intercept it before the collision takes place; and fourth, we must have the ability to deflect or disrupt the impactor.

Most objects hazardous to the earth are on near-Earth orbits (see Chapter 2). To reach most of the 50 m sized objects in 10 years, telescopic surveys would have to operate at around $V = 25$ magnitude (this is based on an extrapolation of data in Morrison *et al.* (2002)). By comparison, the surveys that are operating today have a limiting magnitude near 19.7 mag, i.e., more than a factor of 100 brighter. These rough figures simply mean that *at present* telescopic technology is very far from what would be required to meet the goal of the national mitigation program. However, plans are already afoot that will push the present survey capability to a limiting magnitude of $V = 24$ where most 200 m objects could be found in a ten year period. The proposed Large-aperture Synoptic Survey Telescope (LSST) facility could do this if the requirement is built into the design. The implementation of such a telescope, which is at the edge of present engineering technology, has already been advocated in the reports of two independent committees backed by the National Research Council (Space Studies Board 2001, 2002a). To reach 90% completeness at $V = 25$ in a reasonable amount of time new technological limits would need to be achieved on the ground or space based systems will be required (e.g., Jedicke *et al.* 2002; Leipold *et al.* 2002). As put succinctly by Jewitt (2000) if these, or similar, facilities are not made available, ". . . we will have to face the asteroidal impact hazard with our eyes wide shut."

Detection of NEOs is only a part of the equation. Also essential is the capability for rapid determination of accurate orbits to yield long warning times and accurate calculation of impact location and probability. These are not minor requirements and demand extended post-discovery follow-up observations (see Chapter 2), and advances in astronomical radar systems (see Chapter 3) and in computing technology (Milani *et al.* 2002). While the above discussion indicates that a large increase above today's capability is called for and a considerable amount of telescope building and observational and interpretive work over an extended period of time are implied, there appear, at least in my opinion, to be no fundamental showstoppers to this aspect of a mitigation system. Time and money are the limiting factors.

Detailed knowledge of the general physical properties (mass, spin state, shape, moments of inertia, state of fracture, and a range of surface properties) will be needed for any hazardous asteroid that becomes a target (see Chapter 9). Just the choice of a particular mitigation technology and its operating parameters will obviously be sensitive to the physical and compositional nature of the target. Experience shows that only a few of these parameters can be deduced with any precision from Earth-based observations and *in situ* space missions will need to be flown to determine these parameters. Since this would at least take the time needed to build and launch the spacecraft and to intercept a hazardous target, typically four or five years, it is possible that there will not be enough warning time to accomplish this. In such a case the mitigation system itself may have to determine some of the critical properties (e.g., shape, mass, moments of inertia, internal state of fracture, etc.) when it arrives at the target *while other properties would have to be inferred from a database of properties that has been built up as part of a more general exploration and research program.* The latter will also play a crucial role in developing several new and essential measurement techniques, e.g., radio tomography (see Chapter 10) and seismic assessment (see Chapter 11; also Chapter 12) of the interior structure of small asteroids, and new ways to measure the composition and porosity of surface materials. It seems clear that an aggressive *near-Earth asteroid* space exploration program will need to be integrated within the mitigation program.

The requirement for robotic spacecraft to intercept and to land on a small asteroid is easily within current capability and has already been demonstrated by the NEAR mission at the asteroid Eros (Veverka *et al.* 2001). Mitigation techniques may require more advanced capability for operations around these small, very low mass, objects as discussed elsewhere (see Chapter 14), but, again no serious impediments that could derail a future mitigation project are anticipated.

Our ability to disrupt, or adequately deflect, a rogue asteroid of a particular size headed towards Earth is completely hypothetical at the present time. There are many ideas (for a summary see Chapter 9) on what should be done and there are clearly many serious uncertainties in the application of nuclear devices (see Chapter 6). Similar uncertainties are also latent in the application of a solar concentrator (see Chapter 9). From a purely theoretical point of view it should be possible to find technical solutions these problems. However, it is clear that early *in situ* interaction experiments need to be done on small objects before we can be sure where the problems are and which techniques are viable. The B612 Foundation (http://www.b612foundation.org) has been formed to address the challenge of demonstrating that significant alterations to the orbit of an asteroid can be made in a controlled manner by 2015. Success with this endeavor would also be a major landmark in any mitigation program.

In summary, it would seem that we already have experience with many of the elements needed for mitigation, but that significant development, new capability, and time will be required for success. The lack of a demonstrated technique for deflection or disruption is a particular cause for concern. There are also other serious uncertainties, the chief being whether or not human activities in space (e.g., for the assembly of parts of the system in low Earth orbit, or at the target asteroid) would need to be included. This could strongly affect the ultimate cost of a practical mitigation system and therefore its viability. But overall, though there are many technical areas that need considerable investment in time and money to achieve success, there appear to be no fundamental reasons why a mitigation system could not succeed.

The third condition has to do with the cost of a mitigation system. For costs to be acceptable the mitigation program costs should be comparable (hopefully less) than estimates of the cost of the damage caused by the most probable kind collision, i.e., that of a 50 m asteroid, on the territory of the United States in the lifetime of the current population. The advantage of estimating costs this way is that we can deal with real examples of costs incurred as a result of damage to infrastructure that are provided by historical events.

The United States is a well-developed country and has many large metropolitan areas and valuable, if modestly populated, rural areas. Even its underpopulated desert areas often have valuable resources embedded in them. The economic losses, mainly timber, civil works, and agricultural losses associated with the 1980 Mt. St. Helens event in rural Washington State (approximate energy release: 24 megatons) were estimated at $1.1 billion in a congressionally supported study by the International Trade Commission. In a metropolitan area near Los Angeles, the 1994 Northridge earthquake caused economic loss that was officially estimated at $15 billion with most of the damage within 16 km of the epicentral area, and here the energy release was far less than that which could be released by the kind of impact that we are considering. I believe that these two examples are near the extremes of the economic losses that might be incurred as a result of a localized 10 megaton event occurring at a random place within the United States. On this basis I would argue that a $10 billion cost cap to a mitigation program would not be out of line. In the planning roadmap developed below an investment of approximately $5 billion should cover the costs of the initial preparatory phase of a mitigation program with the expenditures extending over 25 years, i.e., an average funding level of $200 million per year. This is not far from the typical levels invested in major program lines at NASA today, and so the amount is not unusually large. This leaves a further $5 billion that would be available for the implementation of mitigation mission to a specific target. Providing human spaceflight participation is not needed, this is within the expected costs of other extremely large robotic missions that have been flown or proposed. My conclusion is that condition on cost can be met and that the annual budget for a mitigation program will not be too different from costs experienced in existing robotic space programs. If human spaceflight is shown to be an essential element in a mitigation system, then the cost argument made here will need to be substantially modified.

The fourth and final condition has to do with environmental and civil security. Mitigation concepts that depend on even a modest proliferation of explosive nuclear devices in space or on the ground will, in my opinion, be non-starters if this condition is to be met.

4 Mitigation programmatics

Mounting a defense against a sizable incoming object from space will be a complex task. There are national and international issues that need to be resolved; there are issues involving the delegation of responsibility between civil and military authorities; there are science issues; there are political issues involving goal-setting, mission scope, and cost containment; and, finally, there are environmental and civil security issues.

Here I advocate a three-phase process to establish a mitigation capability that roughly separates out strategic, preparatory, and implementation functions. It is probably prudent if these are accomplished sequentially since changes in one can be expected to have large consequences for the phases that follow.

The purpose of the first, or *strategic,* phase is to clarify the overall goal of the program, set up its scope, identify funding, and then assign responsibilities. Because of the significance of the mitigation program to the entire population, it should be initiated by a responsible entity within the federal government, either in the administration or the congress, with, presumably, expert advice from individuals and grass-roots organizations.

The second, or *preparatory,* phase includes all that needs to be done to achieve the scientific and engineering requirements on which the design of a reliable and effective mitigation system will depend. This phase begins once an assignment of responsibility is made and funds are available to proceed. It should ideally be completed before a target on a collision course is identified, but in case we are not this fortunate, it should also include an "amelioration" element that takes care of what to do if an unexpected collision occurs.

The last, or *implementation,* phase can only be pursued efficiently after the preparatory phase is completed and a hazardous target has been identified. In this phase all of the specific requirements of a particular target are addressed and the construction, test, and implementation of an actual mitigation device is carried out. To my knowledge no one has advocated beginning work on this phase at this time. It is probably the most expensive part of the work and may involve elements of human spaceflight.

5 The strategic phase

I have already advocated that the goal of a national program would be to design and implement a system to negate the most probable collision threat to United States territories in the next 100 years: a 50 m or larger near-Earth asteroid. The prime task in the strategic phase, which might take three to five years to accomplish, would be to assess this goal in competition with alternative program concepts and make a definitive selection. Identification of an approximate timeline, suitable programmatic arrangements, and an adequate budget profile, i.e., a roadmap, would follow. Institutional responsibility would need to be assigned. Expert preliminary technical evaluation in the strategic phase is necessary to ensure that the goal is achievable and to obtain a better basis for cost estimation. There are many sources of advice including existing expertise within government agencies, their advisory committees, and committees of the National Academies.

I have placed considerable stress on the idea that the program should start out as a national program rather than one that is international in scope. This is a matter of pragmatism rather than xenophobia. Fostering program growth from existing expertise within the national space program should be more effective and less costly than initiating a brand new top–down international effort. The program may also involve discussion and use of military assets that could be a sensitive issue if placed in an international context. Finally, it is well known that national policies and priorities change on short timescales tied to political cycles, while stable funding and a sustained effort over two or three decades is needed for a mitigation program. I believe that such stability is best obtained in the context of a national program. Cost can also be expected to be an issue in an international program. While it would be beneficial to share development costs, I would expect the total program costs to be enlarged over that of a national program in order to immediately encompass a mitigation system capable of addressing the more difficult goal of combating large near-Earth asteroids that can do global damage. With this said, it is important to recognize that the collision threat is worldwide and that much expertise lies beyond national boundaries. International cooperative projects that contribute to a national program are obviously to be encouraged. For an indication of the level of international interest and direction the reader is referred to the conclusions reached in the Final Report of the Workshop on Near-Earth Objects: Risks, Policies and Actions sponsored by the Global Science Forum (OECD 2003) that suggest actions that could be taken at governmental level.

It should also be understood at the outset that the mitigation program advocated here is aimed at a specific technical goal and is not a scientific or space exploration program. To be sure, the program will have remarkable scientific and exploratory spin-offs, but these are not in any sense the primary goal. This is important because closely allied scientific and exploratory endeavors already have well thought out priorities and widely supported goals that should not be perturbed by the establishment of a mitigation program. This is particularly so in astronomy and astrophysics, in solar system exploration, and in space physics where goals are focused on understanding origins – particularly of life, physical and chemical evolution, and the processes that explain what we experience in space (Space Studies Board, 2001, 2002a, 2002b). It would, in my opinion, be disruptive to try and embed a national mitigation program within one of these scientific endeavors. For mitigation, a separate program with a clear technical goal is required.

6 The preparatory phase

This phase should include at least the following five elements: hazard identification, amelioration, basic research, physical characterization of targets, and what I call interaction system technology.

6.1 Hazard identification

The operational goal of this element would be to locate and determine the orbit of the next 50 m, or larger, near-Earth object that will, if mitigation measures are not taken, collide with the Earth. This goal must be accomplished with sufficient accuracy to determine whether the object will also collide on United States territory. It should also provide a sufficiently long warning time. Initially I propose to set the goal for this warning time as at least ten years, which is the minimum time that I expect it would take to implement a robotic mitigation system that might be capable of deflecting a 50 m object. Astronomical survey systems are expected to yield much longer warning times (~100 year) for collisions with the Earth itself. But these warning times shrink when the impact error ellipse must fit within the area of United States territories (D. Yeomans, personal communication).

This is a distinctly different kind of goal from that associated with the Spaceguard Survey and clearly goes far beyond it. Yet it is, in my opinion, a necessary goal if a national mitigation program is to be justified to the public. To pursue this goal, this element should contain the following components: (1) completion of the Spaceguard Survey; (2) implementation of the Large-aperture Synoptic Survey Telescope project, along the lines recommended in the recent Solar System Exploration Survey (Space Studies Board, 2002a), and a parallel development of the USAF/Hawaii PanStarrs telescope system (http://pan-starrs.ifa.hawaii.edu) to pursue a modified Spaceguard goal which will lead to the detection and orbital properties of 90% of near-Earth objects down to a size of 200 m within about ten years from the start of the survey; (3) design and implementation of a technologically advanced survey system, or possibly a satellite project to take the Spaceguard goal down to the 50 m size range; (4) a ground-based radar component developed from the capabilities that already exist at Goldstone and Arecibo in conjunction with other facilities (see Chapter 3) to provide improved orbits for potentially hazardous objects and to lengthen collision warning times; (5) the final component is a suitably fast computing, data reduction, orbit determination, and archival capability. This capability could be part of the arrangements of one or more of the above telescope projects. To scope the size of the problem there are an estimated 1 million near-Earth space objects down to 50 m in size and, using the results given in Chapter 1, only about

250 of these may be hazardous to the Earth at the present time. However, there are some 210 000 objects in this population that, while not currently Earth impactors, could, through the effects of planetary perturbations, become hazardous to Earth in the relatively short-term future (D. Yeomans, personal communication).

In the roadmap (Fig. 17.1) I show these projects with some overlap stretched out over a period of 25 years. It is envisioned that these telescope systems (and others available to the astronomical community) would provide follow-up observations for each other and, where possible, make physical observations.

6.2 Amelioration element

The goal of the *amelioration* element is to mitigate the effects of unavoidable impacts. There are many community organizations that could fulfill this function throughout the United States and on a national level the new Department of Homeland Security would obviously be involved. However, none of these organizations has, to my knowledge, been tasked on how to respond to an unanticipated impact. As the mitigation program progresses accurate warnings and alerts should become available and the newly invented Torino Scale (Binzel 2000) will be used to communicate the level of danger to the public. Resources in the event of an actual disaster would presumably be allocated as is done today to provide relief from the effects of tsunamis, earthquakes, fires, and other natural disasters and not charged to the mitigation program itself.

6.3 Basic research

There is a need for a small basic research program within the umbrella of the mitigation effort that is unfettered from well-focused goals of the other components. Here a research scientist or engineer would be able to obtain funds to support the investigation of novel theoretical ideas or laboratory investigations that are related, but not necessarily tied, to established mitigation goals. Examples are investigations into the causes of the low bulk densities that are being found for many asteroids (Britt *et al.* 2002; Hilton 2002; Merline *et al.* 2002), or the details of how shocks propagate in macroscopically porous materials. There are already a number of individuals, many at academic institutions or private research facilities, undertaking such investigations in the United States who could form the core of this effort.

6.4 Target characterization

The goals of this element are two-fold: (1) to obtain the information needed so that observations of a hazardous target can be confidently interpreted in terms of the

		Year 1	5	10	15	20	25	Cost ($M)	Subtotal
Hazard Indentification (NASA/NSF/DOD)									**600**
	Spaceguard							20	
	LSST							100	
	PanStars							80	
	Radar							50	
	HIR&A							50	
	SBAS							300	
Amelioration (DHS)									**36**
	Indentification of needs							1	
	Training and operations							12	
Basic Research (NASA/NSF)									**50**
	Risk and disaster management research							23	
	Theory							25	
	Laboratory							25	
Target Characterization (NASA)									**2345**
Reconnaissance line									1050
	1							175	
	2							175	
	3							175	
	4							175	
	5							175	
	6							175	
Interiors line									1200
	1							400	
	2							400	
	3							400	
	Data analysis							80	
	Characterization (R&A) (DOD)							15	
Interception System Technology (DOD)									**2000**
	Interaction Experiments							400	
	Mission 1							400	
Intercept Technology									
	Mission 1							400	
	Mission 2							400	
	Mission 3							400	
Total								**5031**	

Figure 17.1 The elements of the preparatory phase of a national mitigation plan are listed in the left-hand column and the estimated costs to completion in the right-hand columns. This phase would be preceded by a strategic phase and followed by an implementation phase. The goal of the preparatory phase is to accomplish all that needs to be done to lay the scientific and engineering basis for the design of a reliable and effective mitigation system. The approximate phasing of each element within the timeline is shown by a line of Xs. The reasons for the choice of a 25-year timeline are discussed in the text.

surface and interior properties that are of most interest to mitigation; (2) to develop and gain experience with measurement techniques that allow characterization of the state of the interior of a small asteroid and the materials within a few tens of meters of its surface to the level of detail required for mitigation.

To meet these goals the program should provide opportunities to try out novel types of instrumentation and perform detailed characterizations of the physical, compositional, and dynamical properties of a wide sample of the primary asteroidal types with the purpose of creating an archive of such properties. This kind of research, of course, already has a substantial history with considerable advances in understanding spin properties (Pravec *et al.* 2002), multiplicity and bulk density (Britt *et al.* 2002; Hilton 2002; Merline *et al.* 2002) for asteroids as a group and the distribution of taxonomic groups within the NEOs (e.g., Dandy *et al.* 2003). Nevertheless, studies of the physical and compositional properties of these NEOs are being outstripped by their discovery rate. There are three elements that should run in parallel: (1) an Earth-based observational program focused on physical and compositional characterization, including radar studies, that can reach large numbers of objects and sample their diversity (diagnostic spectral features over a broad frequency range should be sought to better characterize the nature of each object); (2) a reconnaissance program of low-cost multiple flyby missions, similar to that advocated by the UK NEO Task Force (Atkinson *et al.* 2000), to sample a wide diversity of objects and to respond quickly to particular hazardous objects so that a first-order characterization of their properties can be accomplished; (3) a program of medium-sized rendezvous missions that can sample their interiors, and get down onto their surfaces to do seismic investigations. I have included four of these relatively costly missions that would include ion drive propulsion and visit at least two targets each.

The final component is a strong, coherent, data analysis and interpretation program. This should cut across all missions and include Earth-based work. Participation beyond the membership of the scientific flight teams would be strongly encouraged. The goal here is to integrate the net experience of the entire suite of investigations and produce the most complete database available on the properties of near-Earth asteroids, a database that can be confidently used to diagnose the properties of a potential Earth impactor.

6.5 Interaction system technology

This element is the most technically oriented part of the preparation phase. Here the goal is to learn how to operate spacecraft and instruments in the close vicinity of the surfaces of very small asteroids, emplace and attach devices to their surfaces, learn their response to the application of various forms of energy and momentum, etc. All of these techniques must be *learned* (see, for example, the advice of Naka

(1997)). Experience must be gained over the full range of surface environments that the various types of asteroids present. Experiments to test the ability and efficiency of candidate techniques to deflect and, possibly, disrupt very small, i.e., otherwise harmless, near-Earth asteroids should be done as part of this element. The history of spaceflight tells us that when the time comes to implement a particular mitigation device we should not trust the first-time application to deliver on its promise. Much can go awry and practice will be needed. It is in this element of the plan that the necessary practice should be acquired.

It is also in this element where it will become clear what, if any, role human spaceflight might play in a mitigation system. A completely robotic approach would presumably be much cheaper if, in fact, such an approach were feasible. But it is possible that human participation may be essential for the effectiveness and reliability of a mitigation system.

7 The implementation phase

The goal of this phase is to safely deviate, disrupt, or otherwise render harmless a 50 m or larger object found to be on a collision course with United States territory in the most reliable manner and at the lowest cost. This goal can be extended to the entire Earth if the hazardous object is found to be above the size that can cause global-scale havoc. If the object is smaller than this critical size and not threatening national territory, the United States may still be involved in the implementation of a mitigation device, but jointly with those nations whose territory is threatened. While this goal is clearly stated, addressing it will have some subtle difficulties due to errors latent in locating the precise impact point. Locating the latter within United States territory is much more difficult than determining that the Earth will undergo a collision. It may be that the implementation phase may have to start before it is determined for sure that United States territory is at risk (I thank D. Yeomans for this insight).

It will not be possible to outline a detailed plan for this phase until the preparation phase is largely complete. Nevertheless, a few essential attributes seem self-evident: (1) it would only begin when a collision threat is confidently identified; (2) it would normally, i.e., if there were enough warning time, involve many of the same components found in the preparatory phase, but with their focus entirely oriented towards the target object itself; (3) it would include the design, construction, and application of the chosen mitigation system.

8 A planning roadmap

Figure 17.1 lays out a crude timeline for the preparatory phase that shows how the different activities that have been described interlace with one another. Estimated dollar costs, without allowance for inflation, are simply based on personal experience in NASA flight programs. The timeline for the preparatory phase is presented over a 25-year period. This time-span is somewhat arbitrary and could have been made shorter by increasing the parallelism of the components. However, there are practical limits to such parallelism. These include the availability of facilities and qualified manpower, as well as acceptable limits on average and peak annual dollar costs. In my experience, average costs of $200–250 million per year with a peak of $300–400 million in any one year are not untypical. The profile for this plan gives an average cost of $200 million per year with a peak of $610 million in year 15. This relatively large peak is due to the confluence of work on six flight missions in a single year. Expert consideration of this plan with more focus on costs could presumably relieve the magnitude of this peak.

Hazard Identification includes the remainder of the Spaceguard program, half of the LSST, and PanStarrs programs, and, towards the end of the phase, a space-based asteroid survey mission (SBAS) for the smaller objects and objects in orbits that are difficult to observe from the ground. In the case of the Spaceguard program, which is underway at the present time, I have assumed that this program would continue until the LSST and PanStarrs survey are well under way. The National Science Foundation (NSF) would presumably support the LSST and part of the PanStarrs program. Also included in this component are provisions for an underlying and continuing research and analysis program. One provision (Hazard Identification Research and Analysis, HIR&A) is focused on providing search software, archiving, orbital analysis, and related tasks; the other is support for an ongoing program of radar observations related to high precision orbital determination. I have assumed that the SBAS mission would be pursued on the scale of a NASA Discovery program.

For the Amelioration component I have assumed that elements of the Department of Homeland Security would undertake this task for a modest cost of $1.5 million per year. This includes approximately $1 million per year for research into such issues as risk control, management, disaster preparation, etc. In the unlikely event that a collision occurs during the preparation period, special disaster relief funds would need to be appropriated as is usually done for unanticipated natural disasters on a case-by-case basis.

The Basic Research component is shown as equally divided between theoretical and laboratory investigations. The correct balance between these lines would have to be judged on the basis of proposal pressure. The program scope is at the modest level of $2 million per year, which should adequately support some 20 independent investigations.

Target Characterization is broken down into four groupings:

(1) A Reconnaissance mission line, which is conceived of a series of low-cost multiple flyby, impact, or multiple rendezvous missions similar to those recommended by the UK NEO Task Force (Atkinson *et al.* 2000). Its purpose is to provide basic physical and compositional data on the wide variety of NEOs that are known to exist. Based on experience with planning proposals, three targets per mission seems feasible with a new start every four years, i.e., six missions seems plausible. To lower costs, I also assume that the basic fight system will be similar in each mission with an average cost of $175 million per mission.

(2) An Interiors mission line consisting of three moderately complex missions with the goal of making a detailed survey of the state of the interior and sub-surface of six different types of asteroids including, if possible, a candidate cometary nucleus. These multiple-rendezvous missions are conceived of as focusing on either radio tomography or seismic investigations and would address at least two targets each. They are expected to fall near the low end of the cost range of the NASA New Frontiers mission line.

(3) A Data analysis line. Here the object is to encourage the larger science community (i.e., beyond the scientific flight teams) to get involved in the interpretation of the return from these missions and ensure that the data from all of the missions are looked at in an integrated way.

(4) A Characterization (Research and Analysis) line which is to primarily to support Earth-based telescopic investigations, including radar, of NEOs and potentially hazardous objects from the point of view of understanding their global physical and compositional properties.

The Interaction System Technology component is, at present, the most poorly defined part of the preparation phase. The necessity and scope of this component is based on the discussion of Naka (1997) and in the roadmap I have broken the tasks down into two broad elements: (1) Interaction experiments, and (2) Intercept technology. It is clear that this element has goals of significant complexity and will need a considerable amount of detailed pre-planning. The lead responsibility for carrying out these missions should lie with the Department of Defense, although some sharing of responsibility with NASA may be required. I have imagined that the tasks in this element could be carried out within the scope of five relatively complex missions with costs similar to those of the Interiors line.

9 Major milestones

In programs of this size it is helpful to identify major accomplishments towards the underlying goal through a series of milestones. In Table 17.1 I list some candidate milestones showing the relative year in which they might be accomplished and the agency that would presumably be responsible.

Table 17.1 *List of milestones*

Milestone	Responsibility[a]	Year
Start of strategic phase	Congress or Administration	1
Assignment of authority and responsibility	Administration	2
Congressional approval for a new program line	Congress	4
Start of preparatory phase	NASA, DOD, DHS	5
Start of reconnaissance line missions	NASA	5
Beginning of LSST survey (objects down to 200 m)	NSF	8
Start of Interiors line missions	NASA	9
Beginning of SBAS survey (objects down to 50 m)	NASA	20
First demonstration of a deflection technique	DOD	21
Determination of need for human participation	DOD	21
Conclusion of preparatory phase	NASA, NSF, DHS, DOD	30

[a] NASA, National Aeronautics and Space Agency; DOD, Department of Defense; NSF, National Science Foundation; DHS, Department of Homeland Security.

10 Summary

I have presented what I believe is a practical approach to a national program to mitigate the threat from asteroidal collisions. It is based on a goal that addresses the most probable threat from an extraterrestrial object to the United States during the lifetime of the current population, i.e., the impact of a 50 m or larger near-Earth object within the territorial boundaries of the United States during the next 100 years. I propose four conditions that would need to be met before the start of a program could proceed. In essence these conditions try to balance a presumed public lack of interest due to the low probability of an impact and the relatively large cost of a program to deal with it, against the typical cost of damage to the social infrastructure that might occur and the bonus in scientific knowledge that the program would produce.

The program itself is constructed from three components that would be pursued sequentially. A strategic phase, which lays the political and programmatic basis; a preparatory phase, which creates the necessary scientific and technical knowledge that is needed to provide a secure foundation for the design and implementation of a mitigation system; and an implementation phase, in which a mitigation system is built and flown with the goal of preventing a collision.

A plan is outlined that accomplishes the strategic and preparatory phases within three decades at a modest annual budgetary level for a total cost of approximately $5 billion dollars. The final implementation phase needs to be accomplished within a cost cap of $5 billion in order for the above argument to hold. It is expected that this can be achieved with a purely a robotic system. If, however, it is determined during the preparatory phase that human presence in space is needed as part of the system, the implementation costs can be expected to be larger than are allowed by the above arguments.

In developing this program, I largely downplay three important issues often associated with mitigation: an impact by comet nucleus, an asteroidal collision by an object that is sufficiently large to cause a civilization-wrecking global catastrophe, and the large number of deaths that could caused by such events. This is done simply because of the rarity of such events, and the lack of any palpable public experience of the destructive force of such an incredible event on the Earth and, finally, what I perceive as a necessity: we must *learn* how to deal with small asteroids before we can expect much success in mitigating a collision involving a large one. Asteroidal collisions will continue to happen and, as our society grows, will have increasingly costly consequences. I would hope that the program that I have sketched out here might be considered as a first step towards the realization of an operational mitigation system in the United States.

Acknowledgments

I would like to thank D. Yeomans, D. Morrison, C. R. Chapman, and W. Huntress for critical reviews of an early draft of this chapter.

References

Atkinson, H., Tickell, C., and Williams, D. 2000. *Report of the Task Force on Potentially Hazardous Near-Earth Objects*. London, UK: British National Space Centre. Available online at http://www.nearearthobjects.co.uk

Belton, M. J. S. 2003. Final Report. *NASA Workshop on Scientific Requirements for Mitigation of Hazardous Comets and Asteroids*. Arlington, VA, Sept. 3–6, 2002. Available online at http://www.noao.edu/meetings/mitigation/

Binzel, R. P. 2000. The Torino Impact Hazard Scale. *Planet. Space Sci.* **48**, 297–303.

Britt, D. T., Yeomans, D., Housen, K., *et al.* 2002. Gravitational aggregates: evidence and evolution. In *Asteroids III*, eds. W. F. Bottke, Jr., A. Cellino, P. Paolicchi, and R. Binzel, pp. 485–500. Tucson, AZ: University of Arizona Press.

Chapman, C. R. 2001. Impact lethality and risks in today's world: lessons for interpreting Earth history. In *Proc. Conf. on Catastrophic Events and Mass Extinctions: Impacts and Beyond*, eds. C. Koerberl and K. G. MacLeod, pp. 7–19. Boulder, CO: Geological Society of America.

Chapman, C. R. and Morrison, M. 1994. Impacts on the Earth by asteroids and comets: assessing the hazard. *Nature* **367**, 33–39.

Dandy, C. L., Fitzsimmons, A., and Collander-Brown, S. J. 2003. Optical colors of 56 near-Earth objects: trends with size and orbit. *Icarus* **163**, 363–373

Harris, A. W. 2002. A new estimate of the population of small NEAs. *Bull. Amer. Astron. Soc.* **34**, 835.

Hilton, J. L. 2002. Asteroid masses and densities. In *Asteroids III*, eds. W. F. Bottke, Jr., A. Cellino, P. Paolicchi, and R. Binzel, pp. 103–112. Tucson, AZ: University of Arizona Press.

Jedicke, R., Morbidelli, A., Spahr, T., *et al.* 2002. Earth and space based NEO survey simulation: prospects for achieving the Spaceguard goal. *Icarus* **161**, 17–33.

Jewitt, D. 2000. Astronomy: eyes wide shut. *Planet. Rep.* **20**, 4–5.

Leipold, M., von Richter, A., Hahn, G., *et al.* 2002. Earthguard-1: a NEO detection space mission. *Proc. Conf. Asteroids, Comets, Meteors ACM2002, ESA SP-500*, pp. 107–110.

Merline, W. J., Weidenschilling, S. J., Durda, D. D., *et al.* 2002. Asteroids do have satellites. In *Asteroids III*, eds. W. F. Bottke, Jr., A. Cellino, P. Paolicchi, and R. Binzel, pp. 289–312. Tucson, AZ: University of Arizona Press.

Milani, A., Chesley, S. R., Chodas, P. W., *et al.* 2002. Asteroid close approaches: analysis and potential impact detection. In *Asteroids III*, eds. W. F. Bottke, Jr., A. Cellino, P. Paolicchi, and R. Binzel, pp. 55–69. Tucson, AZ: University of Arizona Press.

Morrison, D., Harris, A. W., Sommer, G., *et al.* 2002. Dealing with the impact hazard. In *Asteroids III*, eds. W. F. Bottke, Jr., A. Cellino, P. Paolicchi, and R. Binzel, pp. 739–754. Tucson, AZ: University of Arizona Press.

Naka, F. R., 1997. *Report on Space Surveillance, Asteroids and Comets, and Space Debris*, vol. 2, *Asteroids and Comets*. US Air Force Scientific Advisory Board. SAB-TR-96-04.

OECD. 2003. Final Report. *Workshop on Near Earth Objects: Risks, Policies and Actions*. Organization for Economic Cooperation and Development (OECD), Global Science Forum, Frascati, Italy, January 20–22, 2003.

Pravec, P., Harris, A. W., and Michalowski, T., 2002. Asteroid rotations. In *Asteroids III*, eds. W. F. Bottke, Jr., A. Cellino, P. Paolicchi, and R. Binzel, pp. 113–122. Tucson, AZ: University of Arizona Press.

Space Studies Board. 2001. *Astronomy and Astrophysics in the New Millennium*. Washington, DC: National Academy Press.

2002a. *New Frontiers in the Solar System: An Integrated Exploration Strategy*. Washington, DC: National Academy Press.

2002b. *The Sun to Earth – and Beyond: A Decadal Research Strategy in Solar and Space Physics*. Washington, DC: National Academy Press.

Vasilyev, N. V. 1998. The Tunguska meteorite problem today. *Planet. Space Sci.* **46**, 129–150.

Veverka, J., Farguhar, B., Robinson, M., *et al.* 2001. The landing of the NEAR-Shoemaker spacecraft on asteroid 433 Eros. *Nature* **413**, 390–393.

Ward, S. N. and Asphaug, E. 2000. Asteroid impact tsunami: a probalistic hazard assessment. *Icarus* **145**, 64–78.

Weiler, E. J. 2002. In invited remarks at the *NASA Workshop on Scientific Requirements for Mitigation of Hazardous Comets and Asteroids*. Held in Arlington, VA.

Index